Six Sigma Quality Improvement
with Minitab

Six Sigma Quality Improvement with Minitab

Second Edition

G. Robin Henderson

Halcro Consultancy, Midlothian, UK

A John Wiley & Sons, Ltd., Publication

Library of Congress Cataloging-in-Publication Data

Henderson, G. Robin.
  Six Sigma quality improvement with Minitab / G. Robin Henderson. – 2nd ed.
     p. cm.
  Includes bibliographical references and index.
  ISBN 978-0-470-74175-7 (cloth) – ISBN 978-0-470-74174-0 (pbk.)
  1. Process control. 2.   Six sigma (Quality control standard) 3.   Minitab.   I. Title.
  TS156.8.H45 2011
  658.5'62–dc23

                                                             2011012320

A catalogue record for this book is available from the British Library.

The cover image was created using a Minitab macro that simulates the Deming funnel experiments that highlight the dangers of tampering with stable processes. The experiments and macro are referred to in the text.

HB ISBN: 978-0-470-74175-7
PB ISBN: 978-0-470-74174-0
ePDF ISBN: 978-1-119-97533-5
oBook ISBN: 978-1-119-97532-8
ePub ISBN: 978-1-119-97618-9
Mobi ISBN: 978-1-119-97619-6

Set in 10/12pt, Times Roman by Thomson Digital, Noida, India

To Fiona and Iain

Contents

Foreword

If it were possible to add up, on a global basis, all of the benefits that organisations have experienced as a result of deploying Six Sigma techniques, the result would be truly staggering. The place of Six Sigma as an effective methodology for improving quality and performance is very well established.

Six Sigma is often introduced to organisations through training. Six Sigma practitioners gain personal development by attending a spectrum of courses from introductory sessions to expert level that is normally referred to as Six Sigma Black Belt. I personally have led many hundreds of practitioners through this process, and whilst I hope my interventions have been successful, I am only too aware that there is a limit to what participants can be expected to absorb in a classroom.

This is where Robin Henderson's book becomes truly invaluable. As well as providing students new to Six Sigma with a very readable and easy to understand introduction, this publication serves as comprehensive consolidation for those already trained. Furthermore this book extends the knowledge gained by recognised experienced practitioners.

Those looking for relevant and modern case studies from both service and manufacturing environments will be most satisfied to find them in abundance throughout the following pages. Robin Henderson demonstrates the wide applicability and power of these methods with an impressive collection of analyses and improvements drawn from his broad experience of working as a consultant as well as an academic. The combination of many chapter-end exercises, follow up activities, and the accompanying web site, form a wealth of extremely useful resources.

The success of Six Sigma would not have been realised had it not been for the development of statistical software such as Minitab. Minitab brings techniques to all of us that previously were only the domain of statisticians. Robin Henderson's book complements other well-known texts by taking the theory and explaining how to implement these methods in real situations through the use of Minitab software. Starting from a gentle introduction to Minitab, Robin builds our knowledge through detailed yet friendly explanations, and as we practice, gradually leads us on to tackle more sophisticated techniques with confidence.

I have spent much of my career working in the field of quality and performance improvement, utilising these tools and techniques, and teaching the subject to others. Since its publication, I have regularly turned to Robin Henderson's first edition of Six Sigma Quality Improvement with Minitab to check my understanding, to learn a bit more, and to find a way of explaining to my teams, a challenging concept in a straightforward way.

Now we have the benefit of this second edition that keeps us up to date on the latest developments within the Minitab tool and its brand new features. For example, Robin Henderson introduces the new Minitab Assistant that guides users through an analysis process.

There is no doubt that Robin Henderson has helped all of us in our endeavours to improve quality as author of various papers on the subject and through his involvement with the Six Sigma Study Group and the Quality Improvement Section of the Royal Statistical Society.

I warmly welcome the 2nd edition of Six Sigma Quality Improvement with Minitab and thoroughly recommend it to both new students and experienced practitioners of Six Sigma methodologies.

Colin Barr BSc (Hons), Six Sigma Black Belt

Colin Barr is the founder of Colin Barr Associates, providers of training and consultancy in Business Improvement using techniques such as Lean and Six Sigma. He is also founder of Stratile, where he developed FocalPoint, a web based strategy and performance management system. Colin Barr gained a BSc (Hons) Physics from Strathclyde University, and has held director level positions at DEC and Motorola where he was trained as a Six Sigma Black Belt.

Preface

Rationale

The Statistics Division of the American Society for Quality defines *statistical thinking* (http://www.asqstatdiv.org/stats-everywhere.htm, accessed 29 January 2011) as a philosophy of learning and action based on three principles:

- All work occurs in a system of interconnected processes.

- Variation exists in all processes.

- Understanding and reducing variation are key to success.

In a paper entitled 'Six-Sigma: the evolution of 100 years of business improvement methodology', Snee (2004) states that 'The three key elements to statistical thinking are process, variation and data' and that 'Statistical thinking enhances the effectiveness of the statistical methods and tools'. He describes Six Sigma* as a strategy and methodology for the deployment of statistical thinking and methods within an organization. This book aims to explain some of the most important statistical methods and demonstrate their implementation via the statistical software package Minitab® (Release 16). Minitab® and the Minitab logo are registered trademarks of Minitab, Inc. There are many excellent texts available on statistical methods for the monitoring and improvement of quality. In writing this book the author set out to complement such texts by providing careful explanation of important statistical tools coupled with detailed description of the use of Minitab, either to implement the statistical tools or as an aid to understanding them.

In *Six Sigma Beyond the Factory Floor*, Hoerl and Snee (2005, p. 23) wrote:

> Another reason Six Sigma has been effective is the general availability of user-friendly statistical software that enables effective and broad utilization of the statistical tools. The statistical software package most widely used in Six Sigma is Minitab. ... Prior to the availability of such user-friendly software, statistical methods were often the domain of professional statisticians, who had access to, and specialized training in, proprietary statistical software. Specialists in statistical methods have an important role to play in Six Sigma, but practitioners who are not professional statisticians do the vast majority of statistical applications.

The author believes that his book will be of value to such practitioners and to people involved in quality improvement strategies other than Six Sigma, to students of quality improvement and indeed to anyone with an interest in statistical methods and their implementation via software.

---

*Six Sigma is a registered trademark and service mark of Motorola Inc.

Content

Among the features of the book are the following:

- Exposition of key statistical methods for quality improvement – data display, statistical models, control charts, process capability, process experimentation, model building and the evaluation of measurement processes.

- Detailed information on the implementation of the methods using Minitab with extensive use of screen captures.

- Demonstration of facilities provided by Minitab for learning about the methods and the software, including the new Assistant.

- Use of random data generation in Minitab to aid understanding of important statistical concepts.

- Provision of informative follow-up exercises and activities on each topic.

- No prior knowledge of statistical methods assumed.

- No prior knowledge of Minitab assumed.

- Access to Release 16 of the Minitab software is essential.

- An associated website providing data sets for download and answers and notes for the follow-up exercises.

There are eleven chapters and four appendices. In addition to the topics covered in the first edition, this edition includes new material on Pareto charts, cause-and-effect diagrams, the multivariate normal distribution, acceptance sampling, time-weighted and multivariate control charts, tolerance intervals, Taguchi experimental designs, comparison of measurement systems, analysis of categorical data and logistic regression. It also includes material on new features provided in Release 16 of Minitab such as the Assistant. A brief summary of the content of each chapter is as follows:

- Chapter 1 introduces the structured approach to quality improvement provided by Six Sigma via DMAIC – define, measure, analyse, improve and control. It outlines the role of statistical methods in Six Sigma and the capabilities of Minitab for their implementation.

- Chapter 2 provides an introduction to data display, and to Minitab and its features. It also addresses data input, output, storage and manipulation.

- Chapter 3 contains further material on the display and summary of data – exploratory data analysis techniques and techniques for use with multivariate data are introduced. Pareto charts and cause-and-effect diagrams are explained.

- Chapter 4 is devoted to fundamentals of probability and to univariate statistical models for measurements and counts. A brief introduction to the multivariate normal distribution is given and key results concerning means and proportions are presented. An introduction to the application of discrete probability distributions in acceptance sampling is provided.

- Chapter 5 gives a comprehensive treatment of control charts and their application. Shewhart, exponentially weighted moving average (EWMA), cumulative sum (CUSUM) and multivariate control charts are covered. Reference is made to the dangers of tampering with processes and to feedback adjustment.

- Chapter 6 addresses the assessment of process capability via capability indices and sigma quality levels. Tolerance intervals are introduced.

- Chapter 7 deals with process experimentation involving a single factor and essentially addresses the question of whether or not process changes have led to improvement. The question is addressed via statistical inference and estimation.

- Chapter 8 extends the ideas introduced in the previous chapter to process experimentation involving two or more factors. Fundamental aspects of design of experiments are introduced together with the powerful features provided in Minitab for experimental design and the display and analysis of the resulting data. Taguchi experimental designs are introduced.

- Chapter 9 utilizes concepts from previous chapters in order to evaluate the performance of measurement processes for both continuous measurement and attribute measurement scenarios. Reference is made to the comparison of measurement systems.

- Chapter 10 is concerned with model building using simple and multiple regression. Response surface methodology and regression modelling with categorical response variables are introduced.

- Chapter 11 concludes the book by looking at ways in which Minitab can assist the user to learn more about the software and the statistical tools that it implements. An introduction to Minitab macros is provided.

Using the book

This is not a book to be read in an armchair! The author would encourage users to follow the Minitab implementation of displays and analyses as he/she reads about them and to work through the supplementary exercises and activities at the end of each chapter. All but the very smallest data sets referred to in the text will be available on the website http://www.wiley.com/go/six_sigma in the form of Minitab worksheets or Microsoft Excel™ workbooks. It is recommended that you download the files and store them in a directory on your computer. Some of the data sets are real, others have been simulated (using Minitab!) to provide appropriate illustrations. Many of the simulated data sets are set in the context of quality improvement situations that the author has encountered. The website will also provide specimen solutions to, and comments on, the supplementary exercises.

The needs of readers will differ widely. It is envisaged that many will find the first four or five chapters sufficient for their needs. It is important to note that although a brief introduction to the Help facilities will be given in Chapter 2, many readers might find it helpful to read the first section of Chapter 11 immediately. The reference to Help has been encountered in Chapter 2 in order to obtain more comprehensive information facilities that are available.

The reader might wonder why the chapter on control charts is before the one on measurement process evaluation, whilst in DMAIC the order appears to be reversed. The author has endeavoured to order the chapter topics in a sequence that is logical from the point of view of the development of understanding of the applied statistics. For example, one cannot fully understand a gauge R&R measurement process evaluation without knowledge of analysis of variance for data from a designed experiment. Designed experiments are usually associated with the improve phase. Indeed, control charts may be of value during all four of the measure, analyse, improve and control phases of a Six Sigma project. Each chapter will give an indication of the relevance of its content to the DMAIC sequence that lies at the heart of Six Sigma.

There is always a danger that statistical software will be used in black box fashion with unfortunate consequences. Thus the reader is exhorted to learn as much as he/she possibly can about the methods and to take every opportunity to learn from successful, sound applications by others of statistical methods in quality improvement, whether on Six Sigma projects or as part of other strategies.

It is the author's earnest hope that, through using this book, you the reader will acquire understanding of statistical methods for quality improvement and Six Sigma, skill in the application of the Minitab software, and appreciation of just how easy it is to use and of all that it has to offer.

Acknowledgements

Grateful appreciation of help and encouragement is due to the following people: Sandra Bonellie, Colin Barr, Gary Beazant, Isobel Black, Roland Caulcutt, Shirley Coleman, Lorraine Daniels, Ross Davies, Martin Dennis, Jeff Dodgson, Geoff Fielding, Alan Fisher, Wendy Ford, Martin Gibson, Mary Hickey, Iain Henderson, Kevin Hetzler, Tom Johnstone, Graham Leigh, Ron Masson, Bill Matheson, Deborah Macdonald, Mark McGinnis, Gillian Mead, Charles Moncur, Douglas Montgomery, Bill Munro, David Panter, Gillian Raab, David Roberts, David Reed, Fiona Reed, Anne Shade, John Shrouder, William Woodall and Andrew Vickers.

Robert Raeside deserves a major thank-you for all that I learnt from him during shared involvement in the development and delivery of many training courses on quality improvement when we were both members of the Applied Statistics Group at Edinburgh Napier University. Another major influence was John Shade, a statistician with vast experience of quality improvement, with whom it was a pleasure to work, in a highly stimulating environment at Good Decision Ltd.

At John Wiley & Sons, Inc., Richard Davies, Heather Kay, Ilaria Meliconi and Prachi Sinha-Sahay have been a pleasure to work with. The author is most grateful to the copy editor Richard Leigh for his highly professional contribution to the project. Abhishan Sharma at Thomson Digital was most helpful during the typesetting phase of the project. Support from Minitab Inc. has been excellent. In particular, the author wishes to acknowledge the help of Austin Davey in the UK and of Eugenie Chung and Linda Holderman in the USA. Portions of the input and the output contained in this book are printed with permission of Minitab, Inc. Use of data sets included with the Minitab software is gratefully acknowledged. Grateful thanks are due to authors and publishers who have given permission for use of data and material. Specific acknowledgements are made in the text.

Finally, in preparing this second edition my wife Anne has been ignored once again for many, many hours – yet her support, as ever, has been immense.

About the Author

Having studied mathematics and physics at the University of Edinburgh, Robin Henderson embarked on a 35-year career in education. For much of that time he was employed in what is now Edinburgh Napier University, teaching mathematics and statistics, at all levels.

His interest in statistics for quality improvement grew during the 1980s, largely due to involvement with colleagues Professor Robert Raeside and Professor Ron Masson in providing training courses, including courses to prepare engineers to sit the Certified Quality Engineer examinations of the American Society for Quality, and consultancy for local organizations, particularly microelectronics companies. This interest led him to leave the University in 1998 in order to work as a statistical consultant for Good Decision Ltd, Dunfermline, where he was heavily involved in the development and delivery of training courses for industry in process monitoring and adjustment, measurement process evaluation and design and analysis of multifactor experiments.

His interest in and enthusiasm for Minitab also developed during the 1980s when it began to be used at Edinburgh Napier University in the teaching of students on a wide variety of courses. Release 7 of the software, with line-printer graphics, was a far cry from the sophistication of the current version!

Since 2001, Robin has been operating as a sole consultant, trading as Halcro Consultancy, Loanhead, providing training and consultancy in statistics for quality improvement and Six Sigma. He has assisted Colin Barr Associates with the training of Six Sigma Black Belts and with statistical consultancy projects. In 2009 Edinburgh Napier University received the Queen's Anniversary Prize, the highest accolade that can be conferred on a higher or further education institution in the UK, for its pioneering research in innovative construction techniques to improve insulation in new build homes. The author is very pleased to have provided the spin-out company, Robust Details Ltd., that was created to oversee uptake of the construction solutions stemming from the research, with training in both statistical methods and Minitab.

He is also currently employed as coordinator at the Royal Infirmary of Edinburgh for the Scottish National Stroke Audit. On this project he has introduced the use of Shewhart control charts for monitoring aspects of the processes involved in the care of stroke patients. Since the first edition was published he has been principal author of two papers on healthcare applications of control charts and co-author of a paper on the technical details of estimation of process variability. He has acted as secretary to both the Committee of the Quality Improvement Section and the Six Sigma Study Group of the Royal Statistical Society, of which he is a Fellow. In these roles he was responsible for collating the views of colleagues on the draft international standards BS ISO 13053-1/2 *Quantitative methods in process improvement – Six Sigma – Part 1: DMAIC methodology, Part 2: Tools and techniques* and for subsequently preparing the comments submitted by the Society on the drafts. He is a member of ENBIS, the European Network for Business and Industrial Statistics.

1

Introduction

Six Sigma is a strategic approach that works across all processes, products and industries.
(Snee, 2004, p. 8)

Overview

In an overview of Six Sigma, Montgomery and Woodall (2008) state:

> Six Sigma is a disciplined, project-oriented, statistically based approach for
> reducing variability, removing defects, and eliminating waste from products,
> processes, and transactions. The Six Sigma initiative is a major force in today's
> business world for quality and business improvement. Statistical methods and
> statisticians have a fundamental role to play in this process.

This book is about the understanding and implementation of statistical methods fundamental
to the Six Sigma and other approaches to the continuous improvement of products, processes
and services. Release 16 of the widely used statistics package Minitab is used throughout.
Mindful that in the vast majority of situations those applying statistical methods in quality
improvement and Six Sigma are not statisticians, information on the implementation of each
method covered will be preceded by some background explanation, typically employing small
data sets. The role of each method within the define–measure–analyse–improve–control
(DMAIC) framework of Six Sigma will be highlighted.

This chapter deals with quality and quality improvement and in particular with the highly
successful Six Sigma approach to quality improvement. It describes the role of statistical
methods in quality improvement and Six Sigma and outlines how Minitab can be used to
implement these methods.

Six Sigma Quality Improvement with Minitab, Second Edition. G. Robin Henderson.
© 2011 John Wiley & Sons, Ltd. Published 2011 by John Wiley & Sons, Ltd.

1.1   Quality and quality improvement

Definitions of quality abound. Wheeler and Chambers (1992, p. xix) define quality as being 'on-target with minimum variance'. They state that operating 'on-target' requires a different way of thinking about processes and that operating with 'minimum variance' can only be achieved when a process is behaving in a reasonably stable and predictable way. Wheeler and Poling (1998, p. 3) state that continual improvement requires a 'methodology for studying processes and systems, and a way of differentiating between the different types of variation present in processes and systems'. They refer to the cycle of activities involved in continual improvement as the plan–do–study–act (PDSA) cycle – see Figure 1.1 (Wheeler and Poling, 1998, p. 5), reproduced by permission of SPC Press, Inc. This cycle of activities is often referred to as the Shewhart–Deming cycle in honour of two key figures in the development of quality improvement methodology, Dr Walter A. Shewhart and Dr W. Edwards Deming.

In 1988 the author had the pleasure, along with some 400 others, of attending a seminar for statisticians by the late Dr Deming at the University of Nottingham. In his presentation handout he described the cycle as follows (Deming, 1988, p. 33):

- *Plan* – Plan a change or test, aimed at improvement.

- *Do* – Carry it out, preferably on a small scale.

- *Study* – Study the results. What did we learn?

- *Act* – Adopt the change or abandon it or run through the cycle again, possibly under different environmental conditions.

In his book *Out of the Crisis*, Deming (1986, p. 23) listed *14 Points for Management*, including the following:

- *Constancy of purpose* – create constancy of purpose for continual improvement of products and service.

- *Improve every process* – improve constantly and forever every process for planning, production and service.

- *Eliminate targets* – substitute aids and helpful supervision; use statistical methods for continual improvement.

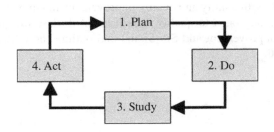

Figure 1.1   The Shewhart–Deming PDSA cycle.

Thus the Deming philosophy of quality improvement advocates never-ending continual improvement of all processes within an organization. The improvement process itself should be structured and employ statistical methods. 'Improvement nearly always means reduction of variation' (Deming, 1988).

1.2 Six Sigma quality improvement

Six Sigma is one of a number of quality improvement strategies based on the Shewhart–Deming PDSA cycle. Truscott devotes a chapter to comparison of Six Sigma with other quality improvement initiatives. He puts Six Sigma in perspective earlier in his book as follows (Truscott, 2003, p. 1):

> Six Sigma focuses on establishing world-class business-performance benchmarks and on providing an organizational structure and road-map by which these can be realized. This is achieved mainly on a project-by-project team basis, using a workforce trained in performance-enhancement methodology, within a receptive company culture and perpetuating infrastructure. Although particularly relevant to the enhancing of value of products and services from a customer perspective, Six Sigma is also directly applicable in improving the efficiency and effectiveness of all processes, tasks and transactions within any organization. Projects are thus chosen and driven on the basis of their relevance to increased customer satisfaction and their effect on business-performance enhancement through gap analysis, namely, prior quantitative measurement of existing performance and comparison with that desired.

Six Sigma originated at Motorola Inc. as a long-term quality improvement initiative entitled 'The Six Sigma Quality Program'. (Six Sigma® is a registered trademark and service mark of Motorola Inc.) It was launched by the company's chief executive officer, Bob Galvin, in January 1987 with a speech that was distributed to everyone in the organization. In the speech Galvin reported on many visits to customers in the previous six months during which desires were expressed for better service from Motorola in terms of delivery, order completeness, accurate transactional records etc. The customers had also indicated that, with better service and an emphasis on total quality, Motorola could expect an increase of between 5% and 20% in future business from them. He therefore challenged employees to respond urgently to make the necessary improvements, emphasized the leadership role of management in the implementation of the programme and announced that Motorola's corporate quality goal had been updated accordingly. The goal included the objective 'Achieve Six Sigma capability by 1992' (Perez-Wilson, 1999, p. 131).

In addition to being a strategy for the continual improvement of quality within organizations, Six Sigma indicates a level of performance equating to 3.4 nonconformities per million opportunities, a level which some regard as being 'world-class performance'. This often leads to confusion. In this book the phrase 'sigma quality level' will be used for this indicator of process performance, as advocated by Breyfogle (2003, p. 3). Thus a sigma quality level of 6 equates to 3.4 nonconformities per million opportunities. The link between sigma quality level and number of nonconformities per million opportunities will be explained in detail in

Table 1.1 Sigma quality levels and nonconformities per million opportunities.

Sigma quality level	Nonconformities per million opportunities
1.5	501 350
2.0	308 770
2.5	158 687
3.0	66 811
3.5	22 750
4.0	6 210
4.5	1 350
5.0	233
5.5	32
6.0	3.4

Chapter 4. Table 1.1 gives some sigma quality levels and corresponding numbers of nonconformities per million opportunities.

The plot in Figure 1.2 (created using Minitab) shows nonconformities per million opportunities plotted against sigma quality level, the vertical scale being logarithmic. Many authors refer to defects rather than nonconformities and state that a sigma quality level of 6 equates to 3.4 defects per million opportunities or 3.4 DPMO.

Imagine a bottling plant where there was concern over the number of bottles of whisky containing contaminant particles. A bottling run of 14 856 bottles had yielded 701 non-conforming bottles in terms of contamination. This corresponds to $(701/14\,856) \times 10^6 =$ 47 186 nonconforming bottles per million. Since 47 254 lies between 22 750 and 66 811, Table 1.1 indicates a sigma quality level between 3.0 and 3.5. The more comprehensive table in

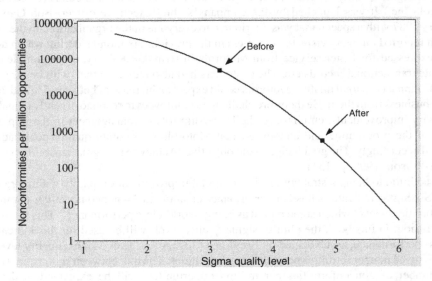

Figure 1.2 Sigma quality levels and nonconformities per million opportunities.

Appendix 1 gives the sigma quality level as 3.17 (47 461 being the entry in the table closest to 47 186).

The main source of contamination was found to be the wax-coated corks used to seal the bottles. Dialogue with the supplier of the corks led to a trial with corks produced using a new method for application of the wax. For the trial there were 8 nonconforming bottles out of 15 841. The reader is invited to verify that this corresponds to 505 nonconforming bottles per million and a sigma quality level of 4.79. Thus the higher the sigma quality level for a process the better is its performance. The points corresponding to the situations before and after the process change are indicated on the curve in Figure 1.2. Montgomery and Woodall (2008) state that 'the 3.4-ppm metric, however, is increasingly recognized as primarily a distraction; it is the focus on reduction of variability about the target and the elimination of waste and defects that is the important feature of Six Sigma'.

Another source of confusion arises from the use of σ, the lower-case Greek letter sigma, as a statistical measure of variation called standard deviation. It will be explained in detail in Chapter 2. Bearing in mind Deming's comment that 'Improvement nearly always means reduction of variation', the consequence is that improvement often implies a reduction in σ. Thus it is frequently the case that improvement corresponds to an increase in sigma quality level on the one hand, and to a decrease in sigma on the other. Therefore it is essential to make a clear distinction between both uses of sigma.

Larry Bossidy, CEO of Allied Signal led the implementation of Six Sigma with his organization (Perez-Wilson, 1999, p. 270). In June1995 Bossidy addressed the Corporate Executive Council of the General Electric Corporation (GE) on Six Sigma quality at 'one of the most important meetings we ever had' (Welch, 2001). The Council was impressed by the cost saving achieved at Allied Signal through Six Sigma and, as an employee survey at GE had indicated that quality was a concern for many GE employees, Welch 'went nuts about Six Sigma and launched it' (Welch, 2001). One of the first steps GE took in working towards implementation of Six Sigma was to invite Mikel Harry, formerly a manager with Motorola and founder of the Six Sigma Academy, to talk to a group of senior employees. In a four-hour presentation 'he jumped excitedly from one easel to another, writing down all kinds of statistical formulas'. The presentation captured the imagination of Welch and his colleagues, the discipline of the approach being of particular appeal to engineers. They concluded that Six Sigma was more than quality control and statistics – 'ultimately it drives leadership to be better by providing tools to think through tough issues' – and rolled out their Six Sigma program in 1996 (Welch, 2001, p. 330). Examples of early Six Sigma success stories at GE, in both manufacturing and non-manufacturing situations, are reported by Welch (2001, pp. 333–334).

Perez-Wilson states that Motorola had looked for a catchy name to shake up the organization when introducing the concept of variation reduction and that in Six Sigma they found it. However, in spite of confusion over the different interpretations of sigma, in his opinion 'It [Six Sigma] reflects a philosophy for pursuing perfection or excellence in everything an organization does. Six Sigma is probably the most successful program ever designed to produce change in an organization' (Perez-Wilson, 1999, p. 195). A Six Sigma process, i.e. a process with a sigma quality level of 6, corresponds to 3.4 nonconformities per million opportunities – 'That's 99.99966 percent of perfection' (Welch, 2001). Harry and Schroeder (2000) refer to a Six Sigma process as the Land of Oz and to the Six Sigma Breakthrough Strategy as the Yellow Brick Road leading there. The Six Sigma 'roadmap' is the subject of Section 1.3.

Antony (2010) and Montgomery and Woodall (2008) refer to the evolution of Six Sigma through three generations of implementations:

- Generation I – focus on elimination of defects and variation reduction, primarily in manufacturing. Spanned the period 1987–1994, with Motorola being a good exemplar.

- Generation II – in addition to the focus in the previous generation there was an emphasis on linking efforts to eliminate defects and reduce variation to efforts to improve product design and reduce costs. Spanned the period 1994–2000, with General Electric a prime exemplar.

- Generation III – since 2000 there has been an additional focus on value creation for both organizations and their stakeholders.

In a keynote presentation at the European Network for Business and Industrial Statistics conference in 2009, Tom Johnstone, CEO of global company SKF, referred to its employment of Six Sigma 'to make it easier and attractive for our customers and suppliers to do business with us'. He referred to four dimensions – 'Standard' Six Sigma, Design for Six Sigma (DFSS), Lean Six Sigma and Six Sigma for Growth. He also referred to the integration of Six Sigma with other improvement initiatives as being a challenge faced by SKF (Johnstone, personal communication, 2010) This concurs with the comment by Montgomery and Woodall (2008) that they 'expect Six Sigma to become somewhat less outwardly visible, while remaining an important initiative within companies'. These comments indicate that Six Sigma is still evolving. In the final section of their paper, on the future of Six Sigma, Montgomery and Woodall (2008) state:

> Six Sigma has become a widely used implementation vehicle for quality and business improvement. It is logical to ask about its future. Some have speculated that 'Six Sigma' is the 'flavour of the month' as management looks for the quick fix to crucial operational problems. However, since Six Sigma is over 20 years old and implementations are growing worldwide, it is difficult to believe that it is simply a management fad. In an ideal implementation, Six Sigma, DMAIC, DFSS and lean tools are used simultaneously in an organization to achieve high levels of process performance and significant business improvement.

Further evidence of the important role of Six Sigma is provided by the current development of two international standards, BS ISO 13053-1/2: *Quantitative methods in process improvement – Six Sigma – Part 1: DMAIC methodology* and *Part 2: Tools and techniques*.

1.3   The Six Sigma roadmap and DMAIC

The ideal roadmap for implementing Six Sigma within an organization is claimed to be as follows (Pande *et al.*, 2000).

1. Identify core processes and key customers.

2. Define customer requirements.

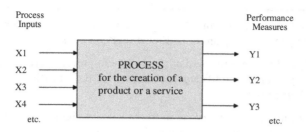

Figure 1.3   Process model.

3. Measure current performance.

4. Prioritize, analyze, and implement improvements.

5. Expand and integrate the Six Sigma system.

A process or system may be represented by the schematic in Figure 1.3, where the Xs represent inputs and the Ys represent performance measures on the process output. For the process of baking biscuits examples of inputs are flour supplier, oven temperature and cooking time. Some inputs are controllable by the baker, such as flour supplier and oven temperature. Others, such as the moisture content of the flour, are not. Examples of performance measures are the proportion of broken biscuits and the height of a stack of 16 biscuits – the number in a standard pack. Problems with the next process of packaging the biscuits may arise because stack height is off target or because there is too much variation in stack height. Xs are referred to as inputs, independent variables, factors and key process input variables. Uncontrollable inputs are often designated noise factors. Ys are referred to as performance measures, dependent variables or responses. Where performance measures are of vital importance to the customer, the phrases key process output variables or critical to quality characteristics are employed.

The process improvement model at the heart of Six Sigma is often referred to as define–measure–analyse–improve–control (DMAIC). The first step is to define the problem or opportunity for improvement that is to be addressed. Writing in *Quality Progress*, a journal of the American Society for Quality (ASQ), Roger Hoerl, who at the time of writing is Manager of the Statistics Lab at General Electric Global Research, described the last four steps as shown in Box 1.1 (Hoerl, 1998). He also Hoerl (2001, p. 403) refers to DMAIC as 'the glue which holds together the individual tools and facilitates solving real problems effectively'.

Some experts in Six Sigma apply the DMAIC improvement model both to process improvement projects and to process design/redesign projects. Others prefer to think in terms of DFSS and use define–measure–analyse–design–verify (DMADV) in the case of process design/redesign. Montgomery (2009) states that 'the I in DMAIC may become DFSS' – in other words, that improvement may only be possible through redesigning a process or creating a new one.

Six Sigma projects are normally selected as having potential in terms of major financial impact through facets of a business such as quality, costs, yield and capacity. Teams working on projects are typically led by employees designated 'Black Belts'. Black Belts will have

Measure – Based on customer input, select the appropriate responses (the *Y*s) to be improved and ensure that they are quantifiable and can be accurately measured.

Analyze – Analyze the preliminary data to document current performance or baseline process capability. Begin identifying root causes of defects (the *X*s or independent variables) and their impact.

Improve – Determine how to intervene in the process to significantly reduce the defect levels. Several rounds of improvement may be required.

Control – Once the desired improvements have been made, put some type of system into place to ensure the improvements are sustained, even though additional Six Sigma resources may no longer be focused on the problem.

Box 1.1   Description of key phases in applying Six Sigma methodology.

been awarded their titles after several weeks of intensive training, including statistical methods, and on completion of a successful Six Sigma project. A Six Sigma organization will invariably have a Six Sigma 'Champion' on the senior management team and may also have Master Black Belts. 'Green Belts' undergo less extensive training than Black Belts and may lead minor projects but normally assist on Black Belt led projects. Some organizations designate employees who have undergone basic Six Sigma training as 'Yellow Belts'. External consultants are often used to train the first group of Black Belts within an organization. As the organization matures in terms of Six Sigma the Black Belts frequently undertake training of Green Belts and may devote 50–100% of their time to project activities. A discussion of curricula for the training of Black Belts is given by Hoerl (2001). He states that 'the BB [Black Belt] role is intended to be a temporary assignment – typically two years' (Hoerl, 2001, p. 394). Many organizations look to Black Belts to progress to senior roles; for example, at SKF it is envisaged that future company leaders will be former Black Belts (Johnstone, personal communication, 2010). Snee (2004) provides a Six Sigma project case study.

1.4   The role of statistical methods in Six Sigma

Measurement is fundamental to Six Sigma and measurement creates data. Many improvement initiatives lead to 'before' and 'after' data for *Y*s – data collected before and after process changes are implemented. In terms of the process model shown in Figure 1.3 the key to improvement is knowledge of how the *X*s influence the *Y*s, i.e. knowledge of the 'formula' linking the *X*s to the *Y*s, represented symbolically as $\mathbf{Y} = f(\mathbf{X})$. (The use of bold symbols indicates that invariably a process has associated with it a set of *Y*s and a set of *X*s so that $\mathbf{Y}$ represents $Y_1, Y_2, Y_3, \ldots, Y_m$ and $\mathbf{X}$ represents $X_1, X_2, X_3, \ldots, X_n$ in the case of a process with m performance measures and n inputs.) Statistics provides a series of tools to aid the search for such knowledge – tools such as design of experiments and regression.

The author recalls the late John Tukey stating, in a Royal Statistical Society presentation in Edinburgh in 1986, that 'display is an obligation' whenever one is dealing with data. Data display can be highly informative, so the topic will be emphasized. Tools from exploratory data analysis (EDA), on which Tukey's (1977) book *Exploratory Data Analysis* is regarded as a classic, are included.

In many cases, process improvement is clearly evident from data display. In order to formally answer the question 'Has improvement been achieved?' the topics of statistical inference and estimation are crucial for evaluation of evidence from 'before' and 'after' type data and for quantifying the extent of any improvement. Much process improvement experimentation effort is wasted through the use of badly planned experiments, particularly experiments in which the effect of only a single X on the Ys is considered. The use of properly designed experiments can be of immense benefit, especially when the effects of the Xs are not independent of one another. Regression and correlation also provide tools for exploring and modelling relationships between Xs and Ys.

Shewhart control charts have a major role to play and many important applications. The example in Figure 1.4 (reproduced by permission of NHS Lothian) displays data from an improvement project at a major hospital. The aim of the project was to reduce delays to assessment at a rapid access transient ischaemic attack and stroke clinic. In the measure and analyse phases of the project, the chart indicated that delay between patient referral and attendance at clinic averaged around 13 days. In early 2007, during the improve phase, a hotline to a consultant was introduced, open 24 hours a day, seven days a week. The consultant on call provided immediate advice and, if appropriate, made an appointment for a clinic visit. The chart indicated a reduction in the average delay from 13 days to around 3 days.

There is a danger that project improvement teams can become involved in measurement without considering the measurement process itself. Statistics provides tools for the evaluation of measurement processes, which 'can and should be continuously improved, just as you would "regular" work processes' (Pande *et al.*, 2000, p. 203).

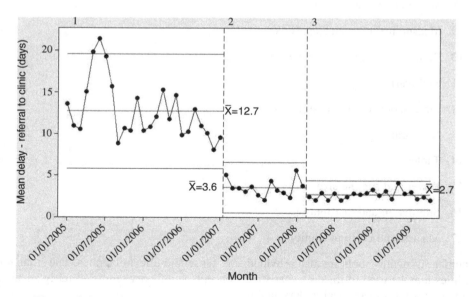

Figure 1.4 A Shewhart control chart demonstrating major improvement.

1.5   Minitab and its role in the implementation of statistical methods

Many involved in Six Sigma projects are busy professional people who welcome rapid and easy access to relevant statistical methods. In a chapter on the history of statistics and quality improvement, Jeroen de Mast (2008, p. 26) refers to thousands of Black Belts and Green Belts worldwide whose applications of statistical methods to quality improvement are 'supported by easy-to-use software'. Minitab is without doubt a statistical software package that is easy to use and its creators claim to be the leading provider of software for statistics education and lean, Six Sigma, and quality improvement projects. Essentially Minitab can provide all the methods referred to in the previous section – and much more.

The author's experience with Six Sigma Black and Green Belt and other trainees has generally been one of their feelings of delight at the ease which statistical methods can be implemented using Minitab. It provides the following capabilities and features:

1. Data and file management

2. Assistant (new in Release 16)

3. Basic statistics

4. Graphics

5. Regression analysis

6. Analysis of variance

7. Design of experiments

8. Statistical process control

9. Measurement systems analysis

10. Reliability and survival analysis

11. Multivariate analysis

12. Time series and forecasting

13. Nonparametrics

14. Tables

15. Power and sample size

16. Simulation and distributions

17. Macros and customizability

A number of quality tools are also provided, e.g. cause-and-effect diagrams. The simulation facilities may be used to illustrate statistical concepts and theory requiring fairly advanced mathematics for a theoretical understanding.

1.6 Exercises and follow-up activities

1. For a familiar process for the creation of a product, list inputs (Xs) and performance measures (Ys). Indicate which inputs you view as being key, and identify those that can be controlled and those that cannot be controlled, or are not controlled, during routine operation, which are noise and which performance measures are critical.

2. Repeat for a process for the creation of a service.

3. Access the Minitab website: http://www.minitab.com/en-GB/theater/default.aspx? video=Minitab16Tour and view the Minitab 16 Statistical Software product tour video. Note the availability of seven other Minitab 16 videos that you might find informative as you progress through the book.

2

Data display, summary and manipulation

In God we trust; all others must bring data! (Attributed to W Edwards Deming)

Overview

Data are essential in the monitoring and improvement of processes and in the measure and control phases of Six Sigma projects. Such data are often obtained in time sequence. For example, a critical dimension might be measured every hour on each member of a sample of machined automotive components during production in a factory, the number of mortgage agreements completed successfully each day might be recorded by a building society. A run chart of such data can frequently be highly informative and forms the basis for some control charts. The construction of run charts using Minitab will be used to introduce the reader to the software and the key features of sessions, projects, worksheets, menus, dialog boxes, graphs, and ReportPad™, etc. The facility for calculation of derived data will also be introduced.

The use of histograms for the display of data will be described, and widely used summary statistics that indicate location and variability defined. The chapter concludes with consideration of a variety of methods for data entry in Minitab, of data manipulation and of the detection of missing and erroneous data values.

2.1   The run chart – a first Minitab session

2.1.1   Input of data via keyboard and creation of a run chart in Minitab

In their book *Building Continual Improvement*, Wheeler and Poling (1998, p. 31) introduce run charts as follows:

Data tend to be associated with time. When was this value obtained? What time period does that value represent? Because of this association with time there is information contained in the time-order sequence of the values. To recover this information you will need to look at your data in a time-ordered plot, which is commonly called a running record or time-series graph.

In order to introduce running records or run charts, consider the time series of weights (g) of glass bottles in Table 2.1. Bottle weight is a key process output variable in the food packaging industry. Each bottle was formed in the same mould of the machine used to produce the bottles and the time interval between sampling of bottles was 15 minutes. The target weight is 490 g and the production run was scheduled to run for a total of 12 hours.

On opening Minitab the screen displayed in Figure 2.1 will appear. Two main windows are visible. The Session window displays the results of analyses in text format; initially it displays the date, time and a message of welcome. (It is also possible to perform tasks by entering commands in the Session window instead of using the Minitab menus.) The Data window

Table 2.1   Initial bottle weight data.

Sample	Weight	Sample	Weight	Sample	Weight	Sample	Weight	Sample	Weight
1	488.1	6	493.1	11	490.5	16	489.7	21	490.2
2	493.4	7	487.4	12	492.2	17	488.5	22	489.8
3	488.7	8	488.4	13	490.6	18	493.6	23	486.1
4	484.4	9	488.6	14	490.8	19	489.1	24	487.0
5	491.8	10	485.9	15	486.7	20	489.4	25	485.4

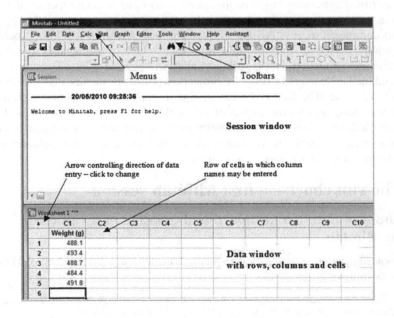

Figure 2.1   Initial Minitab screen.

Figure 2.2   Project Manager toolbar.

displays an open worksheet that has the appearance of a spreadsheet. (It is possible to use multiple worksheets within a single project.) Note that the blue band across the top of the Session window is a deeper colour than that across the top of the Data window, indicating that the Session window is currently active. (Take a moment to click on the blue bands to make the worksheet active and then the Session window active again.) A third component, in addition to the Session and Data windows, of the new Minitab project just opened up, is the Project Manager, which is minimized at this stage. Note the corresponding icon labelled **Proje...** at the foot of the screen.

Figure 2.2 shows the Project Manager toolbar, which has 12 icons, with the mouse pointer located on the icon for the **ReportPad** on the toolbar. Clicking on an icon makes the corresponding component active. Note that the message displayed indicates the project component associated with the icon and gives the keys that may be used to make the component active as an alternative to clicking on its icon. From the extreme left the icons displayed are: Show Session Folder, Show Worksheets Folder, Show Graphs Folder, Show Info, Show History, Show ReportPad, Show Related Documents, Show Design, Session Window, Data Window, Project Manager and Close All Graphs, respectively.

Enter all 25 of the weight data values in Table 2.1 into the first column, C1, of the worksheet in the current Data window, and enter Weight (g) as the column heading to name the variable. Figure 2.1 displays the worksheet with the variable name and the first five data values entered.

In order to access the dialog box for the creation of a run chart first click the **Stat** menu icon (see Figure 2.3) then **Quality Tools** and finally **Run Chart....** (Throughout this book the shorthand **Stat > Quality Tools > Run Chart...** will be used to indicate such sequences of steps.)

The Run Chart dialog box can now be completed as shown in Figure 2.4. Highlight Weight (g) in the window on the left of the dialog box and click on the **Select** button so that Weight (g) appears in the window labelled **Single column:**. (Alternatively highlight

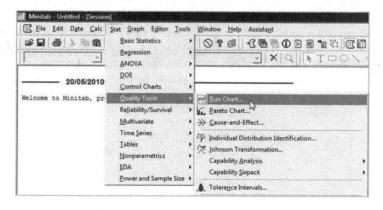

Figure 2.3   Accessing the run chart dialog box.

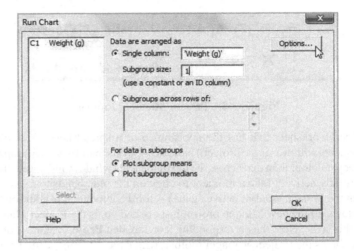

Figure 2.4 Completion of the run chart dialog.

Weight (g) and double-click.) Enter **Subgroup size:** 1 in the appropriate window as each sample consisted of a single bottle. (Had the weights of samples of four bottles been recorded every 15 minutes then **Subgroup size:** 4 would have been entered.)

Clicking the **Options...** button reveals a subdialog box that may be used to create a title for the run chart as indicated in Figure 2.5. Click **OK** to return to the main dialog box and then click **OK** again to display the run chart – see Figure 2.6. (Had the option to create a title not been used then Minitab would have assigned the title 'Run Chart of Weight (g)').

Those involved in running the process can learn about its performance from scrutiny of the run chart. Weight is plotted on the vertical axis, with the weight of each bottle represented by a square symbol and the symbols connected by dotted lines indicating the sequence of sampling. (Had, for example, four bottles been weighed every 15 minutes then Minitab offers the choice of a run chart with symbols corresponding to either the mean or median of the weights of each sample or subgroup of four bottles.) The horizontal axis is labelled Observation – each sample of a single bottle may be thought of as an observation of the process behaviour. The horizontal line on the chart corresponds to the median weight of the entire group of 25 bottles weighed. On moving the mouse pointer to the line a textbox containing the text 'Reference line at 489.1' is displayed. The median weight is 489.1 g, which in this case is the weight of bottle number 19.

In the technical language of statistics the median is a measure of location, which gives an indication of 'where one is at' in terms of process performance. Scrutiny of the run chart reveals 12 points above the median line, 12 points below the median line and the point

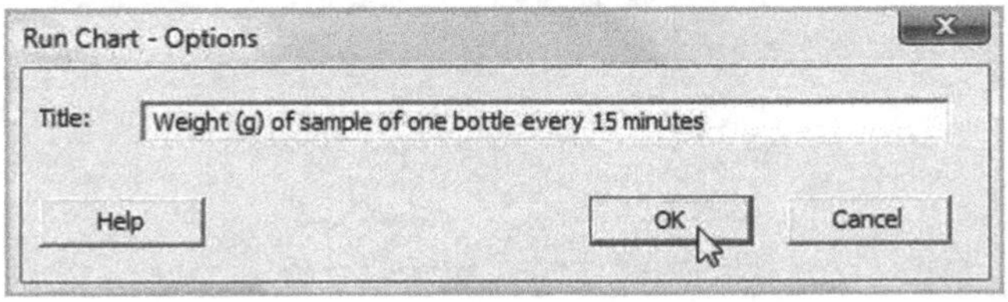

Figure 2.5 Creating a run chart title.

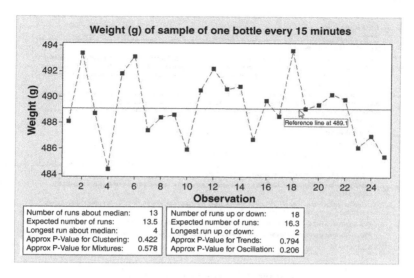

Figure 2.6   Run chart of bottle weight data.

corresponding to bottle number 19 actually on the line. Thus the median is simply the 'middle' of the data set. (The sample of five bottle weights 488.4, 490.8, 486.1, 489.3 and 490.5 may be ordered to yield 486.1, 488.4, **489.3**, 490.5 and 490.8 so the median is the middle value 489.3. The sample of six bottle weights 487.4, 488.1, 493.6, 492.2, 488.6 and 490.6 may be ordered to yield 487.4, 488.1, **488.6**, **490.6**, 492.2 and 493.6. In this case, where there is an even number of bottles in the sample, the median is taken as the value midway between the middle pair of values, i.e. (488.6 + 490.6)/2 = 489.6.)

The median of 489.1 for the sample of 25 bottles is a measure of process performance in relation to the target weight of 490 g. Faced with just this information, should the process owner take action to 'shift' the measure of location closer to 490? This is the sort of question on which statistics can shed light – further discussion of such questions and the tools available in Minitab to answer them appears in Chapter 7.

However, before considering the performance in relation to any target or specification limits, a much more fundamental question should be asked: is the process performing in a stable, predictable manner? The information displayed beneath the run chart is relevant. For an explanation, the Help facility provides the overview of run charts displayed in Figure 2.7. One way to access this information is to click on the **Help** button in the bottom left-hand corner of the Run Chart dialog box – see Figure 2.4. Note also the provision of how to, example, data and see also links to further sources of information to aid the user to learn about run charts. In addition, an explanation of the dialog box items is given and a link to information on the options available in creating a run chart in Minitab via <Options>. Links to related topics are also provided within the text – in the case of the run chart the related topics are subgroup and median.

A process performing in a stable and predictable manner is said to exhibit *common cause variation* only and to be in a state of statistical control. When a process is affected by *special cause variation*, i.e. by variation resulting from causes extraneous to the process, evidence of the presence of such variation may be provided by the tests referred to in the second paragraph of the overview. Data for a process affected only by common cause variation exhibit randomness while data for a process do not. The tests are often referred to as tests for

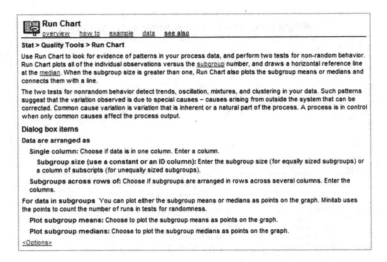

Figure 2.7   Help on run charts.

randomness. In order to conduct these one must scrutinize the *P*-values in the text boxes beneath the run chart. *P*-values will be explained in Chapter 7, but at this stage one need only know that it is generally accepted that any *P*-value less than significance level or α-value of 0.05 provides evidence of the presence of special cause variation, i.e. of the presence of a factor or factors affecting process performance. For the weight data none of the *P*-values is less than 0.05 so it would appear that the bottle production process is behaving in a stable, predictable manner as far as the mould from which the bottles were sampled is concerned.

The run charts in Figures 2.8–2.11 display weight data for moulds where the tests do provide evidence of special cause variation. In the first scenario, displayed in Figure 2.8,

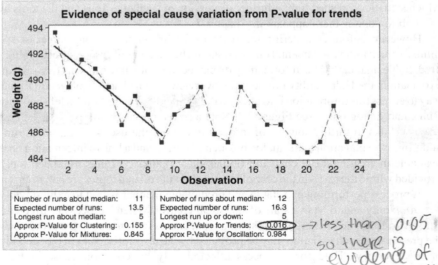

Figure 2.8   Evidence of special cause variation – scenario 1.

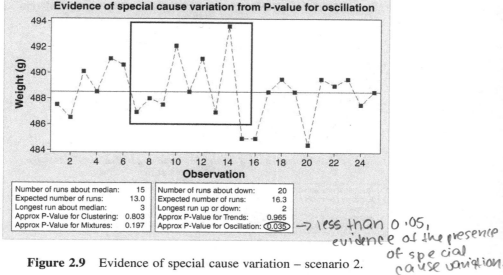

Number of runs about median: 15
Expected number of runs: 13.0
Longest run about median: 3
Approx P-Value for Clustering: 0.803
Approx P-Value for Mixtures: 0.197

Number of runs about down: 20
Expected number of runs: 16.3
Longest run up or down: 2
Approx P-Value for Trends: 0.965
Approx P-Value for Oscillation: 0.035 → less than 0.05, evidence of the presence of special cause variation.

Figure 2.9 Evidence of special cause variation – scenario 2.

the *P*-value for trends is 0.016, which is less than 0.05, so there is evidence of the presence of special cause variation affecting the corresponding mould. The line segment superimposed on the display indicates an apparent initial downward trend in weight.

In the second scenario, displayed in Figure 2.9, the *P*-value for oscillation is 0.035, which is less than 0.05, so there is evidence of the presence of special cause variation affecting the corresponding mould. The rectangle superimposed on the display indicates a period during which weight oscillates rapidly.

In the third scenario, displayed in Figure 2.10, the *P*-value for mixtures is 0.012, which is less than 0.05, so there is evidence of the presence of special cause variation affecting the

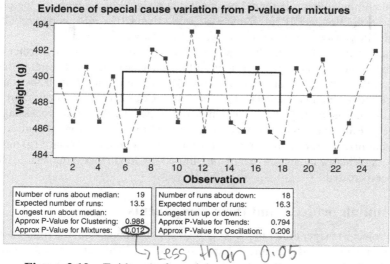

Number of runs about median: 19
Expected number of runs: 13.5
Longest run about median: 2
Approx P-Value for Clustering: 0.988
Approx P-Value for Mixtures: 0.012

Number of runs about down: 18
Expected number of runs: 16.3
Longest run up or down: 3
Approx P-Value for Trends: 0.794
Approx P-Value for Oscillation: 0.206

↳ Less than 0.05

Figure 2.10 Evidence of special cause variation – scenario 3.

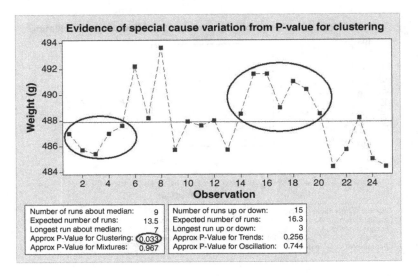

Figure 2.11   Evidence of special cause variation – scenario 4.

corresponding mould. The rectangle superimposed on the display indicates a period during which there is an absence of weight values close to the median weight represented by the reference line – typical when mixtures occur.

In the fourth scenario, displayed in Figure 2.11, the *P*-value for clustering is 0.033, which is less than 0.05, so there is evidence of the presence of special cause variation affecting the corresponding mould. Two clusters – groups of points corresponding to bottles with similar weights – are indicated in the display.

A process team should respond to evidence of special cause variation by taking steps to carry out a root cause investigation in order to determine the extraneous factor or factors affecting process performance. Once any such factor or factors have been identified, steps may be taken to eliminate them. It should be noted that a signal of evidence of the presence of special cause variation from a run chart *P*-value less than 0.05 could arise purely by chance, even when a process is operating in a stable and predictable manner.

The reader is urged to tap the huge Minitab Help resource constantly. Further details are provided in Chapter 11, and the author suggests that it will be beneficial to refer to these details in parallel with study of this and later chapters. Returning to the Minitab session currently being described, it should be noted that use of **Edit** > **Copy Graph** enables a copy of the run chart to be copied and pasted into a document being prepared using word-processing software. Alternatively, **File** > **Save Graph As...** may be used to save the run chart in a variety of formats. Minimize the run chart and note how the text **run chart of Weight (g)** has appeared in the Session Window indicating that in the Minitab session to date a run chart of the weight data has been created.

2.1.2   Minitab projects and their components

View the Project Manager either by clicking on the icon second from the right on the Project Manager toolbar displayed in Figure 2.2, using the Project Manager icon at the bottom left of the screen or by using keystrokes (Ctrl + I). The display in Figure 2.12 results.

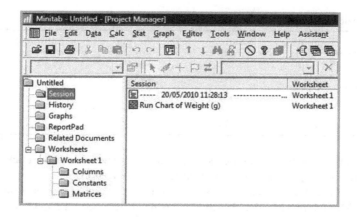

Figure 2.12   Project Manager.

In the top left-hand corner the word 'Untitled' indicates that the Project has not yet been named. The contents of the open Session folder indicate the date and time of the creation of Worksheet1 and the subsequent display of the data in the run chart. The run chart is in the Graphs folder. On opening the ReportPad folder, a report document may be created with appropriate text being entered as shown in Figure 2.13. Subsequently the run chart may be inserted into the ReportPad using **Edit** > **Paste** if the **Edit** > **Copy Graph** option was used

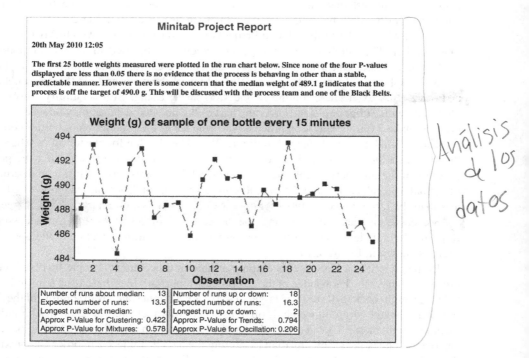

Figure 2.13   ReportPad showing text and run chart.

Table 2.2 Additional bottle weight data.

Sample	Weight	Sample	Weight	Sample	Weight
26	484.5	31	487.0	36	486.4
27	489.9	32	485.1	37	485.3
28	485.0	33	486.3	38	486.9
29	485.9	34	490.0	39	486.7
30	488.1	35	485.2	40	484.6

earlier. Alternatively, on right-clicking the active run chart a menu appears; clicking the option **Append Graph to Report** adds the chart directly to the ReportPad.

The typical final step in such a first session with data from a process is the naming and saving of the Minitab project file. To achieve this use **File** > **Save Project As. . .** to save the project with the name Weight in an appropriate folder. The project file will be created as Weight.MPJ, with the extension .MPJ indicating the file type as a Minitab project. **File** > **Exit** closes down Minitab – you should do just that and take a well-earned rest! To continue working with some other data in a new Minitab project, when one has finished work on a current project and saved it, use **File** > **New** > **Minitab Project**. One may use **File** > **New** > **Minitab Worksheet** to create additional worksheets within a project.

Imagine that a discussion took place with the process manager on concern about bottle weight being on target and that he consults a Six Sigma Black Belt, who does some further analysis of the available data using Minitab and reassures the process team that there is no evidence to suggest that the process is off target. As production of the batch of bottles continues, further data became available which are displayed in Table 2.2.

Launch Minitab. Use **File** > **Open Project** to open the project file Weight.MPJ created and saved earlier. The Toolbar at the top of the screen may be used to access components of the Project as indicated earlier (Figure 2.2). Click on the Current Data Window icon (or on the Show Worksheets Folder icon) to access the only worksheet currently in the project. Add the additional data to the first column of the worksheet. Using: **Stat** > **Quality Tools** > **Run Chart. . .** (with **Subgroup size:** 1), a run chart of the updated data set may now be created in order to make a further check on process performance as the production of bottles continues.

The updated chart is displayed in Figure 2.14. The P-value for clustering of 0.028 is less than 0.05, so therefore there is evidence of a possible special cause affecting the process. Scrutiny of the run chart reveals that the additional data points form a cluster, and scrutiny of the actual data values in Table 2.2 indicates that all but one of the additional bottles had weight less than the target value of 490 g. Thus it would appear that corrective action, to remove a special cause of variation affecting the process, could be necessary. On accessing the ReportPad via its icon one can type appropriate further comments and add the updated run chart as shown in Figure 2.14.

Bottle weight is a key process output variable in this context. People involved with the process will have the knowledge of the key process input variables that can be adjusted in order to bring weight back to the desired target of 490 g. One might state, from scrutiny of the second run chart, that the 'drop' in weight is obvious. There was a 'cluster' of 25 bottles initially with median weight of 489.1 g and a later cluster of 15 bottles with a median weight of 486.3 g. (You can readily check the second median by putting the data in Table 2.2 in order and picking out the middle value; the calculation of medians using Minitab will be covered later in the chapter.) However, the objective evaluation of evidence from data using sound statistical methods is preferable to subjective decision-making.

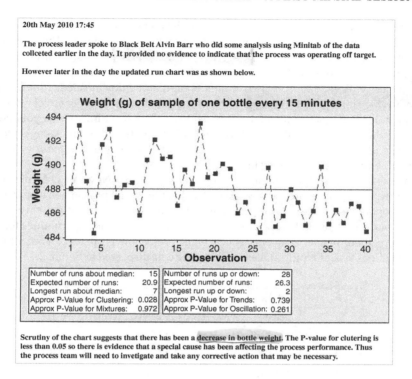

20th May 2010 17:45

The process leader spoke to Black Belt Alvin Barr who did some analysis using Minitab of the data colleted earlier in the day. It provided no evidence to indicate that the process was operating off target.

However later in the day the updated run chart was as shown below.

Scrutiny of the chart suggests that there has been a decrease in bottle weight. The P-value for clutering is less than 0.05 so there is evidence that a special cause has been affecting the process performance. Thus the process team will need to invetigate and take any corrective action that may be necessary.

Figure 2.14   ReportPad showing run chart of the 40 bottle weights and further comments.

Another useful source of information in Minitab is StatGuide™. Right-clicking on an active run chart and then clicking on **StatGuide** opens a window with a display of **Contents** on the left of the screen and with **Index** and **Search** tabs. On the right specific information on run charts is given. Arrows enable navigation around the topics provided, example output is given, and the location of the data used to create it is indicated together with interpretation. The **More** button leads to in-depth details of how the interpretation is made.

The reader will find further details of Minitab StatGuide in Chapter 11, and the author suggests that it will be beneficial to refer to these details in parallel with study of this and later chapters.

Having completed your work with the bottle weight data discussed above, the natural thing to do would be to save the updated Minitab project file Weight.MPJ using **File > Save Project.** Were the project 'for real' then the worksheet could be updated as new data became available and the ReportPad could be used as a dynamic document containing informative displays and analyses of the data and a log of any changes made to the process. Worksheets may be stored independently of projects, in the first instance, using **File > Save Current Worksheet As. . .**, and subsequently updates may be saved using **File > Save Current Worksheet.** It is recommended that you save the worksheet containing the 40 weights using the name Weight1. The worksheet file will be created as Weight1.MTW, with the extension .MTW indicating the file type as a Minitab worksheet.

The response variable, weight, considered above is an example of a *continuous random variable* in the jargon of statistics. When a bottle is weighed on a set of analog scales one can think of the possibility of the pointer coming to rest at any point on a continuous scale of measurement. For a second example of a run chart the number of incomplete invoices per day produced by the billing department of a company will be considered. The daily count of incomplete invoices is referred to as a *discrete random variable* in statistics. It should be noted that both weighing bottles and counting incomplete invoices are examples of measurement.

The majority of the data sets used in this book may be downloaded from the web site http://www.wiley.com/go/six_sigma in the form of Minitab worksheets or Excel workbooks. It is recommended that you download the files and store them in a directory on your computer. The data for this example, available in Invoices1.MTW, are from 'Finding assignable causes' by Bisgaard and Kulachi and are reproduced with permission from *Quality Engineering* (© 2000 American Society for Quality).

In order to create a new project for the invoice data use **File > New**, select **Minitab Project** and click **OK**. (Had you omitted to save the updated bottle weight project you would have been offered the option of doing so on clicking **OK**. It is strongly advised that you save projects as you work your way through this book as many data sets will provide opportunities for analysis using other methods in later chapters.) A new blank project file is opened. In order to save the reader the tedious task of typing in the initial invoice data displayed in Figure 2.15 the data can

↓	C1-D	C2	C3
	Date	No. Invoices	No. Incomplete
1	03/01/2000	98	20
2	04/01/2000	104	18
3	05/01/2000	97	14
4	06/01/2000	99	16
5	07/01/2000	97	13
6	10/01/2000	102	29
7	11/01/2000	104	21
8	12/01/2000	101	14
9	13/01/2000	55	6
10	14/01/2000	48	6
11	17/01/2000	50	7
12	18/01/2000	53	7
13	19/01/2000	56	9
14	20/01/2000	49	5
15	21/01/2000	56	8
16	24/01/2000	53	9
17	25/01/2000	52	9
18	26/01/2000	51	10
19	27/01/2000	52	9
20	28/01/2000	47	10

Figure 2.15   Initial invoice data.

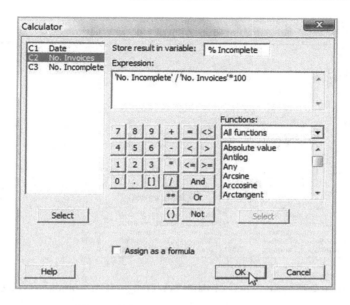

Figure 2.16   Calculator dialog box.

be entered into the project using **File > Open Worksheet**. Select the file Invoices1.MTW from the directory in which you have stored the downloaded data sets and click on **Open**. (You may be asked for confirmation that you wish to add a copy of the content of the worksheet to the current project, in which case it is necessary to click **OK**.)

There are three columns in the worksheet displayed in Figure 2.15. The first is labelled C1-D, indicating that it contains date data. The columns labelled C2 and C3, with no extensions, hold numerical data – the daily number of invoices processed and the daily number of invoices found to be incomplete. A run chart of the number of incomplete invoices per day could be misleading since the number of invoices processed daily varies. Thus there is a need to calculate the proportion of incomplete invoices per day. Using **Calc > Calculator...** gives access to the dialog box displayed in Figure 2.16 for performing calculations in Minitab.

In the **Store result in variable:** window an appropriate name for the new column of percentages to be calculated is entered; %Incomplete was used here. In the window labelled **Expression:** the formula may be created by highlighting the names of columns, using the **Select** button and the calculator keypad. Clicking **OK** implements the calculation. (Note that a menu of functions is available for use in more advanced calculations. Note too that if one checks the **Assign as a formula** box then whenever additional data are entered in the second and third columns the percentage incomplete will be calculated automatically. Columns that have been assigned a formula are indicated by a green cross at their heads.) The run chart of %Incomplete displayed in Figure 2.17 was obtained.

Moving the mouse pointer to the horizontal reference line, representing the median percentage incomplete on the chart, triggers display of a text box giving the median as 16.1% (to one decimal place). Thus the current performance of the invoicing process is such that approximately one in every six invoices is incomplete. The P-value for

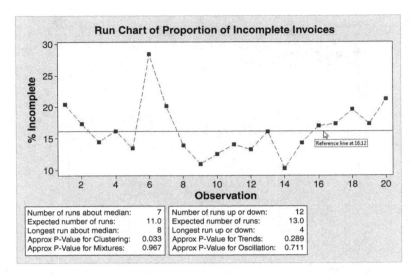

Figure 2.17   Run chart of daily percentage of incomplete invoices.

clustering is less than 0.05, thus providing evidence of the possible influence of a special cause on the process. The percentage for the sixth day appears to be considerably higher than all the other percentages – in fact, a new inexperienced employee processed many of the invoices during that day.

A median of 16.1% was unacceptably high, so a Six Sigma process improvement project was undertaken on the invoicing process, with some process changes being made almost immediately. The run chart in Figure 2.18 shows the data for the year 2000 up to

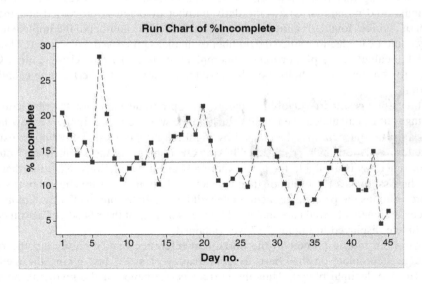

Figure 2.18   Updated run chart of daily percentage of incomplete invoices.

March 3rd. As an exercise the reader is invited to access, in a new project, the updated data stored in the worksheet Invoices2.MTW and re-create this run chart. In doing the calculation of percentage incomplete on this occasion, start to create the **Expression:** required by selecting the Round function under **Functions:**. Choose **All functions** from the menu (if not already on view), scroll down the list, highlight **Round** and click **Select**. The formula that appears in the **Expression:** window is ROUND(number,decimals). By highlighting the word 'number' in the expression, use may be made of highlighting and selection of column names and of the keypad to create the desired formula. The phrase 'decimals' may be highlighted and the digit 1 typed to indicate that rounding of the calculated percentages to one decimal place is required. The final version of the formula in the Expression window is ROUND('No. Incomplete'/'No. Invoices'*100,1). On clicking **OK** the proportions of incomplete invoices as percentages will be calculated and rounded to one decimal place.

The updated run chart in Figure 2.18 was simplified by deleting the two text boxes containing the *P*-values, by clicking on each and then pressing the delete key, and editing the default label for the horizontal axis from the default of **Observation** to **Day no.** in order to make the display less daunting for use in a presentation. Double-clicking an axis label yields an Edit Dialog Label box in which any desired label may be entered in the **Text:** window.

Scrutiny of the updated run chart suggests that the process changes have been effective in lowering the percentage of incomplete invoices. Two of the *P*-values are less than 0.05. This provides formal evidence of a process change having taken place. Alternative data displays, using methods discussed later in the chapter, may be used to highlight the apparent effectiveness of the process changes. The median for the period from 31 January 2000 onwards was 10.7% incomplete invoices per day, compared with 16.1% incomplete invoices per day for the earlier period.

2.2 Display and summary of univariate data

2.2.1 Histogram and distribution

Consider the bottle weight data displayed in the run chart in Figure 2.6. Here we recorded a single variable for each bottle so we refer to *univariate* data. (Had we measured weight and height we would have had *bivariate* data, had we measured weight, height, bore, out of vertical etc. then we would have been dealing with *multivariate* data.) The process appears to have been behaving in a stable, predictable manner during the period in which the data were collected. When a process exhibits this sort of behaviour and the measured response is a continuous variable, such as weight, then display of the data in the form of a histogram is legitimate and can be very informative.

Once a bottle has been weighed, imagine that it is put in one of the series of bins depicted in Figure 2.19 according to its measured weight. The lightest bottle recorded weighed 484.4 g and would be placed in the bin labelled 483.5, 484.5. The second lightest bottle weighed 485.4 g and would be placed in the bin labelled 484.5, 485.5. The next two lightest bottles weighed 485.9 g and 486.1 g and would be placed in the bin labelled 485.5, 486.5. The heaviest bottle weighed 494.5 g and would be placed in the bin labelled 493.5, 494.5. (The convention adopted in Minitab is that a bottle weighing 490.5 g is placed in the bin labelled 490.5, 491.5 unless it

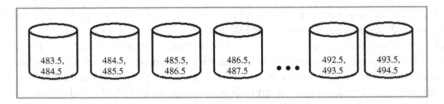

Figure 2.19 Bin concept.

was the heaviest bottle, in which case it would be placed in the bin labelled 489.5, 490.5.) For the complete sample of 25 bottles the number of bottles or observations in each bin is known as its *frequency*. The ranges 483.5–484.5, 484.5–485.5, 485.5–486.5 etc. are referred to as *intervals*. A chart with weight on the horizontal axis and frequency represented on the vertical axis by contiguous bars is a *histogram*.

In order to work through the creation of the histogram with Minitab you require the weights in Table 2.1 in a single column that may be named Weight (g). They are provided in worksheet Weight1A.MTW. To create the histogram with Minitab, use **Graph > Histogram...**. The initial part of the dialog is displayed in Figure 2.20. Accept the default option of **Simple** and click **OK** to access the subdialog box displayed in Figure 2.21.

In the **Graph variables:** window, select the variable to be displayed in the histogram, Weight (g) in this case, and click **OK**. The histogram displayed in Figure 2.22 indicates the distribution of the bottle weights, and three aspects of a distribution may be assessed using a

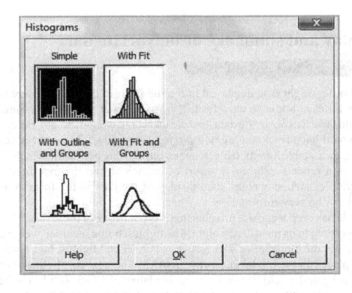

Figure 2.20 Initial part of dialog for creating a histogram.

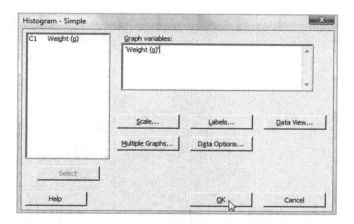

Figure 2.21   Histogram subdialog box.

histogram – *shape*, *location* and *variability*. Note that moving the mouse pointer to a bar of the histogram leads to the bin interval and frequency for that bar being displayed in a text box. In Figure 2.22 the mouse pointer is on the bar corresponding to the bin interval 488.5, 489.5. The frequency for this bin was 5, indicating that five of the 25 bottles in the sample had weight in the interval 488.5 g $\leq$ weight $<$ 489.5 g.

Before reading any further the reader is invited to create a histogram of the bottle weight data stored in the supplied worksheet Weight2.MTW. The histogram will be referred to below and is shown in Figure 2.23.

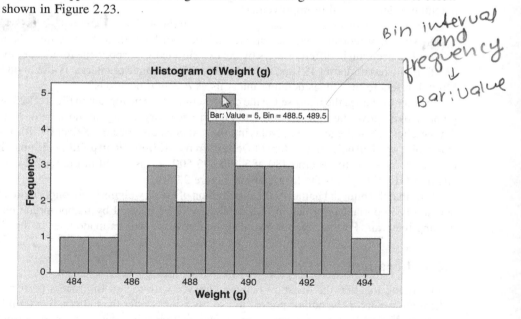

Figure 2.22   Histogram of bottle weight.

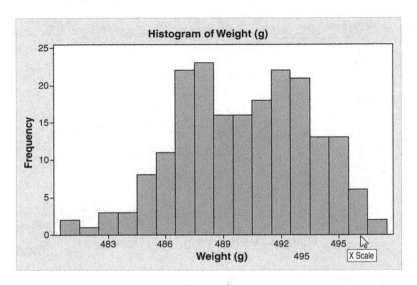

Figure 2.23   Bimodal histogram.

2.2.2 Shape of a distribution

In discussing the shape of a distribution one should consider whether or not the distribution is symmetrical, whether it is skewed to the right (showing upwards straggle) or skewed to the left (showing downwards straggle) or whether there are other features providing insights with potential to lead to quality improvement.

The histogram of weight for a sample of bottles in Figure 2.23 is bimodal – it has two major peaks. This may indicate that the sample includes bottles formed by two moulds, that the process has been run in different ways by the two shift teams responsible for its operation, etc. The third bin has midpoint 483 and the corresponding bin range or interval is 482.5–483.5, and Minitab refers to the values defining intervals as *cutpoint* positions.

In order to change the bins used in the construction of the histogram in Figure 2.23, with the graph active, move the mouse pointer to and, if necessary, along the horizontal X axis and double-click when the text box displaying the text **X Scale** appears. Select the **Binning** tab, **Interval Type Cutpoint** and **Interval Definition** by **Midpoint/Cutpoint positions:**. Editing the list of cutpoints to become 480 485 490 495 500, as displayed in Figure 2.24, and then clicking **OK, OK** yields the histogram in Figure 2.25.

Note that the bimodal nature of the distribution of bottle weights is no longer evident. Thus potentially important information in a data set may be masked by inappropriate choice of binning intervals. Further reference to distribution shape will be made later in the chapter.

2.2.3 Location (Central Tendency)

In discussing location the question being addressed with regard to process performance is 'where are we at?' (The author prefers to use 'location' rather than alternative term 'central tendency'.) Location gives an indication of what is typical in terms of process performance. Suppose that the weight data displayed in Figure 2.23 were actually for 100 bottles produced on each of two different moulding machines, A and B; that it is known which bottles were made

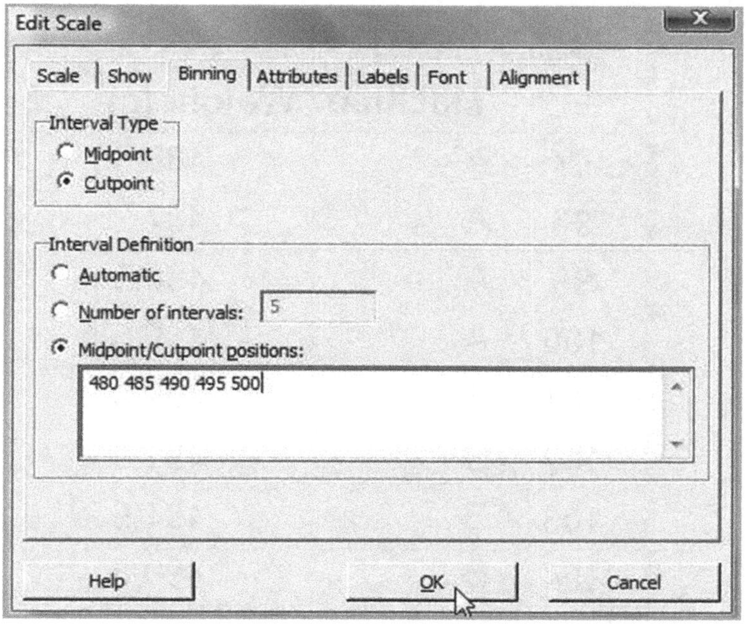

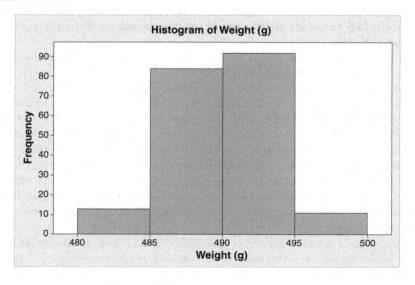

Figure 2.24 Changes to the bin cutpoint positions.

on each machine and that bottle weight for both was stable and predictable during the period when the samples were taken. Part of the supplied worksheet, Weight3.MTW, containing the data is displayed in Figure 2.26. It shows the final four bottle weights for machine A and the first four bottle weights for machine B.

Figure 2.25 Alternative histogram.

↓	C1-T	C2
	Machine	**Weight (g)**
97	A	488.5
98	A	487.2
99	A	486.5
100	A	485.5
101	B	492.0
102	B	491.6
103	B	494.5
104	B	493.8

Figure 2.26 Segment of bottle weight data for two machines.

Column C1 contains text values A and B indicating which of the two machines produced the bottle with weight recorded in column C2. Note the designation of the first column as C1-T, indicating that it stores text values. In order to create a histogram for each machine use **Graph > Histogram...** with the **Simple** option and select the variable to be graphed, **Weight (g)**. Click on **Multiple Graphs...** and **By Variables** and select Machine in the window labelled **By variables with groups in separate panels:** as displayed in Figure 2.27. Finally click **OK, OK**.

The two histograms are shown in Figure 2.28. The histogram for machine A is in the left-hand panel and that for machine B is in the right-hand panel. With the graph active, the **Edit Scale** menu was accessed by moving the mouse pointer to the horizontal X axis and double-clicking when the text box displaying the text **X Scale** appeared. The entries in the window labelled **Positions of ticks:** were changed to 480, 485, 490 and 495. The triangular markers were superimposed using the polygon tool from the Graph Annotation Tools toolbar, but the detail need not concern us here. These marks indicate the mean weight for each machine. The mean will be defined later in this chapter. The markers indicate the horizontal locations of the centroids of the histograms. Cut-outs of the histograms would balance on the knife-edges represented by the upper vertices of these triangles. In terms of the target weight of 490 g for the bottles, it is clear from the data display that both machines are operating off target.

The difference in location for the two machines and in their performance, relative to the target bottle weight of 490 g, may be highlighted as shown in Figure 2.29 with the histograms aligned vertically. Details of how to do this will be the subject of an exercise at the end of this chapter.

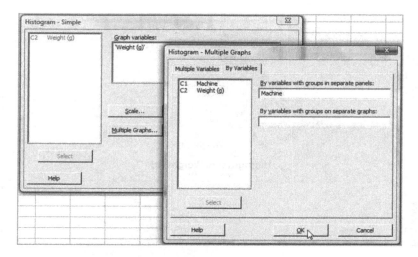

Figure 2.27   Segment of bottle weight data for two machines.

In addition to visual assessment of location from display of the data it is possible to measure location by calculation of descriptive or summary statistics. The median is a widely used measure of location and was referred to in the previous section in relation to run charts. The mean is a second widely used measure of location and is obtained by calculating the sum of data values in a sample and dividing by the sample size, i.e. by the number of data values. In common parlance many refer to the mean as the average. Calculation of the mean with associated statistical notation is given in Box 2.1.

The means of bottle weight for machines A and B are 487.24 g and 492.75 g, respectively. The triangular markers in Figure 2.29 are placed at these values on the appropriate scales. The

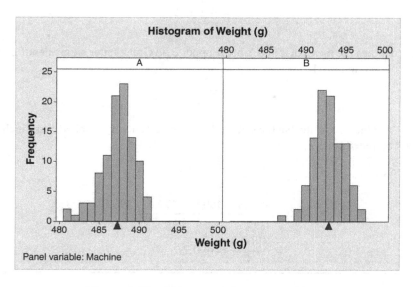

Figure 2.28   Histograms for two machines.

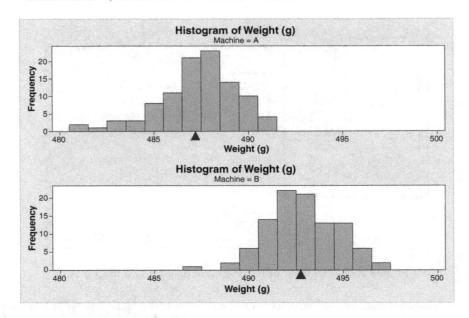

Figure 2.29   Histograms for two machines.

Consider a sample of four bottles with weights (g) 490.3, 489.9, 490.6 and 490.0. The sum of the four data values is 1960.8 and division by the sample size, 4, gives the mean weight 490.2. The mathematical shorthand for this calculation is as follows:

$$\bar{x} = \frac{\sum_{i=1}^{n} x_i}{n}$$

where $\bar{x}$ denotes the mean of x and Σ is the upper case Greek letter sigma denoting 'sum of'. For our sample of 4,

$$\bar{x} = \frac{\sum_{i=1}^{4} x_i}{4}$$

(the $i = 1$ and the 4 indicate that the sum of the measurements, x_i, labelled 1 to 4 inclusive, is to be computed)

$$\bar{x} = \frac{x_1 + x_2 + x_3 + x_4}{4}$$

$$= \frac{490.3 + 489.9 + 490.6 + 490.0}{4}$$

$$= \frac{1960.8}{4} = 490.2.$$

Box 2.1   Calculation of a mean.

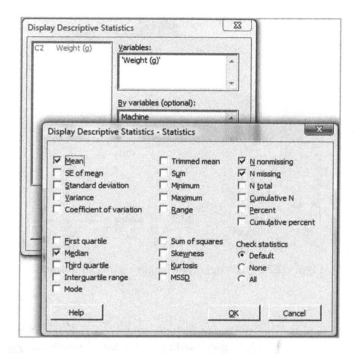

Figure 2.30 Obtaining Descriptive Statistics.

median bottle weights for machines A and B are 487.5 g and 492.7 g, respectively. With an even sample size of 100, the median is the mean of middle two weights when the data have been ordered. Note that the means and medians for each machine are very similar: 487.24 and 487.5 for A and 492.75 and 492.7 for B, respectively. This is typical when distributions (and associated histograms) are fairly symmetrical. In such cases it would not really matter which measure of location is used to summarize the data.

In order to obtain measures of location in Minitab use can be made of the **Stat** menu via the sequence **Stat > Basic Statistics > Display Descriptive Statistics....** The icon to the left of the **Display Descriptive Statistics...** text shows the two symbols $\bar{x}$ and s, representing mean and standard deviation, respectively. The standard deviation is a widely used measure of variability which will be introduced later in this chapter. Weight (g) is entered under **Variables:** and Machine entered in **By variables:**.

In order to obtain the mean and median for the two machines, select Weight (g) in the **Variables:** window and use the **Statistics** button to edit the list of available **Statistics** to the ones shown in Figure 2.30, i.e. **Mean, Median, N nonmissing** and **N missing**. On implementation of the procedure the output in Panel 2.1 appears in the Session Window.

Descriptive Statistics: Weight (g)					
Variable	Machine	N	N*	Mean	Median
Weight (g)	A	100	0	487.24	487.50
	B	100	0	492.75	492.70

Panel 2.1 Session window output from Descriptive Statistics.

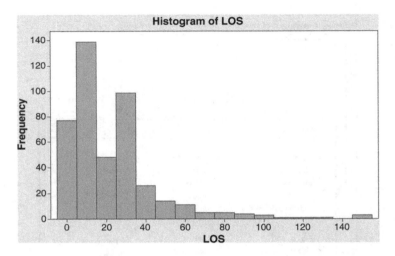

Figure 2.31   Histogram of length of stay in hospital.

The N column indicates that the data set includes values of weight for 100 bottles from each of the machines, A and B. The N* column indicates that no missing values were recorded – missing values are recorded in Minitab as asterisks. The final two columns give the means and medians. Note that, with the mouse pointer located in the Descriptive Statistics section in the Session window, a right click displays a pop-up menu through which access to StatGuide information on Descriptive Statistics may be obtained.

Consider now data on length of stay (days) in hospital (LOS) for stroke patients admitted to a major hospital during a year. The data are available in LOS.MTW. A histogram of the data is shown in Figure 2.31. Such data could be highly relevant during the measure phase of a Six Sigma project aimed at improving stroke care in the hospital.

The histogram is far from symmetrical. With the long tail to the right it exhibits what is known as *positive skewness* or *upward straggle*. (A histogram that had the shape of the mirror image of the one in Figure 2.31 in the vertical axis would be exhibit *negative skewness* or *downward straggle*.) Scrutiny of the bars indicates that the bins used are (−5, 5) (5, 15) (15, 25) etc., with midpoints 0, 10, 20 etc. Of course LOS cannot be a negative number, so a more logical set of bins would be (0, 10) (10, 20) (20, 30) etc. In order to modify the histogram select the **X Scale** as indicated in the previous section, under **Binning** select **Interval Type** as **Cutpoint**, select **Interval Definition** as **Midpoint/Cutpoint positions:** and complete the dialog box as shown in Figure 2.32. This gives the histogram in Figure 2.33.

The default descriptive statistics provided by Minitab for length of stay are shown in Panel 2.2. (SE Mean denotes the standard error of the mean, and Q1 and Q3 denote the first and third quartiles respectively. These statistics are explained later in the book.) As is typical with data exhibiting positive skewness, the median is less than the mean. As the term 'average' is used by some to mean measure of location it is important, in the case of skewness, to ascertain which measure of location is being quoted. In this case the median length of stay is approximately one week less than the mean.

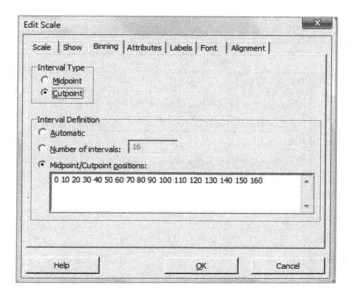

Figure 2.32   Specifying bins for a histogram using cut points.

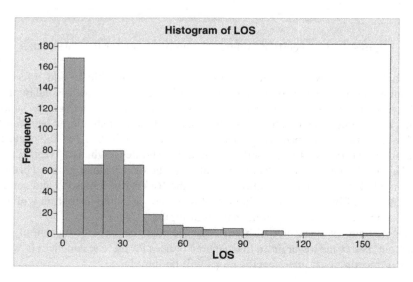

Figure 2.33   Modified histogram of length of stay in hospital.

Descriptive Statistics: LOS

Variable	N	N*	Mean	SE Mean	StDev	Minimum	Q1	Median	Q3	Maximum
LOS	437	0	22.39	1.12	23.35	1.00	6.00	15.00	31.00	154.00

Panel 2.2   Descriptive Statistics for length of stay.

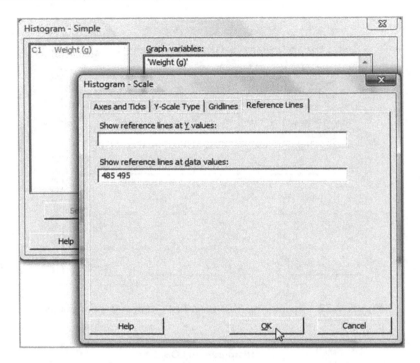

Figure 2.34   Specifying reference line positions for the data scale.

One final facility covered in this section is that of being able to add reference lines corresponding to values on the horizontal scale. This is useful for giving a visual impression of how well a process is performing in terms of customer requirements. For example, suppose that a customer of the bottle manufacturer specifies that bottle weight should lie between 485 and 495 g. Thus the customer is specifying a lower specification limit (LSL) of 485 g and an upper specification limit (USL) of 495 g. For the first sample of bottle weight data (Weight1A.MTW) presented in this chapter, having selected Weight (g) as the variable to be graphed in the form of a histogram, click on the **Scale...** button, then on the **Reference Lines** tab and complete the dialog as shown in Figure 2.34. Clicking **OK, OK** twice yields the histogram with reference lines indicating the specification limits shown in Figure 2.35.

The display indicates that not all bottles met the customer specification limits – there is some 'fall-out' below the lower limit. In Chapter 6 indices for the assessment of how capable a process is of meeting customer specifications will be introduced.

2.2.4 Variability

In discussing variability, or spread, one is addressing the question of how much variation there is in process performance. In order to introduce measures of variability, consider two samples of five bottles from two moulding machines, P and Q. Recall that in order to create a new Minitab project, when one has finished work on a current project and saved it, one may use **File > New > Minitab Project**, and that one may use **File > New > Minitab Worksheet** to create additional worksheets within a project. Set up the data as shown in Figure 2.36. On

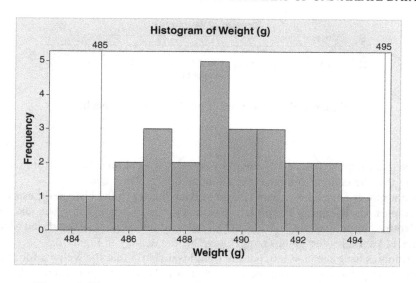

Figure 2.35   Bottle weight histogram with specification limits.

↓	C1-T	C2
	Machine	**Weight (g)**
1	P	488.3
2	P	491.9
3	P	489.6
4	P	487.7
5	P	492.5
6	Q	490.1
7	Q	490.2
8	Q	488.8
9	Q	491.6
10	Q	489.3

Figure 2.36   Bottle weight data from two machines.

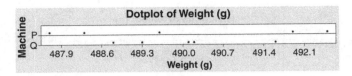

Figure 2.37 Dot plots of bottle weight by machine.

typing the letter P in the first cell of column 1 note that the column changes from C1 to C1-T, indicating that the column contains text as opposed to numeric data.

The dot plot is a useful alternative form of data display to the histogram, especially for small samples. Use **Graph > Dotplot…** to access the appropriate dialog box and select the **With Groups** option for **One Y**. Weight (g) is selected in **Graph variables:** and Machine in **Categorical variables for grouping:**. This yields the display in Figure 2.37.

Both samples have mean 490.0, but the samples differ in that the weights for machine P are more widely dispersed about the mean than are the weights for machine Q. There is greater variability, or spread, for weight in the case of machine P than in the case of machine Q. The reader is invited to verify that the default set of descriptive statistics for the two machines displayed in Panel 2.3 is obtained using **Stat > Basic Statistics > Display Descriptive Statistics…**.

One measure of variability is the *range*, i.e. the difference between the minimum value in the sample and the maximum value in the sample. Using the minimum and maximum values in Panel 2.3 gives the range for machine P to be 4.8 while that for machine Q is 2.8. The greater range for machine P indicates the greater variability in the weight of bottles produced on it than the variability in the weight of bottles produced on machine Q. The range has applications in control charts for measurement data. However, one criticism of the range as a measure of variability is that it only uses two measurements from all the measurements in the sample.

The standard deviations (StDev) given in Panel 2.3 for the two machines are 2.13 and 1.07 respectively. The greater standard deviation for machine P indicates the greater variability of bottle weight for it compared with the variability of bottle weight for machine Q. Detailed explanation of the calculation of standard deviation for machine P is given in Table 2.3.

```
Descriptive Statistics: Weight (g)

Variable    Machine N  N*    Mean SE Mean StDeV Minimun      Q1 Median
Weight (g) P          5   0 490.00   0.954  2.13  487.70 488.00 489.60
           Q          5   0 490.00   0.476  1.07  488.80 489.05 490.10

Variable    Machine      Q3 Maximum
Weight (g) P         492.20   492.50
           Q         490.90   491.60
```

Panel 2.3 Descriptive Statistics for machines P and Q.

Table 2.3 Calculation of measures of variability for weight (Machine P).

Bottle number	Weight x_i	Deviation $x_i - \bar{x}$	Absolute deviation	Squared deviation
1	488.3	−1.7	1.7	2.89
2	491.9	1.9	1.9	3.61
3	489.6	−0.4	0.4	0.16
4	487.7	−2.3	2.3	5.29
5	492.5	2.5	2.5	6.25
Total	**2450.0**	**0**	**8.80**	**18.20**
Mean	**490.0**	**0**	**1.76**	**3.64**

It is widespread practice to use the symbol x_i for a typical data value so that, for example, x_3 represents the weight of the third bottle in the sample, namely 489.6 g. The mean weight, $\bar{x}$, is 490.0 so the deviation of the third bottle weight from the mean is $x_3 - \bar{x} = 489.6 - 490.0 = -0.4$. This indicates that the third bottle had a weight 0.4 g below the mean. Similarly, for example, the fifth bottle had weight 2.5 g above the mean. The deviations from the mean always sum to zero.

The absolute deviation ignores the sign of the deviation and simply indicates by how much each measurement deviates from the mean. The mean absolute deviation (MAD) is 1.76 g for machine P. The reader is invited to verify that the corresponding value for machine Q is 0.76. The greater mean absolute deviation for machine P indicates the greater variability for it compared with that for machine Q.

Although MAD is a perfectly viable measure of variability it has disadvantages from a mathematical point of view. An alternative approach to taking the absolute values of deviations is to square the deviations and to take the mean of the squared deviations as a measure of variability. The mean squared deviation (MSD) for machine P is 3.64 while that for machine Q is 0.91. Once again the greater mean squared deviation for machine P indicates the greater variability for it compared with that for machine Q. However, there are two disadvantages with MSD. First, since the deviations are in units of grams (g) the squared deviations are in units of grams squared (g²). Second, samples are generally taken from populations in order to estimate character- istics of the populations, but statistical theory shows that MSD from sample data underestimates MSD for the population sampled. For example, in the case of a production run of bottles the population of interest would be all bottles produced during that particular run.

An important measure of variability is sample *variance*, which is calculated as the sum of squared deviations divided by the number which is one less than the sample size. Thus for machine P the variance is given by 18.20/4 = 4.55. Finally, in order to get back to the original units, the sample *standard deviation* is obtained by taking the square root of the variance. This yields a standard deviation of 2.13 g for machine P as displayed in the Minitab descriptive statistics in Panel 2.3. The reader is invited to verify that the standard deviation for machine Q is 1.07 g. The main point is that variance and standard deviation are very important measures of variability – the technical details of the underlying calculations are not important.

Table 2.4   Ratios of measures of variability.

Measure of variability	Machine P	Machine Q	Ratio
Range	4.8	2.8	1.7
Mean absolute deviation	1.76	0.76	2.3
Standard deviation	2.13	1.07	2.0

Table 2.4 gives the ratios of the measures of variability that have units of weight for the two machines. In broad terms all three measures indicate that the variability or spread of the weights for machine P is approximately twice that for machine Q. Small artificial samples were used here for illustrative purposes. One would be very wary of claiming that there is a real difference in variability in the weights of bottles produced on the two machines on the basis of such small samples. Readers who wish may examine the mathematics of the standard deviation in Box 2.2.

Consider again the sample of four bottles, referred to in Box 2.1, with weights (g) 490.3, 489.9, 490.6 and 490.0. The mean $\bar{x}$ is 490.2. The mathematical shorthand for the calculation of the sample variance, s^2, and standard deviation s, is:

$$s^2 = \frac{\sum_{i=1}^{n}(x_i - \bar{x})^2}{n-1}$$

$$= \frac{\sum_{i=1}^{4}(x_i - \bar{x})^2}{4-1}$$

$$= \frac{(x_1 - \bar{x})^2 + (x_2 - \bar{x})^2 + (x_3 - \bar{x})^2 + (x_4 - \bar{x})^2}{3}$$

$$= \frac{(490.3 - 490.2)^2 + (489.9 - 490.2)^2 + (490.6 - 490.2)^2 + (490.0 - 490.2)^2}{3}$$

$$= \frac{0.1^2 + (-0.3)^2 + 0.4^2 + (-0.2)^2}{3}$$

$$= \frac{0.01 + 0.09 + 0.16 + 0.04}{3}$$

$$= \frac{0.3}{3}$$

$$= 0.1$$

$$s = \sqrt{0.1} = 0.316.$$

Box 2.2   Calculation of a standard deviation.

The reader is invited to set the data from Box 2.2 up in Minitab and to check the values obtained for the mean and standard deviation. Population mean and population standard deviation are widely denoted by the Greek symbols μ and σ, respectively. The sample mean and the sample standard deviation are generally denoted by $\bar{x}$ and s, respectively. The sample values $\bar{x}$ and s may be considered to provide estimators of the corresponding population parameters μ and σ. In statistics, a fairly general convention is to denote population values (parameters) by Greek letters and sample values (statistics) by Latin letters.

2.3 Data input, output, manipulation and management

2.3.1 Data input and output

Earlier in this chapter we saw the three types of data used in Minitab – *numeric, text* and *date/time*. Data may be stored in the form of columns, constants or matrices. The latter two forms will be introduced later in the book. The key scenario is one of variables in columns and cases in rows stored in a worksheet. The fundamental method of data entry is via the keyboard directly into the Data window. Once data have been entered in this way they may be stored using **File > Save Current Worksheet As...** in the case of a new worksheet or **File > Save Current Worksheet** in the case where further data has been added or changes made to an existing worksheet. Data may also be accessed via Minitab worksheet and project files created previously or from files of other types. One may use **File > Print Worksheet...** to obtain output, on paper, of the data in an active worksheet.

In order to introduce aspects of both data input and manipulation consider the data displayed in Figure 2.38 that are available as the Excel workbook Failures1.xls. It gives, for a period of 6 weeks, the number of units per day that fail to pass final inspection in a manufacturing operation that operates 24 hours a day, 7 days a week. The data set used in this example is relatively small, as are those in examples to follow, and some tasks carried out using Minitab could be done more quickly by simply retyping the data in a new worksheet! However the aim is to use small data sets to illustrate useful facilities in Minitab for the input and manipulation of data.

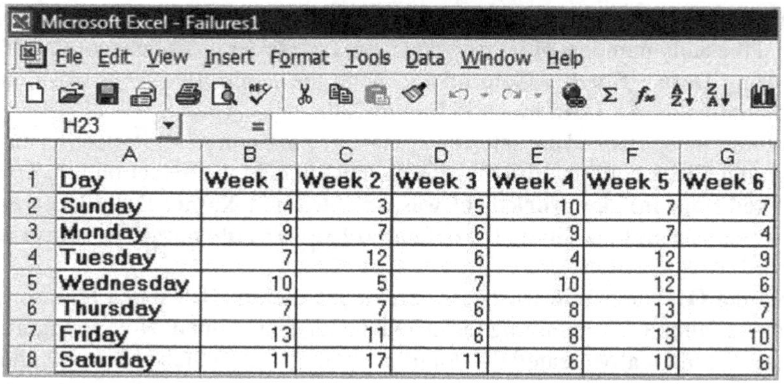

	A	B	C	D	E	F	G
1	Day	Week 1	Week 2	Week 3	Week 4	Week 5	Week 6
2	Sunday	4	3	5	10	7	7
3	Monday	9	7	6	9	7	4
4	Tuesday	7	12	6	4	12	9
5	Wednesday	10	5	7	10	12	6
6	Thursday	7	7	6	8	13	7
7	Friday	13	11	6	8	13	10
8	Saturday	11	17	11	6	10	6

Figure 2.38 Spreadsheet data.

Figure 2.39   Dialog for opening an Excel workbook as a Minitab worksheet.

In order to analyse the data in Minitab the first step is to use **File > Open Worksheet. . ..** As indicated in Figure 2.39, use **Files of type:** to select **Excel(*.xls; *.xlsx)**, and use **File name:** to select the required file. Clicking **Open** enters the data into the Data window. Until **Excel(*.xls; *.xlsx)** has been selected you will not see any Excel workbook files listed. (Note the list of file types catered for. Minitab can directly read and write Minitab portable, Excel, Spreadsheet XML, Quattro Pro, Lotus 1-2-3, and dBase files and text files, with extensions.txt or.csv, or data files with extension.dat.)

2.3.2   Stacking and unstacking of data; changing data type and coding

In order to create a run chart of the number of units per day that fail to pass final inspection it is necessary to stack the blocks of values in columns C2 to C7 of the Minitab worksheet that correspond to columns B to G of the Excel worksheet displayed in Figure 2.38 on top of each other so that the daily numbers of failures appear in time order in a single column. This can be achieved using **Data > Stack > Columns. . ..** Note the descriptive icons positioned beside many of the items on the **Data** menu in Figure 2.40.

The columns to be stacked are selected as indicated in Figure 2.41. Selection may be made by highlighting all six columns simultaneously and clicking the **Select** button. The option to **Store stacked data in:New worksheet** was accepted and **Name:** Daily Failures specified for the new worksheet. In addition, the default to **Use variable names in subscript column** was accepted.

On clicking **OK** the new worksheet is created. Column C1 is named Subscripts by the software and column C2, containing the stacked data, is unnamed. Note in Figure 2.42 how Minitab has automatically created the column of subscripts, which are simply the names of the columns containing the failure counts in the original worksheet. Figure 2.42 shows a portion of the new worksheet with column C1 renamed **Week** and column C2 named

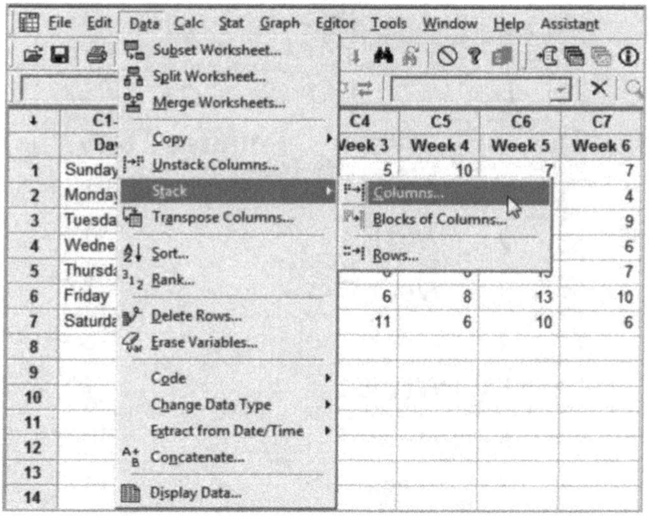

Figure 2.40   Selecting **Stack** > **Columns...** from the **Data** menu.

Failures/Day. (Note that the dialog displayed in Figure 2.41 offers the option of creating the stacked column in the current worksheet, either with or without a column of subscripts.) The data are now arranged in time order and a run chart may be created – this is left as an exercise for the reader.

The Excel workbook Failures2.xls gives the same data in an alternative format as shown in Figure 2.43. Having opened the Excel workbook as a Minitab worksheet, one may create a column with the data in time order in the same worksheet. Use is required of **Data** > **Stack** > **Rows...** with dialog as displayed in Figure 2.44. **Store stacked data in:** C12 indicates that the stacked data are to be stored in column C12. The option **Store row subscripts in:** was checked

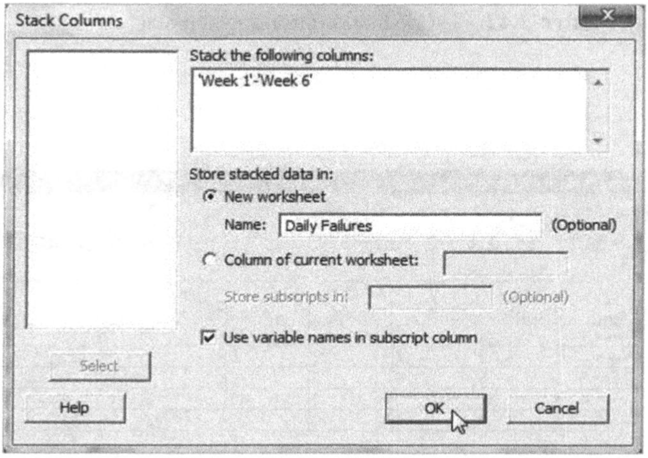

Figure 2.41   Dialog for Stack Columns.

Daily Failures ***	C1-T	C2
↓	Week	Failures / Day
1	Week 1	4
2	Week 1	9
3	Week 1	7
4	Week 1	10
5	Week 1	7
6	Week 1	13
7	Week 1	11
8	Week 2	3
9	Week 2	7
10	Week 2	12
11	Week 2	5
12	Week 2	7
13	Week 2	11
14	Week 2	17
15	Week 3	5

Figure 2.42   Worksheet containing the stacked data.

Microsoft Excel - Failures2

File Edit View Insert Format Tools Data Window Help

Q21

	A	B	C	D	E	F	G	H
1	Day	Sunday	Monday	Tuesday	Wednesda	Thursday	Friday	Saturday
2	Week 1	4	9	7	10	7	13	11
3	Week 2	3	7	12	5	7	11	17
4	Week 3	5	6	6	7	6	6	11
5	Week 4	10	9	4	10	8	8	6
6	Week 5	7	7	12	12	13	13	10
7	Week 6	7	4	9	6	7	10	6

Figure 2.43   Alternative layout for failure data.

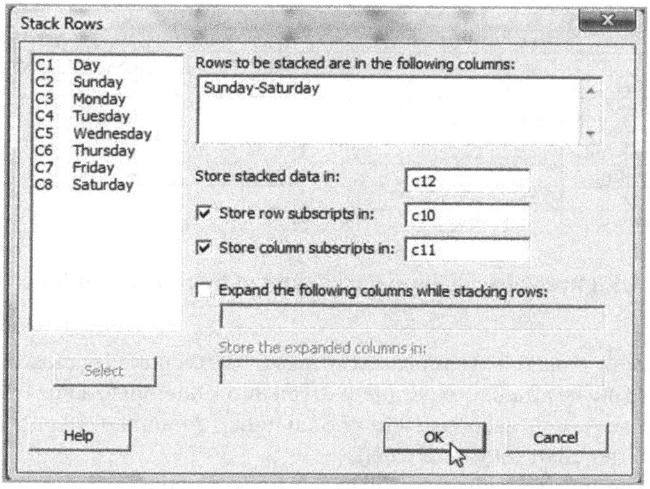

Figure 2.44   Dialog for stacking rows.

and the column C10 specified in the window; the option **Store column subscripts in:** was checked and the column C11 specified in the window.

The new columns require naming. Day is already in use as a column name in the worksheet so the name Day of week was used for the new day column. Column names cannot be duplicated in Minitab. A portion of the stacked data is shown in Figure 2.45. (The allocation of columns for storage of the stacked data and the subscripts in the order C12, C10 and C11 will now appear logical! Column names could have been entered directly during the dialog. Names,

C10	C11-T	C12
Week	Day of week	Failures / day
1	Sunday	4
1	Monday	9
1	Tuesday	7
1	Wednesday	10
1	Thursday	7
1	Friday	13
1	Saturday	11
2	Sunday	3
2	Monday	7
2	Tuesday	12

Figure 2.45   Portion of the stacked data.

	A	B	C	D	E	F	G	H	I	J	K	L	M
1	Day	Wk1	Wk1	Wk2	Wk2	Wk3	Wk3	Wk4	Wk4	Wk5	Wk5	Wk6	Wk6
2	Line	A	B	A	B	A	B	A	B	A	B	A	B
3	Sunday	2	2	1	2	2	3	4	6	2	5	3	4
4	Monday	3	6	3	4	2	4	3	6	3	4	0	4
5	Tuesday	1	6	4	8	1	5	3	1	5	7	3	6
6	Wednesday	4	6	3	2	3	4	4	6	6	6	2	4
7	Thursday	2	5	2	5	4	2	3	5	5	8	3	4
8	Friday	4	9	4	7	2	4	4	4	4	9	4	6
9	Saturday	3	8	4	13	4	7	2	4	5	5	2	4

Figure 2.46   Failure data stratified by production line.

such as Day of week, that are not simple text strings must be entered enclosed in single quotes. The reader should not be afraid to experiment – if an initial attempt to achieve an objective fails then try again or seek assistance via Help or StatGuide.) Again the data are now arranged in time order and a run chart may be created.

Suppose that production actually involves two lines, A and B, and that the data stratifies by line as shown in the Excel worksheet displayed in Figure 2.46. The data are available in the Excel workbook Failures3.xls.

On opening the workbook as a Minitab worksheet the data appear as shown in Figure 2.47. Note how the software names the two columns labelled Wk1 in the Excel spreadsheet as Wk1 and Wk1_1 in order to have unique Minitab column names. We are faced with the problem of having the first row containing the text values A and B and that therefore Minitab 'sees' the columns containing the numerical data we wish to analyse as text columns. This is indicated by C2-T, C3-T, C4-T etc.

The first step in overcoming this is to highlight the entire first row of worksheet entries by clicking on the row number 1 at the left-hand side of the worksheet. On doing this the row will appear as in Figure 2.47. Use of **Edit** > **Delete Cells** will delete the entire row of unwanted text entries. The next step involves use of the facilities for changing data types available via the **Data** menu. The six types of changes that may be made are indicated in Figure 2.48. Here we require a change of data type from text to numeric. **Data** > **Change Data Type** > **Text to Numeric. . .** gives the dialog box displayed in Figure 2.49. The 12 text columns containing the numerical data are specified in **Change text columns:** and the same column names are specified in **Store numeric columns in:**.

Having changed the data type to numeric, C2-T becomes C2 etc. Next we need to use **Data** > **Stack** > **Blocks of Columns. . .** to stack the numeric columns in blocks of two, corresponding to the two production lines, A and B. The completed dialog box is shown in Figure 2.50. Each block of two columns corresponds to one week's data.

↓	C1-T	C2-T	C3-T	C4-T	C5-T	C6-T	C7-T	C8-T	C9-T	C10-T	C11-T	C12-T	C13-T
	Day	Wk1	Wk1_1	Wk2	Wk2_1	Wk3	Wk3_1	Wk4	Wk4_1	Wk5	Wk5_1	Wk6	Wk6_1
1	Line	A	B	A	B	A	B	A	B	A	B	A	B
2	Sunday	2	2	1	2	2	3	4	6	2	5	3	4
3	Monday	3	6	3	4	2	4	3	6	3	4	0	4
4	Tuesday	1	6	4	8	1	5	3	1	5	7	3	6
5	Wednesday	4	6	3	2	3	4	4	6	6	6	2	4
6	Thursday	2	5	2	5	4	2	3	5	5	8	3	4
7	Friday	4	9	4	7	2	4	4	4	4	9	4	6
8	Saturday	3	8	4	13	4	7	2	4	5	5	2	4

Figure 2.47   Stratified data in Minitab.

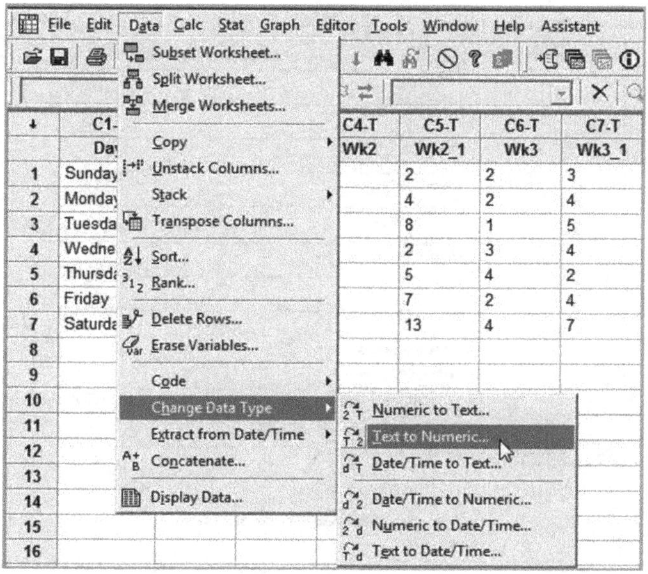

Figure 2.48   Changing data type from text to numeric.

The six blocks of columns to be stacked are specified under **Stack two or more blocks of columns on top of each other:**. In the dialog box shown in Figure 2.50 the default option of storing the stacked data in a new worksheet has been accepted and the name Failures Lines A & B has been specified for it. The default option to **Use variables in subscript column** has also been accepted. This subscript column will appear to the left of a pair of columns, the first of which will contain the sequence of daily failure counts for line A and the second those for line B. These may then be named **Week**, **Line A** and **Line B** respectively.

It would be informative to see a display of run charts for both lines on the same diagram. One may use **Stat > Time Series > Time Series Plot. . .** to create such a display. Choose the **Multiple** option, select the two columns containing the data to be plotted in the **Series:**

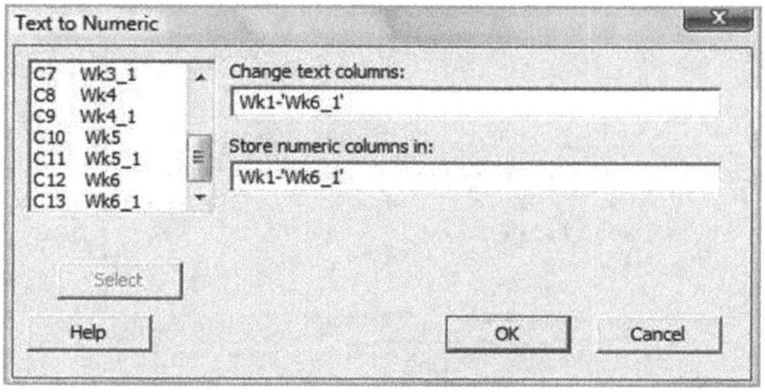

Figure 2.49   Changing text columns to numeric.

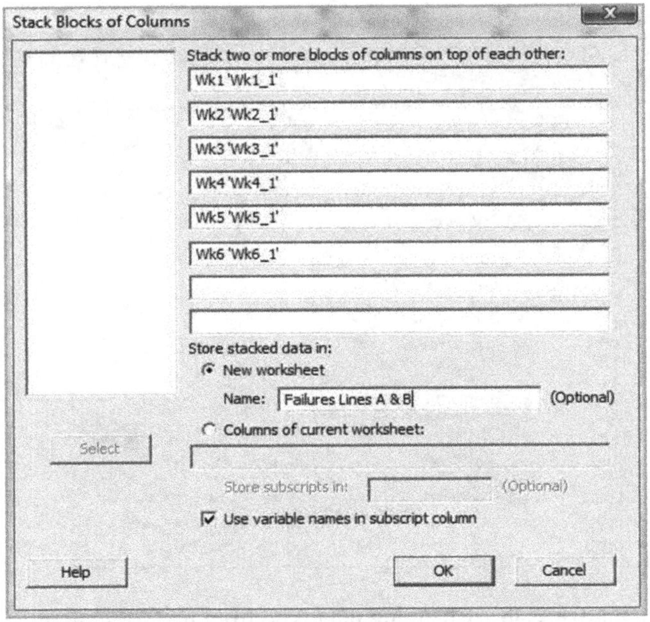

Figure 2.50   Procedure to stack the six blocks of pairs of columns.

window and insert a suitable title in the **Title:** window under **Labels. . . .** Double-clicking on each of the axis labels enables the labels to be edited appropriately. The plot is displayed in Figure 2.51.

Clearly line B has a higher daily rate of failures than line A. You should verify that the medians are 3 and 5 failures per day for lines A and B, respectively. The stratification of the

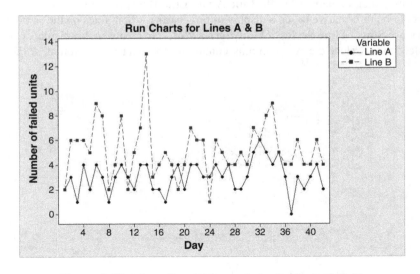

Figure 2.51   Superimposed run charts for lines A & B.

↓	C1	C2	C3	C4	C5	C6	C7	C8
	Pulse1	Pulse2	Ran	Smokes	Sex	Height	Weight	Activity
83	68	68	2	2	2	69.00	150	2
84	72	68	2	2	2	68.00	110	2
85	82	80	2	2	2	63.00	116	1
86	76	76	2	1	2	62.00	108	3
87	87	84	2	2	2	63.00	95	3
88	90	92	2	1	2	64.00	125	1
89	78	80	2	2	2	68.00	133	1
90	68	68	2	2	2	62.00	110	2
91	86	84	2	2	2	67.00	150	3
92	76	76	2	2	2	61.75	108	2
93								

Figure 2.52   Portion of the pulse data set.

daily rate of failures by production line has provided the insight that the performance of line B is worse than that of line A. Given that the lines have identical production capacity, quality improvement could potentially be achieved through investigation of factors contributing to the poorer performance of line B.

A series of data sets that are referred to in examples provided via Help are provided in the Minitab Sample Data folder (typically located in the folder C:\Program Files\Minitab\ Minitab16\English). Alternatively, the folder may be accessed by selecting **File** > **Open Worksheet...** and then clicking on the icon labelled **Look in Minitab Sample Data folder** that appears near the foot of the Open Worksheet dialog box. In order to illustrate further aspects of data manipulation the reader is invited to open the worksheet Pulse.MTW from the Sample Data folder. The worksheet contains data for a group of 92 students. In an introductory statistics class, the group took part in an experiment. Each student recorded their height, weight, gender, smoking habit, usual activity level, and pulse rate at rest. Then they all tossed coins; those whose coins came up heads were asked to run on the spot for a minute. Finally, the entire class recorded their pulse rates for a second time. The data for the final ten students are shown in Figure 2.52.

In this data set codes are used for gender in the column labelled Sex, with 1 representing male and 2 representing female. Before analysing a data set it is always wise to check for any unusual values appearing because of errors in data entry etc. One simple check would be a tally of the values appearing in the Sex column. This can be achieved using **Stat** > **Tables** > **Tally Individual Variables...**. By default **Counts** are given but the Session window output in Panel 2.4 was achieved by also checking the **Percents** box. Thus there were 92 students in the class, of whom 57 were male, i.e. 62% (to the nearest whole number). Had

```
Tally for Discrete Variables: Sex

Sex   Count   Percent
  1      57     61.96
  2      35     38.04
 N=      92
```

Panel 2.4   Counts of males and females.

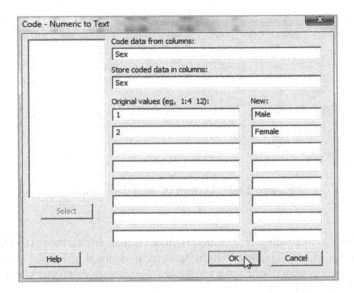

Figure 2.53   Changing numeric codes to text form.

values such as 0 or 3 for Sex appeared then that would have indicated an error in the data or in the data input.

Suppose we wish to replace the numerical codes 1 and 2 with the words Male and Female respectively in the Sex column. This can be achieved using **Data** > **Code** > **Numeric to Text. . ..** One reason for using text rather than numerical values, for example, is that displays of the data can be created having more user-friendly labels. The dialog is shown in Figure 2.53. **Code data from columns:** Sex and **Store coded data in columns:** Sex means that the coded text values will be stored in the same column of the worksheet as the original numerical codes. Under **Original values:** note that 1 and 2 have been entered, with the corresponding replacement codes of Male and Female respectively specified under **New:**.

If a separate worksheet is required of the data for females then this can be achieved using **Data** > **Unstack Columns. . ..** All eight columns were selected as the source of the data to be unstacked in **Unstack the data in:**. (This may be done by highlighting all eight variables and clicking the **Select** button.) The subscripts to be used for unstacking are in the Sex column and this is indicated via **Using subscripts in:**. In the dialog box in Figure 2.54, the default option **Store unstacked data in a new Worksheet** has been accepted with **Name:** Females specified. The default option **Name the columns containing the unstacked data** was accepted. Thus names would automatically be assigned to the columns containing the unstacked data.

The first eight columns of the new worksheet contain the data for the females while the second eight columns contain the data for the males. Note the column headings such as Pulse1_Female in the new worksheet. Minitab has used the subscripts employed for unstacking the original data to extend the original column names appropriately. To delete the data for males from the new worksheet, **Data** > **Erase Variables. . .** may be used, with all names ending in _Male being selected. (Alternatively a left click on the cell containing the text C9, keeping the mouse button depressed and scrolling across to the cell containing the text C16

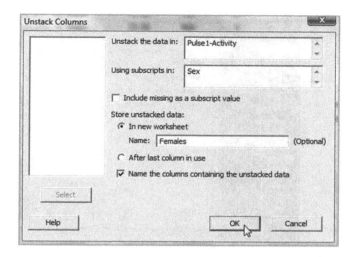

Figure 2.54   Unstacking columns to separate the male and female data.

leads to all the unwanted columns being blacked out. Releasing the mouse button and pressing the delete key completes the operation.) The redundant fifth column may also be deleted by clicking on the cell containing C5-T and using **Edit** > **Delete Cells**. Finally, double-clicking each of the remaining column names in turn enables them to be edited to their original form as shown in Figure 2.55.

Suppose that it is required to have the columns in the order Height, Weight, Activity, Smokes, Pulse1, Ran and Pulse2. The adjacent Height and Weight columns may be moved first as follows. Click on C5 at the head of the Height column, keep the mouse button depressed and drag across to C6 so that the Height and Weight columns are highlighted as shown in Figure 2.56. Choose **Editor** > **Move Columns...**, check **Before column C1** and click **OK**.

Further use of the facility for moving columns yields the worksheet in the desired format. The worksheet may then be saved using **File** > **Save Current Worksheet...**. Under **Save as type:** the default is Minitab. If this option were selected then the worksheet could be saved, for example, as **Females.MTW**. (The reader should note the other available file types and observe, when using Windows Explorer, the subtle difference between the icons used for Minitab project and worksheet files.)

In this section a number of methods of data acquisition in Minitab have been considered. There are situations where the capture of data in real time is of interest. Though Minitab was

Females ***							
↓	C1	C2	C3	C4	C5	C6	C7
	Pulse1	Pulse2	Ran	Smokes	Height	Weight	Activity
1	96	140	1	2	61.00	140	2
2	62	100	1	2	66.00	120	2
3	78	104	1	1	68.00	130	2
4	82	100	1	2	68.00	138	2
5	100	115	1	1	63.00	121	2

Figure 2.55   Section of the data for females.

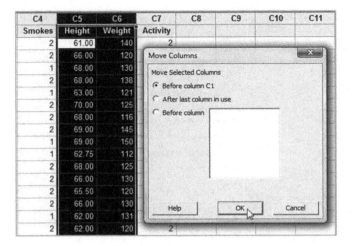

Figure 2.56   Moving columns.

not designed to capture data in real time, software is available to acquire data in real time from devices and transmit it to Minitab. Further information may be obtained from the Minitab website (http://www.minitab.com/en-GB/support/answers/answer.aspx?id=918).

2.3.3   Case study demonstrating ranking, sorting and extraction of information from date/time data

As a final example in this chapter we will consider data for patients referred for a colposcopy at a major hospital. The data cover the period from October 2001 to May 2003. A clinical improvement project was commenced in September 2002 in order to improve waiting times for the patients. The data are stored in the supplied Excel spreadsheet Colposcopy.xls and are reproduced by permission of NHS Lothian and the Colposcopy Services Clinical Improvement Project Team at the Royal Infirmary of Edinburgh, led by Sister Audrey Burnside. On opening the file as a Minitab worksheet it appears as displayed in Figure 2.57.

The first column is the patient reference number, the second gives the date on which the need for an outpatient appointment for a colposcopy was established and the third gives the date on which the procedure was actually carried out. Note that Minitab has recognized the data in the second and third columns as dates; this is indicated by the column headings C2-D and C3-D. The * symbol in the third column for the seventh patient is the missing value code for numeric data in Minitab – the corresponding cell in the Excel spreadsheet is blank. The aim of the example is to demonstrate how to create a run chart of monthly means of waiting times to indicate process performance before and after process changes. Before creating the run chart, screening of the data for anomalous values will be carried out. This process introduces a number of important facilities in Minitab for the manipulation of data.

2.3.3.1   Step 1. Calculation of waiting times

Calculations may be performed on data in the form of dates. Use **Calc** > **Calculator...** to calculate the waiting time in days for each patient as indicated in Figure 2.58. **Assign as a**

↓	C1	C2-D	C3-D
	Patient No.	**Referral Date**	**Colposcopy Date**
1	1	01/10/2001	26/10/2001
2	2	01/10/2001	03/01/2002
3	3	01/10/2001	08/01/2002
4	4	01/10/2001	20/11/2001
5	5	01/10/2001	20/12/2001
6	6	01/10/2001	29/01/2002
7	7	01/10/2001	*
8	8	01/10/2001	18/02/2002
9	9	01/10/2001	11/12/2001
10	10	01/10/2001	06/12/2001
11	11	01/10/2001	24/01/2002
12	12	01/10/2001	04/02/2002

Figure 2.57   Portion of the colposcopy data.

formula was checked – the reason will be explained below. Note how the waiting time for the first patient was 25 days and that the waiting time for the seventh patient is, of course, a missing value. Note, too, the small green cross at the head of the column of Wait values, indicating that the formula used in the calculation has been assigned to the cells in the column. This means that should any dates in the second or third columns be changed, Wait will be automatically

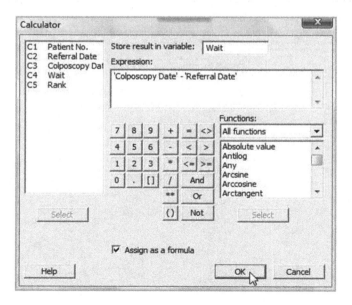

Figure 2.58   Calculating patient waiting time.

recalculated as in spreadsheet software. In addition, if dates for further patients were to be added then their Wait values would be calculated automatically.

2.3.3.2   Step2. Screening the data for anomalous values

Firstly, use can be made of **Data > Rank...** to rank the values of Wait from lowest to highest. In the dialog box enter **Rank data in:** Wait and **Store ranks in:** Rank. Note that the Wait of 25 days for the first patient is ranked 379. When two or more patients have the same value for Wait then the mean of the corresponding ranks is allocated as Rank for these patients.

Next, use can be made of **Data > Sort...** to sort the waiting times in ascending order and to store the sorted data in a second worksheet. The dialog is shown in Figure 2.59. Note that all five columns should be selected in the **Sort columns(s):** window and that sorting **By column:** Rank is specified. Sorting in ascending order is the default so **Descending** is left unchecked. **Store sorted data in:** Ordered data was used to name the new worksheet. The first few rows of the sorted data are shown in Figure 2.60.

Scrutiny of the ranked data reveals errors. For example the lowest Wait value, with Rank value 1, was −4. This is impossible, so the data for patient number 1725 require checking. There were 19 patients with Wait values 0, which means that they underwent the procedure on the day that it was deemed necessary – an occurrence likely in situations where clinicians suspected a serious situation for the patient. These 19 patients would be assigned rank values ranging from 2 to 20 inclusive. These values sum to 209, which on division by 19 yields 11. Thus the rank assigned to each of these 19 patients is 11. The 21st and 22nd ordered Wait values were both 1 day, so the corresponding patients are assigned rank 21.5.

The final section of the sorted data is shown in Figure 2.61. Further relevant information emerges. Five patients have no dates for the procedure recorded. The patient with reference number 1964 had a wait of 1122 days, which exceeds the length of the study period. Suppose that discussion with the project leader reveals that the colposcopy dates for patients 1725 and

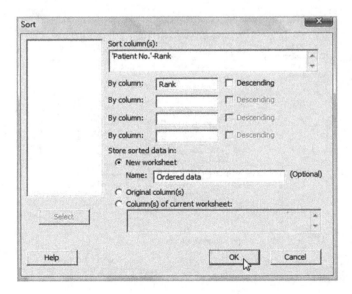

Figure 2.59   Ordering the Wait values.

Patient No.	Referral Date	Colposcopy Date	Wait	Rank
1725	24/01/2003	20/01/2003	-4	1.0
38	04/10/2001	04/10/2001	0	11.0
262	27/11/2001	27/11/2001	0	11.0
434	29/01/2002	29/01/2002	0	11.0
575	11/03/2002	11/03/2002	0	11.0
577	14/03/2002	14/03/2002	0	11.0
789	14/05/2002	14/05/2002	0	11.0
904	06/06/2002	06/06/2002	0	11.0
965	21/06/2002	21/06/2002	0	11.0
1235	23/08/2002	23/08/2002	0	11.0
1348	23/09/2002	23/09/2002	0	11.0
1427	24/10/2002	24/10/2002	0	11.0
1460	04/11/2002	04/11/2002	0	11.0
1530	25/11/2002	25/11/2002	0	11.0
1585	09/12/2002	09/12/2002	0	11.0
1600	10/12/2002	10/12/2002	0	11.0
1708	16/01/2003	16/01/2003	0	11.0
1939	31/03/2003	31/03/2003	0	11.0
2041	28/04/2003	28/04/2003	0	11.0
2042	28/04/2003	28/04/2003	0	11.0
105	17/10/2001	18/10/2001	1	21.5
246	22/11/2001	23/11/2001	1	21.5
153	29/10/2001	01/11/2001	3	25.0

Figure 2.60   Initial section of the sorted data.

1964 were 20/02/2003 and 29/04/2003, respectively. Suppose, too, that she indicates that the patients with reference numbers 7, 1238, 1764 and 1931 should be removed from the data set, as they had moved away from the area served by the hospital. There may still, of course, be further errors. The worksheet of ordered data may be deleted and the appropriate corrections and deletions made in the original worksheet.

	C1	C2-D	C3-D	C4	C5
↓	Patient No.	Referral Date	Colposcopy Date	Wait	Rank
2126	1015	03/07/2002	28/02/2003	240	2126.0
2127	1770	10/02/2003	17/12/2003	310	2127.0
2128	229	19/11/2001	17/10/2002	332	2128.0
2129	93	15/10/2001	24/09/2002	344	2129.0
2130	480	10/02/2002	11/02/2003	366	2130.0
2131	1430	24/10/2002	04/11/2003	376	2131.0
2132	1168	08/08/2002	11/11/2003	460	2132.0
2133	1964	03/04/2003	29/04/2006	1122	2133.0
2134	7	01/10/2001	*	*	*
2135	1238	23/08/2002	*	*	*
2136	1764	07/02/2003	*	*	*
2137	1931	21/03/2003	*	*	*

Figure 2.61   Final section of sorted data.

↓	C1	C2-D	C3-D	C4	C5	C6
	Patient No.	Referral Date	Colposcopy Date	Wait		
1	1	01/10/2001	26/10/2001	25		
2	2	01/10/2001	03/01/2002	94		
3	3					
4	4					
5	5					
6	6					
7	7					
8	8					
9	9					
10	10					
11	11					

Go To

Enter column number or name: 3

Enter row number: 1725

OK   Cancel

Figure 2.62   Locating a cell for correction.

2.3.3.3   Step 3. Correcting the data

The date corrections can be made first. One can scroll around the worksheet in order to locate the cells requiring to be changed or alternatively use **Editor** > **Go To...** (the command with an icon consisting of two footprints) when the worksheet is active. An example of the dialog involved in locating a cell for correction is shown in Figure 2.62 in the case of the patient with reference 1725 whose referral date should be 20/02/2003. Double-clicking the relevant cells enables the edits to be made. Observe how the Wait value for this patient changes automatically from −4 to 27 on making the correction. The change for the patient with number 1964 may be made similarly.

In order to make the deletions use may be made of **Data** > **Delete Rows...** to specify the rows to be deleted, as shown in Figure 2.63. Note that in this case the patient number matches the row number and that the rows have to be deleted from all four of the columns available for selection. The Wait column is not available for selection as it was assigned a formula. The now redundant column Rank may also be deleted using **Data** > **Erase Variables...**.

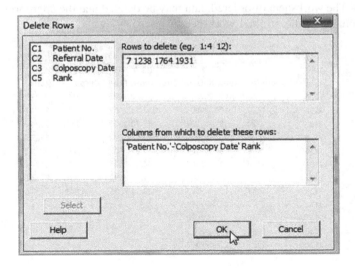

Figure 2.63   Deleting rows.

```
Descriptive Statistics: Wait

Variable      N   N*     Mean  SE Mean    StDev  Minimum       Q1  Median       Q3
Wait       2133    0   65.272    0.951   43.935    0.000   31.000  56.000   96.000

Variable   Maximum
Wait       460.000
```

Panel 2.5   Descriptive statistics for Wait.

For readers who are following the steps 'live', a crosscheck at this point can be made by obtaining descriptive statistics for the revised column of waiting times as shown in Panel 2.5. Note that there are 2133 values of Wait, with none missing, and that the mean was 65.272 days.

2.3.3.4   Step 4. Grouping the patients by month appointment was made

Use can be made of **Data > Extract from Date/Time > To Numeric** to convert the full date a patient's referral was made to a code for the month during which the referral was made. The dialog is displayed in Figure 2.64.

Under **Specify at least one component to extract from date/time** the **Year** was checked, with the **Four Digit** option selected, and **Month** was also checked. With the selection of the four-digit Year component and the Month component in the Minitab dialog box, any referral date in October 2001 will be coded as 200110, any referral date in November 2001 as 200111 etc. Use of **Stat > Basic Statistics > Descriptive Statistics...** for Wait, with **By variables:** Month and with **Mean** checked under **Statistics...**, as the only statistic required, yields the monthly means in the Session window as in Panel 2.6.

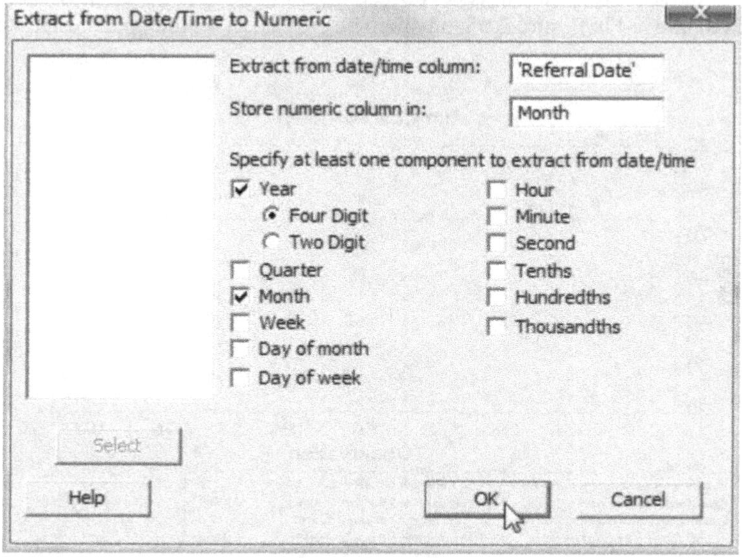

Figure 2.64   Coding dates by month.

Descriptive Statistics: Wait

Variable	Month	Mean
Wait	200110	80.85
	200111	79.68
	200112	77.69
	200201	87.22
	200202	77.98
	200203	64.32
	200204	72.84
	200205	75.31
	200206	75.38
	200207	78.91
	200208	87.16
	200209	75.61
	200210	58.42
	200211	47.48
	200212	53.74
	200301	38.17
	200302	37.07
	200303	34.21
	200304	36.60
	200305	31.80

Panel 2.6   Mean wait by month of referral.

The means can then be copied from the Session window, along with the code for the months, and pasted into a new worksheet as follows. Having highlighted and copied the two columns of 20 numbers from the Session window, click **File > New…**, and with **Minitab Worksheet** highlighted, click **OK**. With the mouse pointer, click on the first cell in column C1 of the new worksheet before pasting, accepting the default setting to **Use spaces as delimiters**. The run chart displayed in Figure 2.65 may then be created. It would appear that there has been

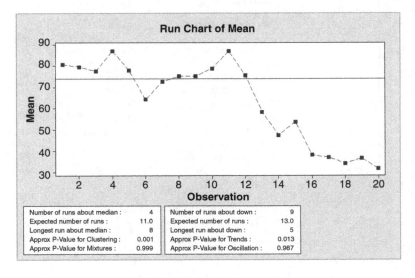

Figure 2.65   Run chart of mean wait by month.

a dramatic reduction in waiting times for patients. Note that the *P*-values provide evidence of the presence of special cause variation.

The author has heard a run chart described as a naked control chart. In Chapter 5 the construction and use of control charts will be introduced. Chapter 4 will be devoted to the introduction of the basic concepts of probability and of statistical models that provides essential underpinning for control charts.

2.4 Exercises and follow-up activities

1. For a familiar process obtain a sequence of measurements and display them in the form of a run chart. If you do not have access to data from a workplace situation then journey durations, your golf scores etc. could be used. Do you consider the process to be behaving in a stable predictable manner? Use ReportPad to note any conclusions you make regarding process performance. Save your work as a Minitab project to facilitate updating the data and to provide a control charting exercise at a later date. You might wish to name the project file Ch2Ex1.MPJ.

2. The Minitab worksheet Scottish_Mean_Annual_Temperatures.MTW gives mean annual Scottish temperatures (degrees Celsius) for the years 1910–2010 (http://www.metoffice.gov.uk/climate/uk/datasets/Tmean/date/Scotland.txt, accessed 30 January 2011.) Display the data in the form of a run chart and comment.

3. In the bottle weight example used in this chapter the sample (subgroup) size was 1. The worksheet Bottles.MTW contains data giving the weights of samples of four bottles taken every 15 minutes from a bottle-forming process. Open the worksheet and create a run chart. Here the data are arranged as subgroups across rows of columns C2, C3, C4 and C5, so this option has to be selected in the Run Chart dialog.

   The default is to plot the means of the subgroups of four weights with reference line placed at the median of the 25 sample means, i.e. 489.638. The means are plotted as red squares and connected by line segments, and in addition the individual weights are plotted as black dots. The alternative option is to plot the subgroup medians, in which case the reference line is placed at the median of the 25 medians, i.e. 489.595.

   Note from the run chart legends that there is no evidence of any special cause variation. Stack the weights into a single column named weight and verify that the mean and standard deviation of the total sample of 100 bottles are 489.75 and 2.09, respectively. Create a histogram of the data with reference lines placed at the specification limits of 485.0 and 495.0 g for bottle weight. Comment on the shape of the distribution and on process performance in relation to the specifications. Save your work as a Minitab project.

4. During the measure phase of a Six Sigma project a building society collected data on the time taken (measured as working days rounded to the nearest day) to process mortgage loan applications. The data are stored in the supplied worksheet Loans1.MTW. Display and summarize the data and comment on the shape of the distribution. Use a stem-and-leaf display to determine the number of times which exceeded the industry standard of 14 working days.

   Following completion of the improvement project a further sample of processing times was collected. The data are stored in Loans2.MTW. Display and summarize the data in order to assess the effectiveness of the project. Save your work as a Minitab project.

5. The file Statin.xls contains monthly numbers of stroke patients admitted to a major hospital for the years 2001–2003 together with the numbers whose medication on admission included a statin. Open the file in Minitab and create a column giving the monthly proportions of stroke patients on a statin at time of admission. You will find that **Data > Transpose Columns...** is useful here. Create a run chart of the proportions and comment. Save your work as a Minitab project.

6. The Minitab worksheet Shareprice.MTW supplied with the software in the Data folder contains monthly share prices for two companies ABC and XYZ. Use **Graph > Time Series Plot...** to create a multiple run chart of these data. Select the option **Multiple**, and enter ABC and XYZ in the **Series:** window. Under **Time/Scale...** select **Calendar** and then choose **Month Year** from the associated menu. For the **Start Values**, accept the default **One set for all variables** and enter 1 for Month and 2009 for Year. Observe how moving the mouse pointer to a point on a plot leads to display of the variable, its value and the corresponding month and year.

7. The histogram was introduced as a type of bar chart in which there are no gaps between the bars, indicating that the variable being displayed is continuous. One can use the histogram facility in Minitab to display discrete random variables and to emphasize the discrete nature of the data by having gaps between the bars. The supplied worksheet AcuteMI.MTW contains daily counts of the number of patients admitted to the accident and emergency department of a major city hospital with a diagnosis of acute myocardial infarction. Create a histogram of the data, making use of **Data View...** to uncheck **Bars** and check **Project lines**. Double-click on one of the project lines, select **Custom** and increase **Size** to, say, 5.

3

Exploratory data analysis, display and summary of multivariate data

Display is an obligation! (Tukey, 1986)

Overview

Two tools of exploratory data analysis (EDA), the stem-and-leaf plot and boxplot, provide versatile and informative displays of process data and enable outliers to be defined and detected. The brushing facility in Minitab is introduced as it enables subsets of data sets displayed in graphs to be readily identified and explored.

There are many situations where two or more performance measurements are made in assessing process performance, so some familiarity with techniques for the display and summary of bivariate and multivariate data is important.

3.1 Exploratory data analysis

3.1.1 Stem-and-Leaf displays

In 1998 when the author worked in Dunfermline, Scotland, he was involved in a process that many of us undertake every day – the process of driving to work. The route from Loanhead, across the Forth Road Bridge, was 25 miles long. A run chart of journey duration (minutes) for 32 journeys undertaken during September and October 1998 is shown in Figure 3.1.

The low P-value for clustering indicates the possible influence on journey duration of some special causes. In addition to journey duration, the weather was recorded as either dry (D) or wet (W). The author had a hunch that this (uncontrollable!) factor influenced journey duration. The data are given in Table 3.1 and are available in the supplied worksheet Travel.MTW. The

Six Sigma Quality Improvement with Minitab, Second Edition. G. Robin Henderson.
© 2011 John Wiley & Sons, Ltd. Published 2011 by John Wiley & Sons, Ltd.

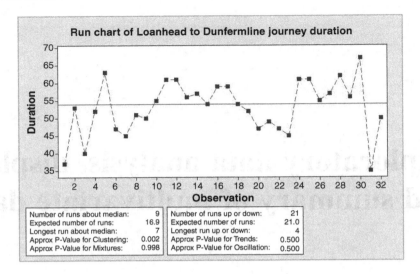

Figure 3.1 Run chart of journey duration.

worksheet also contains a column giving the dates of the journeys and a column named Wcode which contains weather codes 1 for dry and 2 for wet.

In order to compare journey durations under the two categories of weather conditions, dry and wet, and in order to make further exploration of the data easier, stem-and-leaf displays will be constructed. Scrutiny of the data reveals durations ranging from the thirties to the sixties. Stem values of 3, 4, 5 and 6 will be used corresponding to the thirties, forties, fifties and sixties. Each stem will appear twice in the display corresponding to the low thirties, the high thirties, the low forties, the high forties etc. Consider first the durations for dry days: 37, 40, ..., 50. Thus the first dry day duration of 37 minutes is a value in the high thirties, i.e. in the range 35–39 inclusive, and will be considered as a leaf value of 7 attached to the stem value of 3 that corresponds to the upper thirties. The second dry day duration of 40 minutes is a value in the low forties, i.e. in the range 40–44 inclusive, and will be considered as a leaf value of 0 attached to the stem value of 4 that corresponds to the low forties.

In Minitab stem-and-leaf displays are created in the Session window and may be obtained using either **Stat > EDA > Stem-and-Leaf...** or **Graph > Stem-and-Leaf...**. In either case **By variable:** Wcode has to be specified as only numeric variables may be used in this way with the Stem-and-Leaf procedure.

Table 3.1 Journey durations.

Day	1	2	3	4	5	6	7	8	9	10	11	12	13	14	15	16
Duration	37	53	40	52	63	47	45	51	50	55	61	61	56	57	54	59
Weather	D	W	D	W	D	W	D	D	W	W	D	D	D	W	D	W

Day	17	18	19	20	21	22	23	24	25	26	27	28	29	30	31	32
Duration	59	54	52	47	49	47	45	61	61	55	57	62	56	67	35	50
Weather	D	D	W	D	W	D	D	D	D	D	D	D	W	W	W	D

```
Stem-and-Leaf Display: Duration

Stem-and-leaf of Duration  Wcode = 1    N  = 20
Leaf Unit = 1.0


   1    3   7
   2    4   0
   6    4   5577
  10    5   0144
  10    5   5679
   6    6   111123
```

Panel 3.1   Stem-and-leaf display of duration (dry days).

The stems are shown in the second column and the leaves in the third column in Panel 3.1. The first column contains what are known in the jargon of exploratory data analysis (EDA) as the values of depth. The first four depths listed, 1, 2, 6 and 10, correspond to the highest durations on each of the first four stems. Thus, for example, the highest duration of 47 on the third stem is 6 values deep into the ordered data set starting from the minimum. The final two depths listed of 10 and 6 correspond to the lowest durations on each of the final two stems. For example, the duration of 55 on the penultimate stem is 10 values deep into the ordered data set starting from the maximum. The display gives $N = 20$, indicating that the data set included durations for 20 journeys under dry conditions. The median duration corresponds to depth $(N + 1)/2$, which in this case is 10.5, indicating that the median is the mean of the 10th and 11th ordered durations, i.e. $(54 + 55)/2 = 54.5$.

The stem-and-leaf display of the 12 journey durations on wet days is shown in Panel 3.2. The bracketed 4 in the depth column indicates that the median duration lies in the corresponding interval and also that there are four durations in that interval. In Panel 3.1 where the median lies on the boundary between intervals there was no need for a bracketed frequency count. The reader is invited to check that the median in this case is 52.5.

Visual comparison suggests that location and spread are similar for both dry and wet conditions. Thus it would appear that the author's hunch was incorrect! Figure 3.2 indicates how stem-and-leaf displays and histograms are, in essence, equivalent data displays. The stem-and-leaf displays in Panels 3.1 and 3.2 (without the depth values) have been rotated and positioned below the corresponding histograms. The first stem of 3 corresponds to durations in

```
Stem-and-leaf of Duration   Wcode = 2    N  = 12
Leaf Unit = 1.0


   1    3   5
   1    4
   3    4   79
  (4)   5   0223
   5    5   5679
   1    6
   1    6   7
```

Panel 3.2   Stem-and-leaf display of duration (wet days).

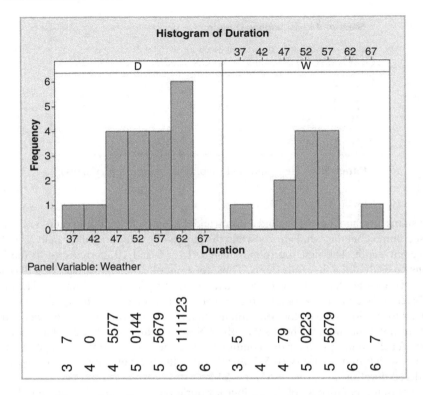

Figure 3.2   Stem-and-leaf displays viewed as histograms.

the upper thirties, i.e. in the range 35–39 inclusive. Thus the midpoint value for the corresponding histogram bin is 37. The second stem of 4 corresponds to durations in the lower forties, i.e. in the range 40–44, the midpoint value for the corresponding histogram bin being 42, and so on. With this choice of binning for the histograms the profiles formed by the stacks of leaves of leaves match the bars of the histograms.

The reader might well ask, therefore, if one needs to know about the stem-and-leaf display in addition to the histogram. An answer is provided by one of the data sets in the Minitab Sample Data folder provided with the software. The worksheet Peru.MTW has data from an anthropological study of the effects of urbanization on people who had migrated from mountainous areas. Pulse rates were recorded for the subjects in the investigation and a stem-and-leaf display is shown in Panel 3.3.

Of the pulse rates for the 39 subjects the only value not divisible by 4 is 74. This suggests that the person recording the pulse rates generally counted heartbeats for a period of 15 seconds and then multiplied by 4 to obtain heart rate. Thus, if the standard procedure required heartbeats to be counted for a full minute, the data appear to provide evidence that the measurement system was not being implemented correctly.

Quartiles are included in the default list of descriptive statistics provided by Minitab – see Panel 3.4 for the results for journey duration by weather. Calculation of the median by 'chopping' the ordered data set in half has already been discussed in Section 2.1.1. The procedure may then be repeated, with the lower half of the data being chopped in half to give a

```
Stem-and-Leaf Display: Pulse

Stem-and-leaf of Pulse   N  = 39
Leaf Unit = 1.0


    1    5  2
    2    5  6
   15    6  0000004444444
   19    6  8888
  (9)    7  222222224
   11    7  6666
    7    8  004
    4    8  888
    1    9  2
```

Panel 3.3   Stem-and-leaf display of pulse rates.

```
Descriptive Statistics: Duration

Variable  Weather  N   N*    Mean  SE Mean  StDev  Minimum    Q1  Median     Q3
Duration  D        20   0   53.30     1.73   7.75    37.00  47.00  54.50  61.00
          W        12   0   52.67     2.21   7.67    35.00  49.25  52.50  56.75

Variable  Weather  Maximum
Duration  D          63.00
          W          67.00
```

Panel 3.4   Descriptive statistics for dry and wet days.

lower quartile Q1 of 47 for journeys on dry days and the upper half chopped in half to give an upper quartile Q3 of 61. The median may also be referred to as the middle quartile, Q2. Essentially the three quartiles split the data set into 'quarters'. The median is also referred to as the 50th percentile, the lower quartile the 25th percentile, and the upper quartile the 75th percentile. Interested readers my access the technical details of the calculation of quartiles in Minitab via the Help facility.

3.1.2   Outliers and outlier detection

One of the benefits of statistical methods is their ability to signal the presence of any unusual values in a data set. The detection of unusual values can lead to insights into how factors (the Xs) affect process responses (the Ys). Unusual values may, of course, simply be due to incorrect data recording or to data input errors. Unusual values or outliers may be detected as follows.

The inter-quartile range (IQR) is the difference Q3 – Q1 and may be used as a measure of variability or spread. In EDA, outlier detection requires the use of lower and upper limits calculated using the formulae given in Box 3.1.

Lower limit = Q1 − 1.5 × IQR
Upper limit = Q3 + 1.5 × IQR

Box 3.1   Formulae for limits used in outlier detection.

```
Stem-and-Leaf Display: Duration (wet days)

Stem-and-leaf of Duration (wet days)   N  = 12
Leaf Unit = 1.0

LO 35

   3    4   79
  (4)   5   0223
   5    5   5679
   1    6
```

Panel 3.5   Outlier detection via Minitab stem-and-leaf.

For the durations on dry days the formulae yield:

$$IQR = 61 - 47 = 14,$$
$$\text{Lower limit} = 47 - 1.5 \times 14 = 26,$$
$$\text{Upper limit} = 61 + 1.5 \times 14 = 82.$$

Outliers are defined as values falling either below the lower limit or above the upper limit. For dry days the minimum and maximum durations were 37 and 63 so therefore there are no outliers.

For the durations on wet days we have

$$IQR = 56.75 - 49.25 = 7.5,$$
$$\text{Lower limit} = 49.25 - 1.5 \times 7.5 = 38,$$
$$\text{Upper limit} = 56.75 + 1.5 \times 7.5 = 68.$$

Thus the duration of 35 minutes on one of the wet days is an outlier as it falls below the lower limit of 38, i.e. the duration of 35 minutes is being 'flagged' as the result of possible special cause variation. On checking his diary for that date, the author discovered that he had delivered a training course and had left home earlier than usual in order to set up the training room and had thereby driven in lighter traffic than normal.

Outliers may be obtained via Minitab by checking the **Trim outliers** option in the Stem-and-Leaf dialog box. Note in Panel 3.5 how the value of 35 appears beside the text LO indicating that it is a low outlier: high outliers appear with the text HI. (For this facility to work the data for dry days has to be in a separate column from the data for wet days.)

3.1.3 Boxplots

The use of the median as a measure of location, the use of IQR as a measure of variability and the outlier detection procedure described above can be combined in a very powerful display called a *boxplot*. In order to display the journey duration data in this way for each weather condition use is made of **Graph > Boxplot...**, with the **One Y, With Groups** options being selected and the dialog box completed as shown in Figure 3.3. Duration is specified under **Graph variables:**. Either Weather or Wcode may be used as **Categorical variable for grouping:** here. The **Labels...** button was used to create the title. Click on **OK** and the boxplots are created as displayed in Figure 3.4.

Date	Duration	Weather	WCode					
01/09/98	37	D	1					
02/09/98	53	W	2					
03/09/98								
04/09/98								
05/09/98								
08/09/98								
09/09/98								
10/09/98								
11/09/98								
12/09/98								
15/09/98								
16/09/98								
17/09/98								
18/09/98								
19/09/98								
22/09/00								
22/09/98								
23/09/98								

Boxplot - One Y, With Groups

C1 Date
C2 Duration
C3 Weather
C4 WCode

Graph variables:
Duration

Categorical variables for grouping (1-4, outermost first):
Weather

Scale...   Labels...   Data View...

Select   Multiple Graphs...   Data Options...

Help   OK   Cancel

Figure 3.3   Dialog for creating boxplots.

Boxplots are sometimes known as box-and-whisker plots, for obvious reasons. The ends of a box correspond to the quartiles, the line across the box corresponds to the median. The medians indicate location for the samples and the box lengths (IQRs) indicate variability. Outliers are denoted by asterisk symbols. The adjacent values are defined as the lowest and highest data values in a sample, which lie within the lower and upper limits. For dry days the adjacent values are 37 and 63. For wet days the adjacent values are 47 and 67. The whiskers extend from the ends of the boxes to the adjacent values.

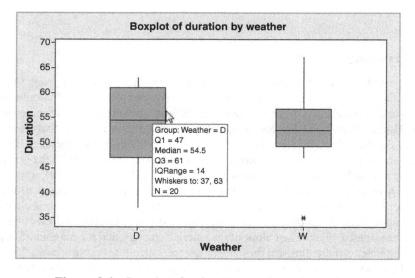

Boxplot of duration by weather

Group: Weather = D
Q1 = 47
Median = 54.5
Q3 = 61
IQRange = 14
Whiskers to: 37, 63
N = 20

Figure 3.4   Boxplots for duration on dry and wet days.

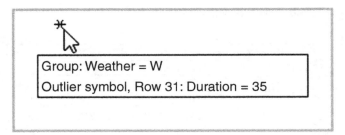

Figure 3.5 Identifying an outlier.

Scrutiny of data displayed as boxplots gives an immediate visual impression of location in terms of the position of the box(es) relative to the vertical measurement scale and of variability in terms of the length(s) of the box(es). The reader may have observed that the default for a histogram in Minitab is to have the measurement scale horizontal whilst the default for a boxplot is to have the measurement scale vertical. The **Scale. . .** button in the dialogs for both displays enables the alternatives to be selected by checking **Transpose Y and X** in the case of the histogram and **Transpose value and category scales** in the case of the boxplot.

In Figure 3.4 the mouse pointer is shown located on the outline of the box part of the display for dry days. This triggers display of a summary of the data in the group, i.e. the sample size, the quartiles and the IQR, together with the group identity and the adjacent values.

On moving the cursor to an asterisk denoting an outlier the corresponding row number and value of duration are displayed, as shown in Figure 3.5. With a large data set this is clearly a very useful facility to have in the search for knowledge of special causes affecting processes, knowledge that may be used to achieve improvement. The brushing facility in Minitab may also be used to identify and view information on plotted points in graphs of interest to the user.

3.1.4 Brushing

Brushing provides a way of investigating data corresponding to a point or group of points of interest to the user in a Minitab graph e.g. outliers on a boxplot. As an example, open the Minitab worksheet Lake.MTW that is supplied with the software and which gives measurements of 71 lakes in northern Wisconsin. Create a boxplot of area and note that there are nine outliers in the plot. Suppose that we wish to create a worksheet that includes, for further analysis, data for those lakes that are flagged as outliers on the boxplot. With the graph active, use **Editor** > **Brush** to activate the brushing facility. (Alternatively click on the brush icon $\boxed{\mathscr{d}}$.) A window, referred to as the Brushing Palette, appears on the left of the screen with the heading Row. Move the mouse pointer over the plot and note how it changes to a pointing finger shape. With the tip of the finger located on the asterisk representing the lake with the greatest area, click. The colour of the asterisk changes and the number 55 appears in the Brushing Palette, indicating that the data for the corresponding lake are in row 55. Keep the shift key depressed, move to the next asterisk and click again. The number 5 will now appear in the window together with the number 55. Continue in this way, keeping the shift key depressed,

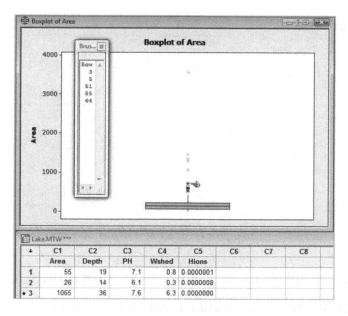

Figure 3.6   Brushing outliers in a boxplot.

until all nine outliers have been identified. The Brushing Palette should now display the row numbers 3, 5, 37, 40, 51, 55, 60, 62 and 64. The row numbers are automatically sorted as each new point is brushed. Figure 3.6 displays the boxplot, the brushing palette and the worksheet after five outliers have been brushed.

Note that in the worksheet there are black dots to the left of the row number for each of the brushed rows. Thus variables other than area could readily be scrutinized for lakes classified as outliers on the basis of area. If the Brushing Palette is closed, by clicking the cross in its top right-hand corner, then the brushed points revert to their original colour and the black dots disappear from the worksheet. The reader is invited to do this, to activate the brushing facility and to brush the nine outliers again by clicking on the graph so that the pointing finger shape appears and, while keeping the mouse button depressed, dragging the mouse to create a dotted rectangle enclosing the nine points of interest. On releasing the mouse button the Brushing Palette should now include the row numbers 3, 5, 37, 40, 51, 55, 60, 62 and 64. Select **Editor > Create Indicator Variable...** and a menu will appear. Enter **Column:** Large, select **Update now**, click **OK** and close the Brushing Palette. Examination of the worksheet reveals that a column named Large has been created with value 1 in rows that have been brushed and value 0 in other rows. Next **Data > Subset Worksheet...** enables the dialog in Figure 3.7.

The default name for the new worksheet may be changed to one that is appropriate, e.g. Large_lake.MTW. The default **Specify which rows to include** was accepted under **Include or Exclude**. Under **Specify Which Rows to Include**, clicking on **Condition...** provides a subdialog box in which 'Large' = '1 is entered under **Condition:**. Clicking **OK** completes the creation of the new worksheet for the nine outlier lakes. Note the availability of the Boolean operators **And, Or** and **Not** so that more complex conditions may be created.

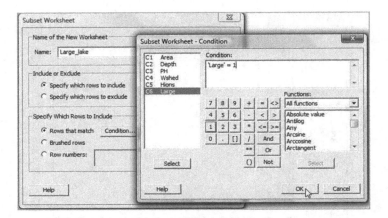

Figure 3.7 Dialog for creation of a subset of a worksheet.

3.2   Display and summary of bivariate and multivariate data

3.2.1   Bivariate data – scatterplots and marginal plots

The Minitab data set Pulse.MTW was referred to in Chapter 2. Consider the random variables Height and Weight for the students, denoted by X and Y, respectively. The first student listed in the worksheet had height 66 inches and weight 140 pounds, so the point $(x_1, y_1) = (66, 140)$ in a two-dimensional coordinate system can be used to represent this student. Use of **Graph > Scatterplot... > Simple**, selection of Weight as **Y variable** (vertical axis) and Height as **X variable** (horizontal axis), and acceptance of defaults otherwise, yields the scatterplot or scatter diagram of the data displayed in Figure 3.8.

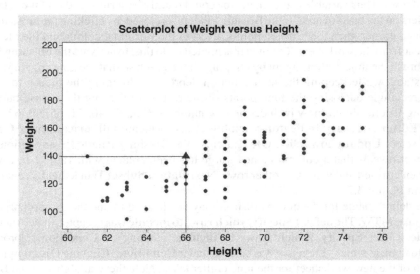

Figure 3.8 Scatterplot of weight versus height.

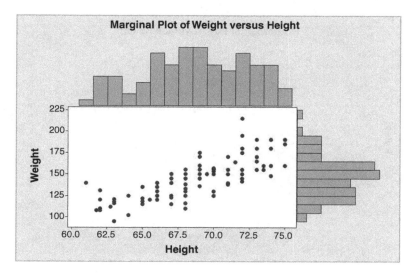

Figure 3.9 Marginal plot of weight versus height.

The scatterplot has been modified so that the point representing the first student is represented by a triangle with lines drawn parallel to the axes to indicate the weight of 140 and the height of 66 for that student. The upward drift in the points as one scans from left to right across the diagram indicates a positive relationship between weight and height. It simply reflects what is common knowledge – taller people tend to be heavier than shorter people. (The vertical stacks of points are due to the rounding of height measurements to the nearest half of an inch.) Here we are investigating two random variables at the same time, i.e. we are exploring a bivariate data set.

Use of **Graph > Marginal Plot...** enables one to explore the univariate aspects also either via **With Histograms**, **With Boxplots** or **With Dotplots** constructed on the margins of the scatterplot. Having selected the **With Histograms** option, the display in Figure 3.9 was obtained. The histogram of Height appears in the upper margin of the scatterplot, that of Weight appears in the right-hand margin.

A scatterplot of the first pulse rate recorded (Pulse1) versus height is shown in Figure 3.10. In this second plot there is no apparent relationship between these two random variables as might be expected.

Scatterplots are a useful tool for exploring relationships between Ys and Xs. The diagram in Figure 3.11 shows the diameter (mm) of a machined automotive component plotted against the temperature (°C) of the coolant supplied to the machine at the time of production. The data for this example, available in Diameters.MTW, are copyright (2000) from 'Finding assignable causes' by Bisgaard and Kulachi and are reproduced by permission of Taylor & Francis, Inc., http://www.taylorandfrancis.com.

Given that the target diameter is 100 mm, this plot indicates the possibility of improving the process through controlling the coolant temperature (an X) to be more consistent, thus leading to less variability in the Diameter (a Y) of the components. Use of **Graph > Scatterplot...** and the **With Regression** option (with all defaults accepted) yields the scatterplot in Figure 3.11 with the addition of a straight line modelling the linear relationship between diameter and

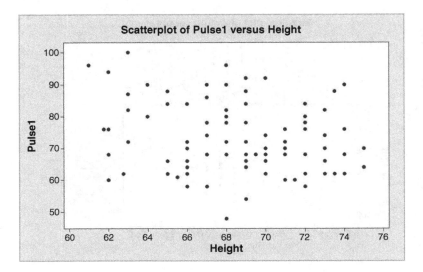

Figure 3.10   Scatterplot of pulse rate versus height.

temperature. The modelling of linear relationships using linear regression will be covered in Chapter 10. The dotted reference lines added to the plot indicate that it appears desirable to maintain the coolant temperature at around 22 °C.

3.2.2   Covariance and correlation

In order to measure the 'strength' of the linear relationship between two random variables the concept of **covariance** is required. Four small data sets will be used to introduce the concept.

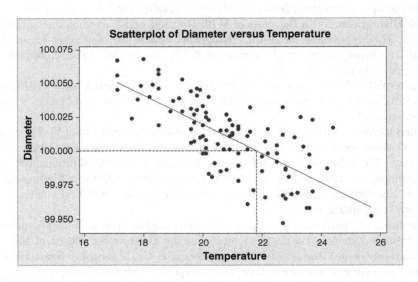

Figure 3.11   Scatterplot of diameter versus temperature of coolant.

The sample covariance between x and y, denoted by $\mathrm{cov}(x, y)$, is given by

$$\mathrm{cov}(x, y) = \frac{\sum_{i=1}^{n}(x_i - \bar{x})(y_i - \bar{y})}{n - 1}.$$

Box 3.2   Formula for covariance.

Table 3.2   Calculation of covariance.

		Data set 1		
x_i	y_i	$x_i - \bar{x}$	$y_i - \bar{y}$	$(x_i - \bar{x})(y_i - \bar{y})$
0	0	-12	-6	72
4	2	-8	-4	32
6	3	-6	-3	18
8	4	-4	-2	8
12	6	0	0	0
14	7	2	1	2
16	8	4	2	8
22	11	10	5	50
26	13	14	7	98
	Total	0	0	288

The formula for calculating sample covariance is given in Box 3.2. It is similar in structure to the formula for variance given in Chapter 2. In fact the covariance between x and x is simply the *variance* of x.

Data set 1, for which the means of x and y are 12 and 6 respectively, is displayed in Table 3.2 together with the results of calculations required to obtain the covariance. The covariance between x and y is

$$\mathrm{cov}(x, y) = \frac{\sum(x_i - \bar{x})(y_i - \bar{y})}{n - 1} = \frac{288}{8} = 36.$$

Having entered the data into two columns named x and y, the result can be verified in Minitab using **Stat > Basic Statistics > Covariance. . . .** The output in the Session window is displayed in Panel 3.6. The covariance of 36 calculated via Table 3.2 is confirmed by the

```
Covariances: x, y

           x          y
x   72.0000
y   36.0000   18.0000
```

Panel 3.6   Covariance from Minitab.

```
Covariances: x, y

           x          y
x    72.0000    36.0000
y    36.0000    18.0000
```

Panel 3.7   Variance–covariance matrix.

Session window output and, in addition, the variances of x and y are given: 72 and 18 respectively. In reading such a table the covariance between x and y is located vertically below the x which appears in the header row of variable names and horizontally across from the y which appears in the left-hand column of variable names. The creators of Minitab could have chosen to present the output as shown in Panel 3.7. The additional value of 36 in the top right-hand corner of the table is the covariance between y and x according to the rule for reading the table given above. But since the covariance between x and y is identical to that between y and x the additional information is redundant. The full square array of values is referred to as the variance–covariance matrix for the variables x and y.

For data sets 2, 3 and 4, displayed in Table 3.3, you are invited to verify, both by grass-roots calculation and using Minitab, that the covariances are 21.5, 0 and − 32 respectively. (The x values are the same in all four cases. The data are available in FourSets.MTW.) Annotated scatterplots for all four data sets are shown in Figure 3.12. The notation y1*x1 is a shorthand used by the software to label the scatterplot of the first y versus the first x, i.e. for data set 1 etc.

For data set 1 a perfect linear relationship with a positive slope is evident; the covariance is positive. For data set 2 there is an upward drift in the points; the covariance is positive. For data set 3 the points appear to be scattered at random; the covariance is zero. For data set 4 there is a downward drift in the points; the covariance is negative. In the case of data sets considered earlier, for weight and height the covariance is 68.18, for pulse rate and height − 8.53 and for diameter and temperature − 0.035. Note how the signs of the covariances are in accord with the appearance of the corresponding scatterplots in the cases of weight and height (positive, see Figure 3.8) and diameter and temperature (negative, see Figure 3.11).

Table 3.3   Further bivariate data sets.

Data set 2		Data set 3		Data set 4	
x	y	x	y	x	y
0	2	0	4	0	11
4	8	4	3	4	13
6	0	6	8	6	8
8	6	8	6	8	4
12	3	12	7	12	7
14	4	14	13	14	6
16	13	16	2	16	3
22	7	22	11	22	2
26	11	26	0	26	0

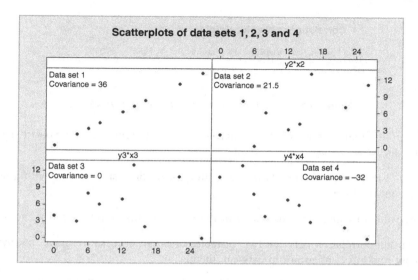

Figure 3.12 Scatterplots of data sets 1, 2, 3 and 4.

Had the weights and heights of the students been given in kilograms and metres then the covariance would have been 0.786 (compared with 68.18 for inches and pounds). Thus a problem with covariance as a summary statistic for bivariate data is that its value depends on the units of measurement used. This problem can be overcome by reporting the covariance between the standardized values of the random variable. In this context, standardization means transforming each data value by subtracting the mean from the original value and then dividing by the standard deviation. The standardized values have mean 0 and standard deviation 1. The covariance between the standardized variables is referred to as the *correlation coefficient*. Formally it is the Pearson product moment correlation coefficient between the original variables and is usually denoted by the symbol r. The **Calc** menu in Minitab enables one to standardize variables and then the **Covariance** facility could be used to obtain the correlation coefficient. However, they are so important that the **Stat** > **Basic Statistics** menu provides a **Correlation...** submenu enabling them to be obtained directly. You should verify that the correlation between weight and height is 0.785 and that between diameter and temperature is $- 0.712$.

Alternatively the correlation coefficient (the Pearson product moment correlation to give it its full title) may be calculated from the covariance using the following formula:

$$r = \frac{\text{cov}(x, y)}{s_x s_y},$$

where s_x and s_y are the standard deviations of x and y, respectively.

Some properties of the product moment correlation coefficient, r, are as follows:

- r is independent of the units of measurement

- r is not dependent on which of the two variables is labelled x and which is labelled y.

- $-1 \leq r \leq 1$, i.e. the value of r must lie between -1 and 1 inclusive.

```
Correlations: Weight, Height

Pearson correlation of Weight and Height = 0.785
P-Value = 0.000
```

Panel 3.8   Weight–height correlation with P-value.

- $r = 1$ only if all the points in the scatterplot lie on a straight line with a positive gradient or slope.

- $r = -1$ only if all the points in the scatterplot lie on a straight line with a negative gradient or slope.

- The value of r measures the extent of the *linear* relationship between x and y.

- A strong correlation between two variables does not necessarily mean that there is a cause-and-effect relationship between them.

- A small value of the correlation coefficient does not necessarily mean that there is no relationship between the variables – the relationship might be nonlinear.

The square of the correlation coefficient, r^2, is known as the *coefficient of determination*. It can be interpreted as the proportion of the variation in y attributable to its linear dependence on x. For the diameter and temperature data $r^2 = 0.51$, so just over half the variation in diameter can be attributed to its linear dependence on temperature. The Greek letter ρ (rho) is used for the population correlation coefficient.

Panel 3.8 shows the Session window output obtained for the correlation between Weight and Height obtained when **Display p-values** is checked in the **Stat > Basic Statistics > Correlation...** dialog box. If the P-value is less than 0.05 then it is generally accepted that there is evidence that the population correlation ρ is nonzero. Thus the student data provides evidence of a nonzero correlation between weight and height in the population from which the sample was drawn, provided that the sample can be regarded as a random one from that population.

So far the only data structure we have considered in Minitab is the column. The matrix is another data structure provided and this is a convenient point at which to introduce it. With the Session window active, enable commands via **Editor > Enable Commands**. Note how the Session window now has the Minitab Command prompt MTB > displayed. Use **Stat > Basic Statistics > Correlation...** and select Weight and Height but check the **Store matrix (display nothing)** option. On implementation the Session Window contains the output in Panel 3.9. The software has created a matrix m1, named it CORR1 and stored in it the correlations involving

```
Results for: Pulse.MTW

MTB > Name m1 "CORR1"
MTB > Correlation 'Weight' 'Height' 'CORR1'.
MTB >
```

Panel 3.9   Storage of correlation matrix.

```
MTB > Print CORR1

Data Display

 Matrix CORR1

 1.00000   0.78487
 0.78487   1.00000
```

Panel 3.10   Printing a correlation matrix.

the variables weight and height. Typing the command Print CORR1, or simply prin m1, after the prompt in the Session window and pressing the enter key yields the Session window output in Panel 3.10.

Alternatively one could use **Data** > **Display Data...**, select the matrix CORR1 and click **OK** in order to display it. Note how all columns are available for selection to be displayed in the Session window if desired.

3.2.3   Multivariate data – matrix plots

A matrix is a rectangular array of numbers, symbols or objects. The **Graph** menu provides a matrix plot as shown in Figure 3.13 for weight and height. The sequence **Graph** > **Matrix Plot...** > **Matrix of plots – Simple**, with the variable names entered in the order Weight Height was used to create the plot.

The four panels in the matrix plot match the four elements of the correlation matrix displayed in Panel 3.12 as indicated by the annotation that includes the correlation coefficients, rounded to three decimal places. The correlation between weight and height is the same as that between height and weight. Had the pulse rate Pulse1 been included in addition to weight and height then a correlation matrix with three rows and three columns would have been obtained –

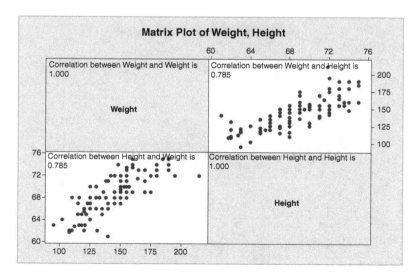

Figure 3.13   Matrix plot of weight and height data.

```
MTB > Name m2 "CORR2"
MTB > Correlation 'Pulse1'  'Weight' 'Height' 'CORR2'.
MTB > Print m2

Data Display

 Matrix CORR2

  1.00000   -0.20222   -0.21179
 -0.20222    1.00000    0.78487
 -0.21179    0.78487    1.00000
```

Panel 3.11   Correlation matrix for pulse rate, weight and height.

see Panel 3.11. For example, the entry in the top right-hand corner, -0.21179, is the correlation between pulse rate and height. The corresponding matrix plot is shown in Figure 3.14. It has been annotated with the correlations, rounded to three decimal places. Note that the pattern of correlations in the correlation matrix matches that in the matrix plot – this is the case since the variables were listed in the same order in the corresponding Minitab dialog boxes.

Scrutiny of the matrix of scatterplots reveals that the strongest relationship is that between weight and height. In order to identify the variables for a particular plot, look to the left or right to find the variable plotted on the vertical axis and then look above or below to find the variable plotted on the horizontal axis.

Matrix plots can be very useful for exploring relationships between the Ys and the Xs for a process. Minitab provides a version of the matrix plot that enables one to specify which variables in the data set are Ys and which are Xs. Consider data collected in an *ad hoc* way by microelectronic fabrication engineers on a plasma etching process. The two response variables of interest were etch rate, Y_1 (Å/m), and uniformity, Y_2 (Å/m). The four factors of interest were gap, X_1 (cm), pressure, X_2 (m Torr), flow, X_3 (standard cubic centimetres per minute), and

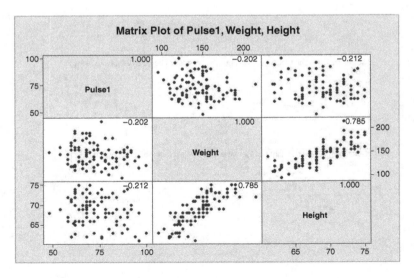

Figure 3.14   Matrix plot of pulse rate, weight and height data.

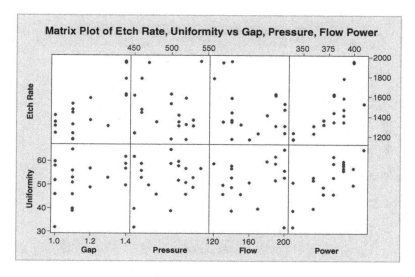

Figure 3.15   Matrix plot of plasma etching process data.

power, X_4 (W). The data are stored in the supplied worksheet Etch.MTW. Use of **Graph > Matrix Plot. . . > Each Y versus each X – Simple** may be used to create the display of the data in Figure 3.15. Both etch rate and uniformity are entered as **Y variables** and gap, pressure, flow and power as **X variables**.

Each horizontal strip displays scatterplots of one Y versus each of the Xs in turn. Examination of the plots suggests that power has an effect on both etch rate and uniformity and that gap influences etch rate and possibly uniformity. There certainly does not appear to be any relationship between the two Ys and the two Xs pressure and flow. The engineers had the problem of maintaining the etch rate within acceptable limits while at the same time maintaining uniformity below a specified maximum value. The above display indicates that gap and power could be factors worth investigating in a designed experiment with the aim of determining appropriate settings for gap and power to be used in running the process.

Before concluding this section it has to be emphasized that the correlation coefficient measures the strength of the *linear* relationship between two random variables. Consider a sales manager who obtained the correlation between annual sales (£000) and years of experience for his sales team using Minitab as displayed in Panel 3.12. The data are available in Sales.MTW.

A P-value well in excess of 0.05 might lead the manager to think that there was no relationship between sales and experience. But a scatterplot of sales versus experience reveals

Panel 3.12   Correlation between sales and experience.

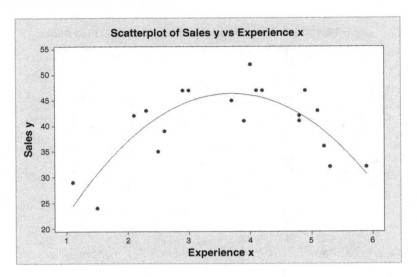

Figure 3.16   A nonlinear relationship.

that there is an apparent relationship. The quadratic regression model curve added to the basic scatterplot was obtained using **Graph** > **Scatterplot...** > **With Regression**. Sales was specified as the Y variable and Experience as the X variable. The **Data View...** option was used for selection of **Quadratic** via the **Regression** tab. The output is displayed in Figure 3.16.

The display suggests that sales increase with experience initially and then decline. This example underlines the importance of displaying bivariate data in the form of scatterplots, and not merely scrutinizing correlation coefficients and corresponding *P*-values.

3.2.4 Multi-vari charts

The multi-vari chart is a graphical tool that may be used to display multivariate data. The author and his wife used a micrometer to make two measurements of the height (mm) of each one of a sample of 10 parts. Experiments of this type are frequently used in the evaluation of measurement processes and will be referred to fully in Chapter 9. For the moment the data, available in ARGageR&R.MTW, will be displayed in a multi-vari chart in order to see what insights, if any, may be gained.

A segment of the data set is shown together with the dialog required to create the chart in Figure 3.17. The data in the first row indicate that Anne obtained a measurement of 1.445 for the height of the first part on the first occasion that she measured it. The data in the eleventh row indicate that Anne obtained a measurement of 1.255 for the height of the first part on the second occasion that she measured it. The chart may be created using **Stat** > **Quality Tools** > **Multi-Vari Chart...**. **Factor 1**: was specified as Trial, **Factor 2**: was specified as Operator and **Factor 3**: as Part. Under **Options...**, **Display Options**, **Display individual points** was checked, **Connect means for...** was checked for all three factors and a title was added. The chart is displayed in Figure 3.18.

The main insight to be gained here was that Anne's measurements were more variable than Robin's. For example, in the third panel of the display the four circular symbols represent the

C1	C2-T	C3	C4	C5	C6	C7	C8	C9	C10	C11
Part	Operator	Measurement	Trial							
1 Anne		1.445	1							
2 Anne		1.819	1							
3 Anne		2.320	1							
4 Anne		2.252	1							
5 Anne		1.579	1							
6 Anne		2.966	1							
7 Anne		3.200	1							
8 Anne		2.507	1							
9 Anne		1.624	1							
10 Anne		2.174	1							
1 Anne		1.255	2							
2 Anne		1.750	2							
3 Anne		1.955	2							

Multi-Vari Chart

C1	Part
C2	Operator
C3	Measurement
C4	Trial

Response: Measurement

Factor 1: Trial

Factor 2: Operator

Factor 3: Part

Factor 4:

Options...

Select

Help

OK

Cancel

Figure 3.17   Dialog for multi-vari chart.

four measurements obtained on the third part, indicated by the 3 in the legend above the panel. The circular symbols with the crosses represent the second measurement made and this is indicated in the legend to the right of the plot. The names of the two operators of the micrometer are indicated in the legend beneath the panel. The two circles for Anne's pair of measurements on the left have a much greater vertical separation then the two circles for Robin's measurements on the right. The fact that this pattern appears for the majority of the parts provides the insight. The square symbols on the line segments joining the pairs of circular symbols represent the means of the pairs of measurements. In the majority of cases Anne's mean for a part was less than Robin's. As the components were made of wood this raises the possibility that Anne was compressing the parts to some extent when taking measurements. The insights indicated a need for further investigation of the measurement process.

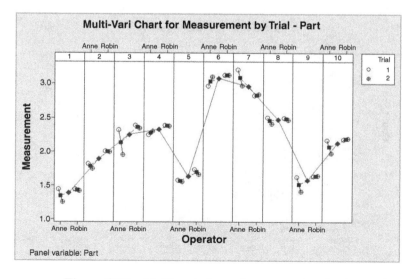

Figure 3.18   Multi-vari chart of measurement data.

The diamond symbols on the line segments joining the pairs of square symbols represent the mean of all four measurements on each of the parts. In order to complete the display the diamond symbols are connected by line segments.

Multi-vari charts may be used to display a response and up to four factors, i.e. one Y and up to four Xs – five-dimensional data displayed in two dimensions! It can be informative to experiment with different ordering of the factors.

3.3   Other displays

3.3.1   Pareto charts

Having collected information on, and established the nature and the level of nonconformities for a process in the measure phase of a Six Sigma project, it is clearly of interest to be able to identify the major contributors in terms of category of nonconformity for the improve phase of the project. Resources can then be focused on reduction of the impact of these contributors. Pareto analysis is a graphical tool for the display of the relative contributions of the categories of nonconformity that can occur.

As an example, consider the data in the worksheet Lenses1.MTW for a lens coating process. Each row of the table corresponds to a nonconforming lens identified at final inspection of the lenses produced during a production run of 2400 lenses of a particular type. Column C1 indicates category of nonconformity, and column C2 gives a reference number for the run. A Pareto chart of the data may be created using **Stat > Quality Tools > Pareto Chart...** with Nonconformity entered in the **Defects or attribute data in:** window and **Combine remaining defects into one category after this percent:** specified as 99.9. The resulting output is shown in Figure 3.19.

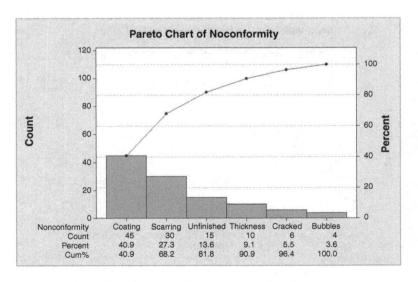

Figure 3.19   Pareto chart of nonconforming lenses.

```
┌─────────────────────────────────────────────────┐
│ Tally for Discrete Variables: Nonconformity      │
│                                                   │
│ Nonconformity   Count                             │
│       Bubbles       4                             │
│       Coating      45                             │
│       Cracked       6                             │
│       Scarring     30                             │
│     Thickness      10                             │
│    Unfinished      15                             │
│            N=     110                             │
│                                                   │
│                                                   │
└─────────────────────────────────────────────────┘
```

Panel 3.13   Frequency of nonconformity by category.

Summation of the six counts below the bar chart indicates that there were 110 non-conforming lenses for the run. Of these 45 were nonconforming due to a coating problem, representing 40.9% of the total number of nonconforming lenses. The first bar in the chart corresponds to coating nonconformities. The left-hand scale gives the count and the right-hand scale gives the corresponding percentage of the total count. The second highest number of nonconforming lenses arose from scarring, of which there were 30, representing 27.3% of the total. Coating and scarring together account for 68.2% of the nonconforming lenses. This cumulative percentage is plotted vertically above the middle of the bar in the chart corre-sponding to scarring. The points that correspond to the cumulative percentages are connected using line segments. (By clicking on **Options...** this cumulative plot can be omitted from the display.) The reader is invited to verify that setting **Combine remaining defects into one category after this percent:** specified as **95** leads to the cracked and bubbles categories being combined into a single category labelled Other by the software.) It is frequently the case that relatively few categories of nonconformity account for a relatively large proportion of the nonconforming product items. In this case it would appear potentially fruitful to seek ways to reduce the impact of coating and scarring problems with the lens coating process.

Use of **Stat > Tables > Tally Individual Variables...** leads to the display in the Session window shown in Panel 3.13.

The reader is invited to verify that by setting up Nonconformity and Count in two separate columns in a worksheet the Pareto chart in Figure 3.19 can be created using **Stat > Quality Tools > Pareto Chart...** with Nonconformity entered in the **Defects or attribute data in:** window, Count entered in the **Frequencies in:** window and **Combine remaining defects into one category after this percent:** specified as 99.9.

The supplier of the coating fluid used during the process was identified as a likely factor leading to nonconforming lenses in the coating category. A switch was made to a new supplier and the data in worksheet Lenses2.MTW was collected for a production run of 2400 lenses manufactured using the coating fluid from the new supplier. The corresponding Pareto chart is shown in Figure 3.20.

For this run there were 80 nonconforming lenses as opposed to 110 for the earlier run. As will be demonstrated formally in Chapter 7, this represents a significant process improvement. Secondly, coating is no longer the most important category so the change of supplier has been of real benefit. Visual comparison of the two charts can be facilitated by having the first two columns in the two worksheets stacked into a single column with a second column indicating the number of the production run. The display shown in Figure 3.21 may be created using **Stat > Quality Tools > Pareto Chart...** with Nonconformity entered in the **Defects or attribute data in:** window plus **BY variable in:** specified as Run with **Default (all on one**

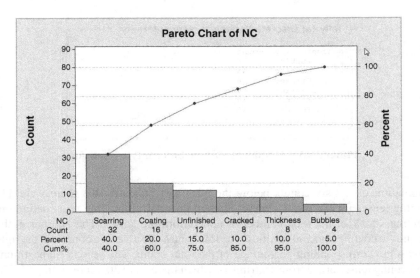

Figure 3.20   Pareto chart of nonconforming lenses after process change.

graph, same ordering of bars) checked and **Combine remaining defects into one category after thispercent:** specified as 99.9.

The display in Figure 3.21 has much more impact in colour! In the example considered above the display was of the counts for the different categories of nonconforming lenses. In some situations it may be desirable to display costs rather than counts. An exercise will be provided on this aspect of Pareto analysis.

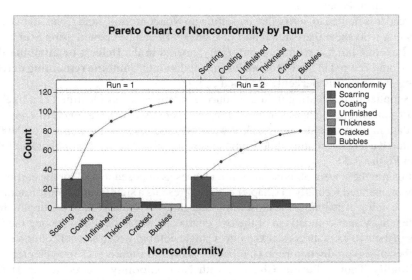

Figure 3.21   Pareto charts in a single display.

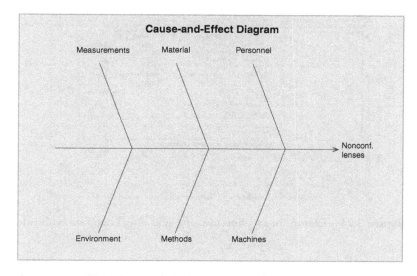

Figure 3.22   Skeleton of Cause-and-Effect diagram.

3.3.2   Cause-and-effect diagrams

Cause-and-effect, Ishikawa or fish-bone diagrams may be created using Minitab. Imagine that the Six Sigma project team involved in the lens coating improvement project followed up scrutiny of the Pareto chart in Figure 3.19 with a structured brainstorming session aimed at determining potential causes leading to lenses with coating that does not conform to requirements. The facilitator of the session could start by displaying the diagram in Figure 3.22 on a flipchart and inviting team members to 'flesh out' the bones of the 'fish' through suggesting additions that could be made.

The diagram may be created in Minitab using **Stat > Quality Tools > Cause-and-Effect...** with **Nonconf. lenses** specified in the **Effect:** box in the dialog. Once the discussion has been completed the possible causes identified in each of the six major cause categories can be listed in a worksheet as shown in Figure 3.23.

Use of **Stat > Quality Tools > Cause-and-Effect...** with Nonconf. lenses specified in the **Effect:** window in the dialog and the lists of potential causes specified as shown in Figure 3.24 yields the diagram in Figure 3.25. Note that in this case the names given to the columns used for the lists of potential causes matched the default category labels offered by Minitab

It may be desirable to subdivide potential causes into subcauses, e.g. under settings for the machines the factors temperature, coat time and agitation speed might all have been

↓	C1-T	C2-T	C3-T	C4-T	C5-T	C6-T
	Measurements	**Material**	**Personnel**	**Environment**	**Methods**	**Machines**
1	Temperature sensor	Quality of uncoated lenses	Insufficient training	Variable ambient temperature	Poor SOPs	Old
2		Supplier of coating fluid	Heavy workload	Dust contamination	Shift variations	Inadequate maintenance
3						Settings
4						

Figure 3.23   Lists of potential causes in the six major categories.

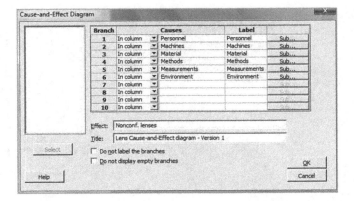

Figure 3.24   Dialog for creation of cause-and-effect diagram with title.

raised during the project team's discussion. These three subcauses can be listed in an additional column, named Settings, and the dialog shown in Figure 3.26 leads to the second version of the diagram in Figure 3.27. It is necessary to click on **Sub...** to the right of Machines in the main dialog box and to enter the column name Settings under the heading **Causes** in the row that contains the **Label** Settings in the subdialog box. Note that the name for the column containing the subcauses was chosen to match the name used for the cause, but this is not necessary.

Montgomery states that cause-and-effect analysis is an extremely powerful tool and that 'a highly detailed cause-and-effect diagram can serve as an effective trouble-shooting aid' (Montgomery, 2009, p. 204).

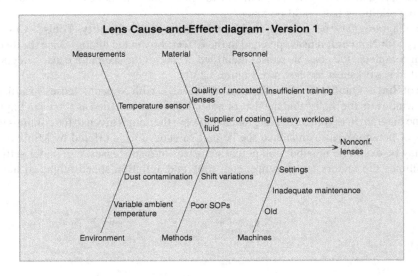

Figure 3.25   Revised cause-and-effect diagram – first version.

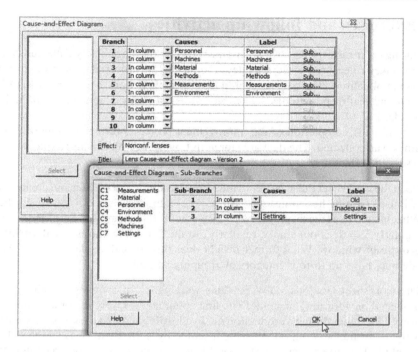

Figure 3.26   Dialog for creation version of cause-and-effect diagram with subcauses.

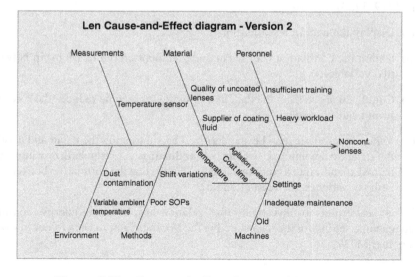

Figure 3.27   Cause-and-effect diagram with subcauses.

3.4   Exercises and follow-up activities

1. During the measure phase of a Six Sigma project a building society collected data on the time taken (measured as working days rounded to the nearest day) to process mortgage loan applications. The data are stored in the supplied worksheet Loans1.MTW. Use a stem-and-leaf display to determine the number of times which exceeded the industry standard of 14 working days.

   Following completion of the improvement project a further sample of processing times was collected. The data are stored in Loans2.MTW. Create boxplots in order to assess the effectiveness of the project.

2. Open the worksheet Pulse.MTW in the Minitab Sample Data folder and create a boxplot of Pulse1 (resting heart rate, beats per minute) rate versus sex (1 = Male, 2 = Female). What insight does the display provide? Re-edit the dialog and add Smokes (1 = Smokes regularly, 2 = does not smoke regularly) to the list in the **Categorical variables for grouping:** window. What further insight does this second display provide? Note that one may have up to four categorical grouping variables.

3. The worksheet Goujons.MTW contains weights (g) of samples of chicken goujons produced on a forming machine. The first column gives weights for a goujon selected every 15 minutes during a production run by Shift Team A and the second a similar set of weights for the next run on the same machine by Shift Team B. Given that the target weight is 17 g, use run charts, stem-and-leaf displays and boxplots to explore the data and comment on the performance of the two teams.

4. The supplied worksheet Wine.MTW gives annual wine consumption (litres of alcohol from wine, per capita) and annual deaths from heart disease (per 100 000 people), reported by Criqui and Ringel (1994) and discussed by Moore (1996, p. 314), for a set of countries. The data are reproduced with permission from Elsevier (*Lancet* 1994, 344, pp. 1719–1723).

   (i) Display the data in a scatterplot.

   (ii) Obtain the correlation coefficient and comment on the relationship between the two variables.

   (iii) Could you argue from the data that drinking more wine reduces the risk of dying from heart disease?

   (iv) Values in a sample may be standardized by subtracting the mean and dividing by the standard deviation. Use **Calc > Standardize...** to standardize x into a column named z_1 and y into a column named z_2. Verify that the correlation between x and y is the covariance between z_1 and z_2.

5. Use a scatterplot to investigate the relationship between energy consumption and machine setting for the data stored in the Minitab Sample Data folder in worksheet Exh_regr.MTW.

6. Engineers responsible for rectifying faults on automatic telling machines recorded repair duration, number of months since last service of machine and classified the faults as either hardware or software. The data are supplied in worksheet Faults.MTW.

   (i) Create a scatterplot of duration versus months and comment.

   (ii) Use the categorical variable fault type to stratify the scatterplot. This requires use of **Scatterplot...** > **With Groups** and selection of Fault Type in the **Categorical variables for grouping:** box. Try also the option **With Regression and Groups**. What insight do you gain from the displays?

7. As part of a project on solar thermal energy, total heat flux was measured for a sample of houses. You wish to examine whether total heat flux can be predicted from insolation, by the position of the focal points in the east, south, and north directions, and by the time of day. The data are stored in the Minitab Sample Data folder in columns C3 to C8 inclusive of the worksheet Exh_regr.MTW. Obtain the correlations between heat flux (Y) and the other five variables (Xs) and display the data in an appropriate matrix plot. Which of the Xs appear to influence heat flux?

8. In March 1999, *Quality Progress*, the monthly magazine of the American Society for Quality (ASQ), included an article by two ninth-grade students, Eric Wasiloff and Curtiss Hargitt, entitled *Using DOE to Determine AA Battery Life*. As one-tenth scale electric RC model-car racing enthusiasts they were interested in the theory that a high-cost battery with gold-plated connectors at low initial temperature will result in superior performance during a race. The results of an experiment they performed are displayed in Table 3.4. The data are reprinted with permission from the *Journal of Quality Technology* (© 1999 American Society for Quality).

   Here the response variable of interest was Y, the life (minutes) of the battery and the factors considered were X_1, the battery cost (low or high), X_2, the temperature at which the battery was stored prior to use (6 °C or 25 °C), and X_3, the connector type (standard or gold). Set up the data in Minitab and display them as a multi-vari chart. Explore different orderings of the factors in the dialog. What overwhelming conclusion would

Table 3.4 Battery life data.

Cost	Temperature	Connector	Life
Low	6	Standard	72
High	6	Standard	612
Low	25	Standard	93
High	25	Standard	489
Low	6	Gold	75
High	6	Gold	490
Low	25	Gold	94
High	25	Gold	493

you make? (You are advised to save the data in a worksheet as it will be referred to again in Chapter 8.)

9. The worksheet Shifts.MTW gives data for the number of units produced daily during eight-hour shifts by three shift teams over a three-week period. The teams were the same size and had identical facilities and supplies of materials available. Plot the data in a multi-vari chart. Find a solution to the problem with the ordering of the days and experiment with the ordering of the factors in the dialog box. What insights do the charts provide?

4

Statistical models

All models are wrong, but some are useful. (Box, 1999, p. 23)

Overview

Control charts are versatile statistical tools with a role to play in all four of the measure, analyse, improve and control phases of Six Sigma projects. In order to develop control charts from run charts, some understanding of statistical models for both discrete and continuous random variables is required, in particular of the normal or Gaussian statistical model. The normal distribution is also fundamental in understanding of the concept of sigma quality level referred to earlier in Section 1.1. Brief reference will also be made to the multivariate normal distribution. An understanding of statistical models in turn necessitates some fundamental knowledge of probability.

Finally, knowledge of the statistical properties of sums of independent random variables yields important results concerning means and proportions – results that are vital for the assessment of whether or not changes made during the improve phase of a Six Sigma project have been effective, for an appreciation of the way in which the various sources of errors in measurement processes contribute to the overall measurement error, and for an understanding of the construction of control charts.

Unlike other chapters, this one includes a number of exercises at various points, some of which do not require the use of Minitab. They are included to help the reader understand the topics of probability and statistical models as they are developed. Solutions to these exercises are provided on the book's website.

4.1 Fundamentals of probability

4.1.1 Concept and notation

Probability theory developed in the seventeenth century due, to some extent, to the dialogue between gamblers and mathematicians concerning the odds for games of chance involving

Table 4.1   Relative frequencies of sound tees.

Number tested, n	Number sound, f	Relative frequency, f/n
10	9	0.900 000
100	85	0.850 000
1 000	780	0.780 000
10 000	7 952	0.795 200
100 000	80 042	0.800 420
1 000 000	799 631	0.799 631

dice and cards. Further impetus came from the development of astronomy in the eighteenth and nineteenth centuries and the work of mathematicians Gauss and Legendre on problems such as the development of models for the orbits of comets. Subsequently the normal or Gaussian probability distribution provided Walter Shewhart with the foundation for the control chart in 1924 and was central to the development of inferential statistics by William Gosset and Ronald Fisher around the same time.

Consider an injection-moulding process for the manufacture of plastic golf tees that is behaving in a stable, predictable manner and where we record successive tees as either sound or defective. Suppose Table 4.1 summarizes findings as the recording of data progresses.

A plot of the relative frequency of sound tees against the logarithm to base 10 of the number of tees tested is shown in Figure 4.1. (The sequence of numbers tested is 10, 100, 1000, etc., which may be written as 10^1, 10^2, 10^3, etc. The sequence of indices 1, 2, 3, etc. is the sequence of logarithms to base 10 of the sequence of numbers tested in the first column.)

For small numbers tested, the relative frequency is unstable but, as the number tested becomes large, the relative frequency stabilizes at around 0.8. For this test there are two outcomes; the tee is either sound (S) or defective (D). The set of possible *outcomes* $\{D, S\}$ is called the sample space for the testing of a tee. The value 0.8 can be assigned as the *probability*

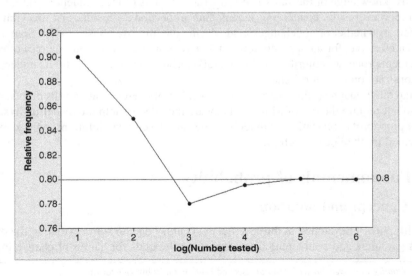

Figure 4.1   Relative frequency of sound tees.

of the outcome that a tee is sound. One can write in shorthand $P(S) = 0.8$. A little reflection should convince the reader that the relative frequency of defective tees would stabilize at 0.2. Therefore, $P(D) = 0.2$. The total of the probabilities assigned to the outcomes in a sample space is 1.

If a conventional cubic die is rolled then the sample space is $\{1, 2, 3, 4, 5, 6\}$, where the integers represent the number showing on the uppermost face of the die when it comes to rest. It is natural to assign probability 1/6 to each of the six outcomes since one would expect the relative frequency of each outcome to stabilize at the same value and the total of the six probabilities is 1. An *event* is defined as a subset of the sample space. For example, the subset $A = \{2, 4, 6\}$ of the sample space above is the event of rolling an even number. The probability of an event is the sum of the probabilities of the constituent outcomes. (Note that an event may consist of a single outcome.) Thus $P(A) = 1/6 + 1/6 + 1/6 = 1/2$. This means that in a long sequence of rolls of a die one would expect the relative frequency of an even result to stabilize at 0.5. Probability is a measure on a scale of 0 to 1 inclusive and may be considered as 'long-term' relative frequency of occurrence. If the probability of an event is 0 then it is impossible for the event to occur. If the probability of an event is 1 then the event is certain to occur.

Let us denote by $P(A \cup B)$ the probability of *either* event A *or* event B (*or* both) occurring is denoted $P(A \cup B)$, and by $P(A \cap B)$ the probability of *both* event A *and* event B occurring. (The 'cup' $\cup$ and 'cap' $\cap$ symbols are the union and intersection symbols used in the mathematics of sets. An informal way to remember that the $\cap$ symbol corresponds to *both* ... *and* ... is to think of fish $\cap$ chips!)

If for the rolling of a die the event B is defined as the score being a multiple of 3, then $B = \{3, 6\}$ so that $A \cup B = \{2, 3, 4, 6\}$ and $A \cap B = \{6\}$. The *complement* of event A, denoted by $\bar{A}$ and referred to as the event 'not-A', is the *non-occurrence* of A. Here $\bar{A} = \{1, 3, 5\}$. The event $B \mid A$ ('B given A') is the event that B occurs, knowing that A has occurred. Thus, for the example, it is the event that the score is a multiple of 3, given the information that it is an even number. Knowing that A has occurred, we are dealing with the reduced sample space consisting of the outcomes 2, 4 and 6, so we assign revised probabilities of one-third to each of these. Hence, $P(B|A) = 1/3$ since, of the three outcomes in the reduced sample space, only one is a multiple of 3. You should verify that $P(A|B) = 1/2$.

Exercise 4.1   Let C be the event that the score is a prime number, so that $C = \{2, 3, 5\}$. Write down $P(C), P(A|C), P(B|C), P(C|A)$ and $P(A \cap C)$.

4.1.2 Rules for probabilities

There are various rules for combining probabilities. Three fundamental ones will be considered.

Rule 1. $P(\bar{E}) = 1 - P(E)$.

For the example of the die above, $P(B) = 1/3$, so Rule 1 yields $P(\bar{B}) = 1 - P(B) = 2/3$. This means that the probability of a result that is not a multiple of three is $2/3$. This can be taken as an indication that out of every, say, 300 rolls of a die, one could expect two thirds of the rolls, i.e. 200 rolls, to yield results that are not multiples of three.

If the event $E_1 \cap E_2 = \{ \ \}$, the empty set, then the events E_1 and E_2 are said to be *mutually exclusive*. In this case $P(E_1 \cap E_2) = 0$, i.e. when E_1 and E_2 are mutually exclusive it is impossible that both occur simultaneously.

Students in an introductory statistics course participated in a simple experiment. Each student recorded his or her height, weight, gender, smoking preference, usual activity level, and resting pulse. Then they all flipped coins, and those whose coins came up heads ran in place for one minute. Then the entire class recorded their pulses once more.

Column	Name	Count	Description
C1	Pulse1	92	First pulse rate
C2	Pulse2	92	Second pulse rate
C3	Ran	92	1 = ran in place 2 = did not run in place
C4	Smokes	92	1 = smokes regularly 2 = does not smoke regularly
C5	Sex	92	1 = male 2 = female
C6	Height	92	Height in inches
C7	Weight	92	Weight in pounds
C8	Activity	92	Usual level of physical activity: 1 = slight 2 = moderate 3 = a lot

Panel 4.1   Description of the Pulse.MTW data set.

Rule 2.   If the events E_1 and E_2 are mutually exclusive then

$$P(E_1 \cup E_2) = P(E_1) + P(E_2).$$

Thus when dealing with an *either ... or* situation for **mutually exclusive** events one must *add* probabilities.

In order to introduce the third rule reference will be made again to the Pulse.MTW data set encountered in the previous chapter. The description of the data set displayed in Panel 4.1 may be accessed using **Help > Help**, clicking on the **Search** tab, entering Pulse in the **Type in the word(s) to search for:** window, clicking the **List Topics** button and then double-clicking on PULSE.MTW in the list of topics.

With the worksheet PULSE.MTW open, use of **Stat > Tables > Tally Individual Values...**, with Ran entered under **Variables:** and both **Counts** and **Percent** checked, yields the Session window output displayed in Panel 4.2. This indicates that 35 of the 92 students in the class ran. As the decision whether or not to run was meant to be based on the outcome of the toss of a coin, one would expect the relative frequency of heads to have been approximately 0.50 while in fact it was 0.38, to two decimal places. (Did some students cheat through not running in place in spite of obtaining a head on their coins? This type of question will be addressed formally in Chapter 7.)

```
┌─────────────────────────────────────────┐
│  Tally for Discrete Variables: Ran        │
│                                           │
│  Ran  Count  Percent                      │
│    1     35   38.04                       │
│    2     57   61.96                       │
│   N=     92                               │
└─────────────────────────────────────────┘
```

Panel 4.2   Tally of variable Ran.

Use of **Stat > Tables > Cross Tabulation and Chi-Square...** with **Categorical variables:** specified as **For rows:** Sex and **For cols:** Smokes, with **Counts** and **Row percents** checked, yields the table in Panel 4.3. The value 1 for the variable Sex indicates a male and the value 1 for the variable Smokes indicates a regular smoker. Thus, to the nearest whole per cent, 30% of the students smoked regularly, while 23% of the female students smoked regularly and 35% of the male students smoked regularly. Is smoking gender-dependent? If $P(E_2|E_1) = P(E_2)$ then we say that the events E_1 and E_2 are *independent*. Let S denote the event that a student is a regular smoker and let F denote the event that a student is female. If we regard the 92 students as a sample from a population of students then we have the estimates $P(S) = 30\% = 0.30$ and $P(S|F) = 23\% = 0.23$. The fact that these two estimates differ suggests that smoking might be gender-dependent in the student population. Formal assessment of dependence will be considered in Chapter 10.

Rule 3.   If the events A and B are **independent** then

$$P(A \cap B) = P(A) \times P(B).$$

Thus when dealing with a *both ... and* situation for independent events one must *multiply* probabilities.

Consider blood groups in the UK population. The mutually exclusive groups are O, A, B and AB with respective probabilities 0.46, 0.42, 0.09 and 0.03 (British Broadcasting Corporation, 2004). Thus, for example, the probability that a randomly selected member of the population has blood of group either A or B is $0.42 + 0.09 = 0.51$. The rhesus factor is

```
┌──────────────────────────────────────────────┐
│  Tabulated statistics: Sex, Smokes             │
│                                                │
│  Rows: Sex    Columns: Smokes                  │
│                                                │
│              1       2      All                │
│                                                │
│  1          20      37       57                │
│          35.09   64.91   100.00                │
│                                                │
│  2           8      27       35                │
│          22.86   77.14   100.00                │
│                                                │
│  All        28      64       92                │
│          30.43   69.57   100.00                │
│                                                │
│  Cell Contents:      Count                     │
│                      % of Row                  │
└──────────────────────────────────────────────┘
```

Panel 4.3   Cross-tabulation of Sex and Smokes.

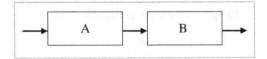

Figure 4.2   System with two subsystems.

present in the blood of 85% of the UK population and absent from the blood of the remainder of the population. Those with the factor present are said to be rhesus positive, while those with the factor absent are said to be rhesus negative. The presence or absence of the rhesus factor is independent of blood group. The author is 'O rhesus negative'. We have P(Rhesus positive) = 0.85, so by Rule 1,

$$P(\text{Rhesus negative}) = 1 - 0.85 = 0.15.$$

By Rule 3,

$$P(\text{Both O and rhesus negative}) = P(\text{O}) \times P(\text{Rhesus negative}) = 0.46 \times 0.15 = 0.069.$$

Thus just less than 7% of the UK population are O rhesus negative.

The multiplication rule for the combination of probabilities of independent events can be applied to systems made up of a series of subsystems or processes that function independently of each other. Figure 4.2 depicts a system comprising two subsystems, A and B. Let the probability that each of the subsystems functions correctly be 0.9. Assuming independence, the multiplication rule yields probability $0.9 \times 0.9 = 0.81$ that the overall system functions correctly. This probability of 0.81 is often referred to as the reliability of the system. With three such subsystems the reliability would be $0.9 \times 0.9 \times 0.9 = 0.9^3 = 0.729$.

Reference was made in Chapter 1 to the concept of sigma quality level (or sigma) for a process. Consider a system consisting of 10 independent subsystems where each of the subsystems is created by a process operating with a sigma quality level of 3. As can be seen from Appendix 1, a sigma quality level of 3 corresponds to 66 811 nonconformities per million, which equates to a probability of $1 - \frac{66\,811}{1\,000\,000} = 0.933\,189$ that each subsystem functions correctly. The reliability of the system would be $0.933\,189^{10} = 0.5008$, to four decimal places. Thus only 50% of the systems would function correctly. (This calculation may be performed in Minitab using **Calc > Calculator** with **Expression:** 0.933 189**10 and **Store result in variable:** Answer. The result may then be read from the column named Answer created in the current worksheet.)

Exercise 4.2   A system consists of 1000 subsystems each of which is produced by a Six Sigma process, i.e. a process with sigma quality level 6. Verify that the relative frequency of system failure is 3 in 1000.

A key contributor to the development of Six Sigma at Motorola was senior quality engineer Bill Smith. One product he worked on had a much higher failure rate than predicted, despite great care having been taken during its design.

Smith came to realize that it was the accumulation of a lot of little defects made during the manufacturing process – not inherent design flaws – that caused the high rate of early-life failures. Eliminating the source of those defects was therefore the only way the company could deliver higher quality to its customers.

(Reynard, 2007, p. 23)

Thus appreciation of the relevance of probability calculations such as that in Exercise 4.2 to the need for low failure rates amongst the large number of individual components in complex systems, such as a cellular telephone, is key.

4.2   Probability distributions for counts and measurements

4.2.1   Binomial distribution

Consider a situation where there is constant probability p that an item produced by a process is nonconforming. Let D denote a nonconforming item and S denote a conforming item. Thus we have $P(D) = p$ and $P(S) = 1 - p$ (by Rule 1), which will be denoted by q. Suppose that samples of $n = 2$ items are selected at random from the process output. The sample space consists of the four sequences SS, SD, DS, DD where, for example, the sequence DS represents the outcome that the first item selected was nonconforming and the second was conforming. Rule 3 gives the probabilities of the above four outcomes as qq or q^2, qp, pq and pp or p^2, respectively.

The number of nonconforming items in samples of $n = 2$ items is a count or, more formally, a discrete random variable. It is conventional to use an upper-case letter to denote a random variable and the corresponding lower-case letter to denote a specific value that random variable may take. Let the number of nonconforming items in samples of two items be denoted by X. Thus specific values of X will be denoted by x. Table 4.2 demonstrates the calculation of the probability that a random sample of two items includes precisely x nonconforming items for $x = 0$, 1 and 2. Thus $P(X = 0) = q^2$, $P(X = 1) = 2qp$ (note use of Rule 2) and $P(X = 2) = p^2$. The probability function for a discrete random variable is typically denoted by $f(x)$ and is defined as:

$$P(X = x) = f(x).$$

The probability function in Table 4.3 is for the specific case of samples of $n = 2$ and for probability $p = 0.2$, i.e. where 20% of items are nonconforming.

Exercise 4.3   Tabulate the probability functions for the cases where $n = 2$ and $p = 0.5$ and where $n = 2$ and $p = 0.7$.

Note that in all cases the sum of the probabilities is 1. This must always be the case for a probability function for a discrete random variable. In general the fact that the probabilities

Table 4.2   Derivation of probabilities.

No. of nonconforming items, x, in sample of two	Outcomes yielding x nonconforming items	Probability that sample includes x nonconforming items
0	SS	$qq = q^2$
1	Either SD or DS	$qp + pq = 2qp$
2	DD	$pp = p^2$

Table 4.3 Probability function.

x	$f(x)$
0	0.64
1	0.32
2	0.04

sum to 1 may be demonstrated as follows, use being made of the fact that since $q = 1 - p$ then $q + p = 1$:

$$q^2 + 2qp + p^2 = (q + p)^2 = 1^2 = 1.$$

Since the two-term expression $q + p$ is referred to as a binomial expression in mathematics, this type of discrete probability distribution is referred to as a *binomial distribution*. In order to specify the distribution, the number of items tested or number of trials, n, and the constant probability, p, that an individual item is nonconforming are required. The numbers n and p are referred to as the two *parameters* of the distribution. The short-hand $B(n, p)$ is used for the binomial distribution with parameters n and p and the short-hand $X \sim B(n, p)$ is used to indicate that the random variable X has the specified binomial distribution.

Minitab provides a facility for the calculation of binomial and other widely used probability functions. Having set up and named columns C1 and C2 as indicated in Figure 4.3, **Calc > Probability Distributions > Binomial...**, with the **Probability** option selected, may be used to reproduce the probabilities in Table 4.3. **Number of trials:** was specified as 2 and **Event probability:** as 0.2. **Input column:** specifies the column, named x, containing the values of the random variable for which probabilities are to be calculated and **Optional storage:** 'f(x)' indicates where the probabilities are to be stored. (Recall that the column named f(x) may be selected for storage by highlighting and left clicking or by typing the column name enclosed in single quotes.) The probability function is displayed as a bar chart in Figure 4.4.

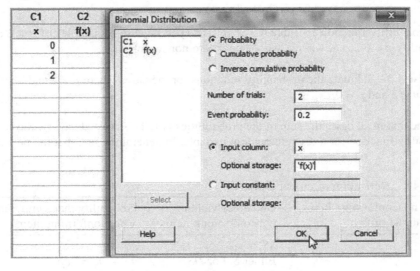

Figure 4.3 Obtaining binomial probabilities.

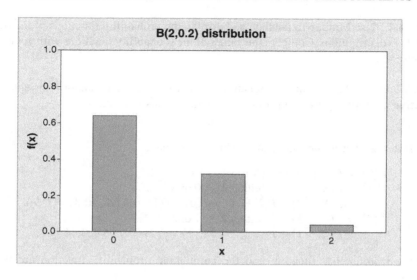

Figure 4.4   Display of $B(2, 0.2)$ probability function.

When the parameter p is less than 0.5 the distribution is positively skewed as in Figure 4.4. When $p = 0.5$ the distribution is symmetrical and when p is greater than 0.5 the distribution is negatively skewed. The display in Figure 4.4 was created using **Graph** > **Bar Chart...** – see the dialog in Figure 4.5. **Bars represent:** Values from a table was selected initially followed by **Simple** under **One column of values**. The column of probabilities f(x) was selected under **Graph variables:** and **Categorical variable:** x for the horizontal axis. **Labels...** was used to create the title B(2, 0.2) distribution. The vertical scale was altered to have maximum value 1, since a probability cannot exceed 1. (The alteration was made by double-clicking on the vertical scale, selecting the **Scale** tab and, under **Scale Range**, unchecking **Auto** for **Maximum** and entering the value 1 in the appropriate window.)

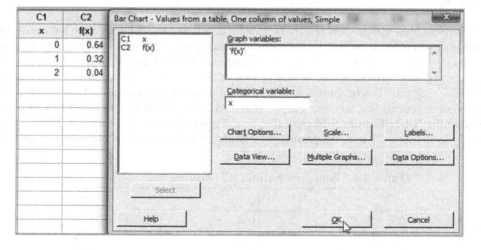

Figure 4.5   Creating display of $B(2, 0.2)$ distribution.

Exercise 4.4   Use Minitab to tabulate and display the probability functions of the $B(2, 0.5)$ and $B(2, 0.7)$ distributions. Compare the values of the probability functions with your answers to Exercise 4.3.

In addition to the probability function, $f(x)$, for a discrete random variable, X, the **cumulative probability function** $F(x)$ is important. This is given by

$$F(x) = P(X \leq x).$$

For the $B(2, 0.2)$ distribution tabulated Table 4.2 we have:

$$
\begin{aligned}
F(0) &= P(X \leq 0) = P(X = 0) = f(0) = 0.64, \\
F(1) &= P(X \leq 1) = P(\text{either } X = 0 \text{ or } X = 1) \\
&= P(X = 0) + P(X = 1) = f(0) + f(1) = 0.64 + 0.32 = 0.96, \\
F(2) &= P(X \leq 2) = P(\text{either } X \leq 1 \text{ or } X = 2) \\
&= P(X \leq 1) + P(X = 2) = F(1) + f(2) = 0.96 + 0.04 = 1.
\end{aligned}
$$

Both the probability function and the cumulative probability function are tabulated in Table 4.4.

You should check the distribution function, $F(x)$, using Minitab by selecting the **Cumulative probability** option in the dialog box displayed in Figure 4.3. The software does not distinguish between f(x) and F(x) as column names so one could give the name F.(x) to column C3 and then use it for storage of the results. In plain English, $f(1)$ gives the probability that a sample includes *precisely one* nonconforming item whereas $F(1)$ gives the probability that a sample includes *one or fewer* nonconforming items. In general the probability function, $f(x)$, gives the probability that a sample includes *precisely x* nonconforming items, while the cumulative probability function, $F(x)$, gives the probability that a sample includes *x or fewer* nonconforming items.

Exercise 4.5   From the tables created in answering Exercise 4.3 tabulate the cumulative probability function $F(x)$ for the $B(2, 0.5)$ and $B(2, 0.7)$ distributions. Check your answers using Minitab.

Suppose that the process referred to above continues to produce items with constant probability of 0.2 that an item is nonconforming and that it is decided to monitor the process by taking samples of $n = 25$ items at regular intervals. Grass-roots calculation of the probability function in this case would be very tedious indeed. Mathematical formulae are available but will not be introduced here. If required, the probability function and the cumulative probability function can be obtained using Minitab. One would use **Calc > Probability Distributions > Binomial...** as in Figure 4.3 with the values $0, 1, 2, \ldots, 24, 25$ set up in column C1 and the **Number of trials** specified as 25 in order to tabulate the

Table 4.4   Functions for $B(2, 0.2)$ distribution.

x	$f(x)$	$F(x)$
0	0.64	0.64
1	0.32	0.96
2	0.04	1.00

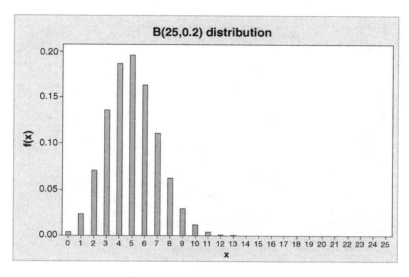

Figure 4.6   Display of $B(25, 0.2)$ distribution.

probability function. This function is displayed in Figure 4.6, with the minimum set to 0 on the vertical scale.

Note that it would be rare to obtain a sample including more than 10 nonconforming items. The most likely number of nonconforming items is 5, and this value is known as the *mode* of the distribution.

What might a run chart of a typical series of counts of numbers of nonconforming items in samples of 25 look like? Minitab enables random data from a wide variety of distributions to be generated. Having assigned No. N-C as the name of column C1, one can use **Calc > Random Data > Binomial...** as indicated in Figure 4.7. By specifying **Number of**

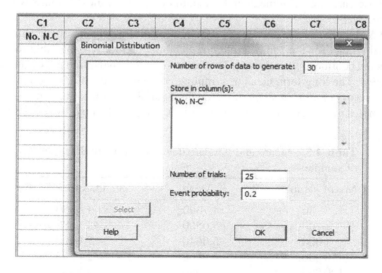

Figure 4.7   Generating data from the $B(25,0.2)$ distribution.

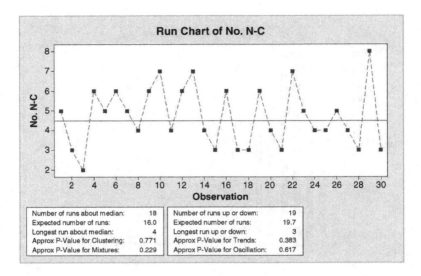

Figure 4.8   Run chart of the data.

rows to generate: 30 we are indicating that we wish to simulate the selection of 30 samples of $n = 25$ items from a population of items in which there is probability $p = 0.2$ that an item is nonconforming. No. N-C is selected for **Store in column(s):** and the distribution parameters $n = 25$ and $p = 0.2$ are specified as before.

A run chart of a data set generated by the author in this manner is displayed in Figure 4.8. Since data have been simulated here for a stable, predictable process with constant probability 0.2 of an item being nonconforming it should be no surprise that none of the P-values are less than 0.05 and that therefore there is no evidence of any variation other than common cause variation. The mean and standard deviation of the 30 counts were 4.700 and 1.535, respectively.

Table 4.5 gives the means and standard deviations for longer and longer series of samples of simulated counts for the same stable, predictable process. It is possible to imagine the sampling being continued *ad infinitum*. The resulting conceptual infinity of counts is known as the *population*. The long-term mean as sampling continues is denoted by μ and is called the *expected value* or *population mean* of the random variable X, the number of nonconforming items in samples of 25. The long-term mean of $(X - \mu)^2$ is denoted by σ^2 and is called the

Table 4.5   Means and standard deviations for simulated series of samples.

No. of samples	Mean	Standard deviation
30	4.7000	1.5350
100	5.0500	1.8880
1 000	5.0640	2.0100
10 000	4.9968	2.0067
100 000	4.9996	1.9997

variance of the population; σ is the *standard deviation* of the population. In standard statistical notation:

Population mean μ = Expected value of $X = E[X]$;
Population variance σ^2 = Expected value of $(X - \mu)^2 = E[(X - \mu)^2]$.

Statistical theory provides important results for the $B(n, p)$ distribution:

$$\mu = np, \quad \sigma^2 = npq, \quad \sigma = \sqrt{npq}.$$

Substitution of $n = 25$, $p = 0.2$ and $q = 1 - p = 0.8$ into the above formulae yields 5 and 2 for the population mean and standard deviation, respectively. This is in accord with the sequences of sample means and standard deviations obtained from the simulations and displayed in Table 4.5. The column of means appears to be 'homing in' on the expected value 5 and the column of standard deviations on the expected value 2. The binomial distribution provides the basis for two widely used control charts to be introduced in Chapter 5.

Exercise 4.6 A public utility company believes that currently 10% of billings of private customers are posted out after the scheduled date. Suppose that samples of 250 of the accounts posted each working day are checked for delay in posting. Denote by X the number of accounts in the sample which are posted after the scheduled date.

   (i) State the distribution of X and its parameters.

  (ii) Calculate the mean and standard deviation of X.

 (iii) Simulate data for 50 days, display it in a run chart, obtain the mean and standard deviation of the 50 daily counts and compare with your answers to (ii).

 (iv) Repeat (iii) for 500 days and for 5000 days.

  (v) Obtain the probability that a sample contains fewer than 15 delayed accounts.

4.2.2 Poisson distribution

Scrutiny of the menu under **Calc > Probability Distributions** reveals a total of 24 distributions available to provide statistical models! Distributions that model count data – in other words, that model discrete random variables – are the binomial, hypergeometric, discrete, integer and Poisson. The hypergeometric distribution has important applications in acceptance sampling and is referred to later in this chapter. From the point of view of quality improvement the two most important discrete distributions are the binomial and Poisson.

    The Poisson distribution is named in honour of the French mathematician Siméon Denis Poisson and may be used to model situations where a discrete random variable X may take any of the integer values 0, 1, 2, 3, ..., with no upper limit on the range of possible values. The Poisson distribution has a single parameter, usually denoted by the Greek letter λ (lambda). Both the mean and variance of the distribution are equal to λ. Thus the standard deviation is $\sqrt{\lambda}$.

Table 4.6 Frequency of V2 bomb hits.

No. hits	0	1	2	3	4	5	6	7	≥8
No. squares	229	211	93	35	7	0	0	1	0

The Poisson distribution provides an important model for counts of random events in both time and space. During the Second World War the city of London was attacked by German V2 flying bombs. In order to assess the ability of the Germans to aim at specific targets, British scientists divided the city into a set of 576 squares, each of side 0.5 km. The number of V2 bomb hits per square was counted, yielding the following frequency table displayed in Table 4.6 (data reproduced by permission of the Institute and Faculty of Actuaries from Clarke, 1946, p. 481).

In order to fit the Poisson model to this data set an estimate of the model parameter is required. The Poisson parameter λ is equal to the mean, so the mean number of hits per square will provide an estimate of this parameter. A convenient way to input the data into Minitab is to first name column C1 in a new worksheet No. Hits, make the Session window active, select the **Editor** menu and check **Enable Commands**. The software responds by presenting the prompt MTB > in the Session window. The user is now able to 'drive' using session commands as well as menu commands. The SET command now be used to input the data as indicated in Panel 4.4. Note, for example, that the notation 93(2) indicates that 93 values of 2 for the number of hits were obtained. The single value of 7 for the number of hits could have been indicated by 1(7) but the solitary 7 is sufficient.

```
MTB > set c1
DATA> 229(0) 211(1) 93(2) 35(3) 7(4) 7
DATA> end
MTB > Tally 'No. Hits';
SUBC>    Counts.
```

Tally for Discrete Variables: No. Hits

```
 No.
Hits   Count
   0     229
   1     211
   2      93
   3      35
   4       7
   7       1
  N=     576
```

```
MTB > Describe 'No. Hits';
SUBC>    Mean;
SUBC>    Count.
```

Descriptive Statistics: No. Hits

```
             Total
Variable   Count    Mean
No. Hits     576   0.9323
```

Panel 4.4 Input and analysis of V2 bomb data.

Table 4.7 Observed and expected frequency of V2 bomb hits.

No. Hits	0	1	2	3	4	5	6	7	≥8
Observed No. Squares	229	211	93	35	7	0	0	1	0
Expected No. Squares	226.7	211.4	98.5	30.6	7.1	1.3	0.2	0.0	0.0

Having input the data, the author then used the **Stat** menu and **Stat > Tables > Tally Individual Values...** to check the input. The session commands corresponding to the menu actions appear in the Session window. Finally, **Basic Statistics > Display Descriptive Statistics...** from the **Stat** menu was used, with **Mean** and **N total** selected via **Statistics...**, to obtain the mean number of hits.

The mean of 0.9323 can be used to compute Poisson probabilities via **Calc > Probability Distributions > Poisson....** The Poisson model gives the probability of no hits in a square to be 0.393 647, so the expected frequency of squares with no hits is $576 \times 0.393\,647 = 226.7$ correct to one decimal place. This is very close to the observed frequency of squares with no hits, 229. The expected frequencies for 1, 2, 3, 4, 5, 6 and 7 hits per square may be calculated in a similar fashion. Use of the **Cumulative probability** facility with 7 entered in the **Input constant:** box yields $F(7) = P(X \le 7) = 0.999\,99$ so $P(X \ge 8) = 1 - 0.999\,99 = 0.000\,01$. Hence the expected frequency of squares with 8 or more hits is $576 \times 0.000\,01 = 0.0$ correct to one decimal place. Table 4.7 gives the summarized results.

The bar chart in Figure 4.9 highlights how well the Poisson distribution models the situation. (A follow-up exercise will indicate how such charts may be created using Minitab.) The good fit of the Poisson distribution with parameter 0.9323, i.e. of the P(0.9323) distribution in shorthand, indicated that the V2 impacts were occurring at random locations within the city and thus provided evidence to the scientists that the flying bombs were not equipped with a sophisticated guidance system.

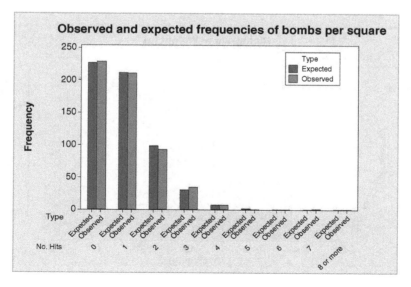

Figure 4.9 Comparison of observed and expected frequencies.

The Poisson distribution often provides a statistical model for counts of events occurring at random in either space or time, e.g. the number of nonconformities on a printed circuit board or the number of line stoppages per month in a factory. As with the binomial distribution, it provides the basis for two very important types of control chart to be introduced in Chapter 5.

4.2.3   Normal (Gaussian) Distribution

The normal distribution is central in the application of statistical methods in quality improvement and in understanding the concept of sigma quality level within Six Sigma programmes. The name 'normal' is unfortunate in that it suggests that such distributions are to be expected as some sort of norm. Some authors refer to it as the Gaussian distribution in honour of the German mathematician, Karl Friedrich Gauss. The normal distribution provides a model for continuous random variables. To introduce the normal distribution use will be made of the bottle weight data that is available in Weight1A.MTW and was displayed in Figure 2.22 in Chapter 2.

In the histogram in Figure 4.10 a different set of bins was used from that used in Figure 2.22. The bins for the histogram in Figure 4.10 were bounded by the values 483.0, 485.0, 487.0, 489.0, 491.0, 493.0 and 495.0. The first bin range from 483.0 to 485.0 has a range of 2 and there was a single bottle from the sample of 25 with weight in this range. The relative frequency of weight in this range for the sample was therefore $1/25 = 0.04$. This relative frequency provides an estimate of the probability that a bottle from the population of bottles sampled has weight in the corresponding bin range. Thus a probability of 0.04 spread over a range of 2 is equivalent to a probability of 0.02 spread over a range of 1; the *probability density* corresponding to the first bin is $0.04/2 = 0.02$. You should verify that the probability densities corresponding to the remaining bins are 0.08, 0.14, 0.16, 0.04 and 0.06. Minitab offers the option of displaying a density histogram of the weight data, in which the height of the bars corresponds to probability density rather than frequency. This histogram is shown in

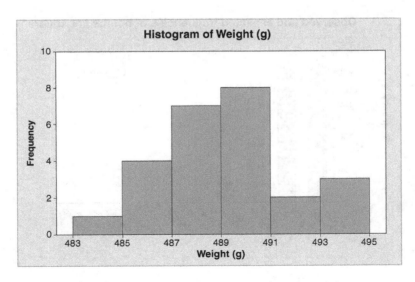

Figure 4.10   Frequency histogram of bottle weights.

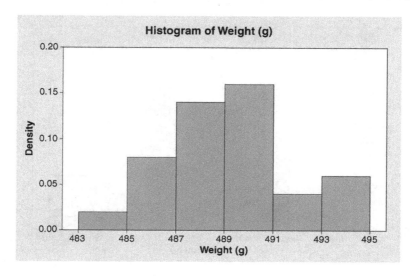

Figure 4.11   Density histogram of bottle weights.

Figure 4.11. Note that the vertical scale now represents density as opposed to frequency. Some properties of the density histogram are given in Table 4.8. The areas of the bars were obtained by multiplying width by height. It should be borne in mind that the histograms are not drawn to scale!

If the probability that a bottle weight is less than or equal to, say, 487 g is required then it may be estimated from the data via the histogram as follows:

$P(\text{Weight} \leq 487)$
$= P(\text{Either weight lies in bin 483, 485 or weight lies in bin 485, 487})$
$= P(\text{Weight lies in bin 483, 485}) + P(\text{Weight lies in bin 485, 487})$
$= 0.04 + 0.16$
$= 0.20.$

Thus the estimated probability that a bottle weight is less than or equal to 487 g is the shaded area indicated in the density histogram in Figure 4.12.

Table 4.8   Properties of the density histogram of weight.

Weight bin range	Histogram bar width	Histogram bar height	Histogram bar area
483–485	2	0.02	0.04
485–487	2	0.08	0.16
487–489	2	0.14	0.28
489–491	2	0.16	0.32
491–493	2	0.04	0.08
493–495	2	0.06	0.12
Total area of histogram			1.00

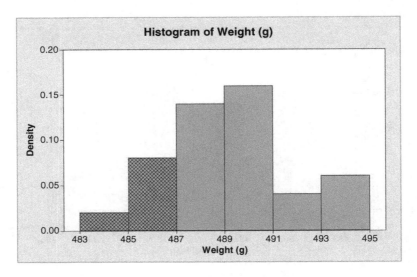

Figure 4.12   Shaded area gives $P(\text{Weight} \leq 487)$.

A further option available for histograms in Minitab is the fitting of a normal distribution. The curve in Figure 4.13 is the probability density function of the fitted normal distribution. The parameters of the fitted distribution are the sample mean and sample variance. It appears that the normal distribution provides a reasonable model for the random variable bottle weight.

The probability density function $f(x)$ for a continuous random variable, X, can never be less than 0 and is such that the total area it encloses with the horizontal axis is 1. This value of 1 represents total probability. The cumulative probability function $F(x)$ gives $P(X \leq x)$. A

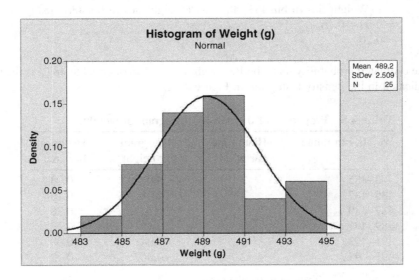

Figure 4.13   Normal distribution fitted to bottle weight data.

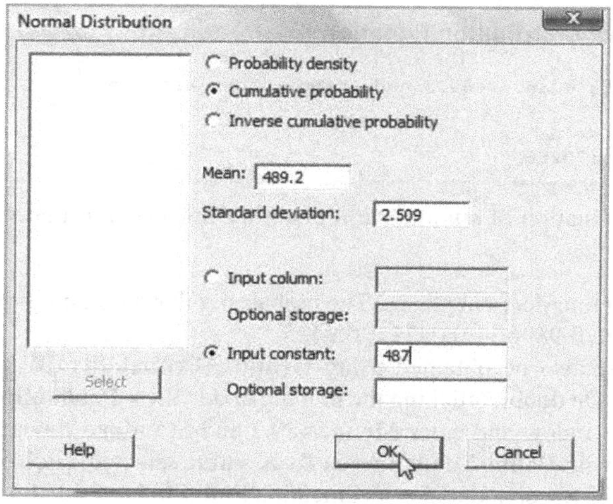

Figure 4.14 Evaluation of cumulative probability function for a normal distribution.

normal distribution is specified by two parameters: the mean and the variance. The normal distribution with mean 0 and variance 1 is referred to as the *standard normal distribution*. Because of the central role in statistics of the standard normal distribution its probability density function and its cumulative probability function are denoted by the special functions $\phi(x)$ and $\Phi(x)$, respectively. The letters ϕ and Φ are the lower- and upper-case Greek letters phi. Tables of the function $\Phi(x)$ are available but with Minitab you will have no need for them.

The sample mean and sample variance for the bottle weight data are 489.2 and 2.509^2 respectively, and these are the parameters used for the fitted normal distribution shown in Figure 4.13. In shorthand this distribution is denoted by $N(489.2, 2.509^2)$. (As with the binomial and Poisson distributions, the letter in front of the brackets indicates the distribution type and the numbers within the brackets are the parameters.) In order to obtain from the model the probability $P(\text{Weight} \leq 487) = F(487)$, the cumulative probability function has to be evaluated for 487. This may be obtained using **Calc > Probability Distributions > Normal. . . .** Note that the cumulative probability function and the standard normal distribution are the defaults. The value of interest, 487, is entered in the **Input constant:** window. The appropriate mean and standard deviation must be specified in the **Mean:** and **Standard deviation:** windows respectively. The dialog is completed as shown in Figure 4.14.

The Session window output is shown Panel 4.5. This indicates that the fitted normal distribution model gives $F(487) = 0.190\,286$, i.e. the probability of a bottle weight of 487 g or less is 0.19 to two decimal places. This agrees quite closely with the empirical probability of 0.20 represented by the shaded area in Figure 4.13. In fact the evaluation of the cumulative probability function involves calculation of the area under the probability density function lying to the left of 487. This area is shaded in Figure 4.15.

In Chapter 2 specification limits of 485 and 495 g for bottle weight were indicated. The fitted model may be used to estimate the proportion of bottles meeting those requirements. You should verify, using dialogs similar to that in Figure 4.14, that $F(485) = 0.0471$ and that

Cumulative Distribution Function

```
Normal with mean = 489.2 and standard deviation = 2.509

   x  P( X <= x )
 487    0.190286
```

Panel 4.5 Evaluation of cumulative probability function for a normal distribution.

$F(495) = 0.9896$ to four decimal places. The probability that a bottle weight lies between 485 and 495 is therefore $0.9896 - 0.0471 = 0.9425$.

This result may also be obtained using **Graph > Probability Distribution Plot... > View Probability**. On double-clicking the graphic under **View Probability**, select Normal in the **Distribution:** window and enter **Mean:** 489.2 and **Standard deviation:** 2.509. On the **Shaded Area** tab, with **Define Shaded Area By X Value** selected, click on the graphic under **Middle** and enter the lower and upper specification limits 485 and 495 for bottle weight in the **X_value 1:** and **X_value 2:** boxes, respectively. The graph in Figure 4.16 results, confirming the required probability as 0.9425.

Thus the model predicts that 94.25% of bottles from the process would conform to requirements on weight. The proportion nonconforming is therefore estimated to be 6.75% or 67 500 per million. Reference to Appendix 1 indicates the sigma quality level to be approximately 3. (The sample size of 25 is small, so in practice estimates based on such samples should be viewed with some caution.)

Earlier it was stated that the normal distribution appears to provide a reasonable model for the random variable bottle weight in view of the manner in which the curve (the model) in Figure 4.13 fitted the data (the histogram). In order to make a formal assessment whether or not a normal distribution provides a satisfactory model, the normality test provided in Minitab may be used. The points in the associated plot should be reasonably linear and

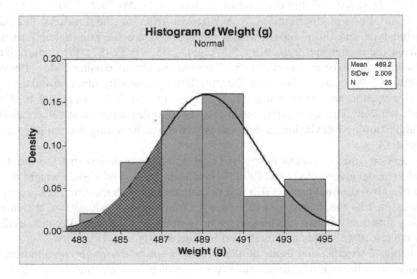

Figure 4.15 Area representing cumulative probability function.

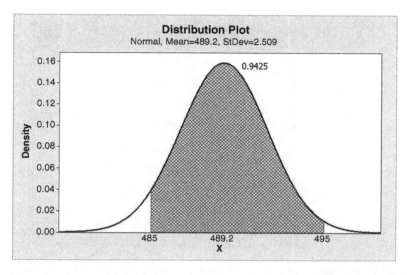

Figure 4.16   Area representing proportion of bottles conforming to weight specifications.

a P-value in excess of 0.05 is usually taken to mean that a normal distribution may be accepted as a satisfactory model. The normality test is available via **Stat > Basic Statistics > Normality Test....** The output obtained with the default options is shown in Figure 4.17. With P-value well in excess of 0.05 the normal distribution model is clearly acceptable. (The AD value quoted is the Anderson–Darling test statistic. Test statistics will be explained in Chapter 7.)

Consider now the conduct of interviews by a researcher undertaking a customer satisfaction survey. Suppose that the duration, in minutes, of interviews can be adequately modelled by the $N(40, 8^2)$ distribution, i.e. by the normal or Gaussian distribution with mean 40 and

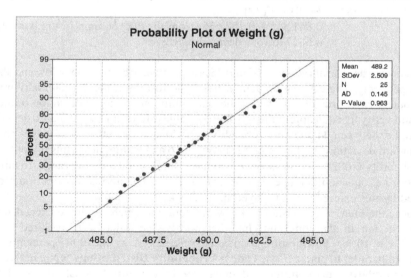

Figure 4.17   Normality test of bottle weight data.

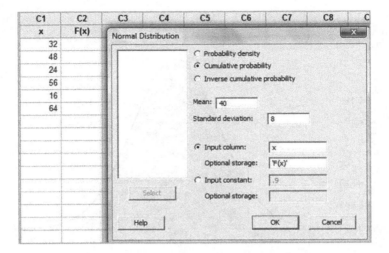

C1	C2	C3	C4	C5	C6	C7	C8	C
x	F(x)							
32								
48								
24								
56								
16								
64								

Normal Distribution

○ Probability density
● Cumulative probability
○ Inverse cumulative probability

Mean: 40

Standard deviation: 8

● Input column: x

Optional storage: 'F(x)'

○ Input constant: .9

Optional storage:

Select   Help   OK   Cancel

Figure 4.18   Evaluation of a series of normal cumulative probability function values.

variance 8^2, standard deviation 8. Minitab can be used to obtain the probabilities that interviews last between 32 and 48 minutes, between 24 and 56 minutes, and between 16 and 64 minutes.

Here the cumulative probability function is required for six different values of the random variable of interest, the duration of an interview. These six values can be entered into a column in Minitab and the corresponding cumulative probabilities stored in a second column, named F(x) in advance of the calculations being performed. The dialog is displayed in Figure 4.18.

The probability that duration lies between 32 and 48 minutes is

$$P(32 < X \le 48) = F(48) - F(32)$$
$$= 0.841\,345 - 0.158\,655 \text{ (from the F(x) column in the worksheet)}$$
$$= 0.6827 \text{ to four decimal places,}$$

or about two-thirds. The durations of 32 and 48 are one standard deviation below and above the mean respectively. The calculation demonstrates the feature of the normal distribution that approximately two thirds of observed values are within one standard deviation of the mean. You should verify that that approximately 95% of values are within two standard deviations of the mean, i.e. between 24 and 56 (probability 0.9545), and that 99.73% of values are within three standard deviations of the mean, i.e. between 16 and 64 minutes. Thus it would be very rare (probability 0.0027) for an interview to have duration outside the range 16 minutes to 64 minutes, i.e. outside the range $\mu - 3\sigma$ and $\mu + 3\sigma$. The normal distribution properties illustrated by this example are displayed in Figure 4.19.

Suppose the interviewer wishes to know the duration which would be exceeded for one interview in ten in the long run. In other words, the value d is required such that $P(X \le d) = 9/10 = 0.9$. Duration d may be referred to as the 90th percentile of the distribution. In order to obtain d, use **Calc > Probability Distributions > Normal...**, select **Inverse cumulative probability** and enter 0.9 in the **Input constant:** window. The appropriate mean and standard deviation must be specified in the **Mean:** and **Standard deviation:** windows, respectively. On execution the Session window displays the text in

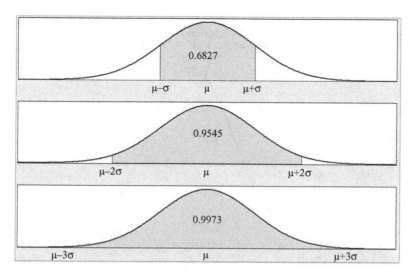

Figure 4.19   Probabilities of values lying within 1, 2 and 3 standard deviations of the mean for a normal distribution.

Panel 4.6. This indicates that d is 50.25 and that approximately one interview in ten would have duration in excess of 50 minutes.

Alternatively the result may be obtained using **Graph > Probability Distribution Plot... > View Probability**. On double-clicking the graphic under **View Probability** select Normal in the **Distribution:** window and enter **Mean:** 40 and **Standard deviation:** 8. On the **Shaded Area** tab, with **Define Shaded Area By Probability** selected, click on the graphic under **Right Tail** and enter **Probability:** 0.1. The graph in Figure 4.20 results, confirming the required duration d to be 50.25 minutes.

In Section 1.1 the concept of sigma quality level was introduced. Minitab can be used to compute the entries in Table 1.1 and in Appendix 1. In order to demonstrate the calculations involved, consider a bottle manufacturing process which is producing bottles with weights (g) which are $N(493, 2^2)$ and for which the specification limits are 486 and 494 g. The target weight can be considered to be 490 g, the weight that is midway between the specification limits. Thus the specification limits here are two standard deviations away from the target and the process is off target, on the high side, by 3 g – equivalent to 1.5 standard deviations. The situation is illustrated in Figure 4.21,created using the **Graph > Probability Distribution Plot... > View Probability** facility, as described earlier for the creation of Figure 4.16, and Graph Annotation Tools.

```
Inverse Cumulative Distribution Function

Normal with mean = 40 and standard deviation = 8

P( X <= x )        x
          0.9  50.2524
```

Panel 4.6   Evaluation of inverse cumulative probability function.

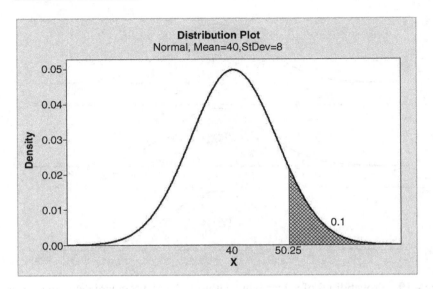

Figure 4.20   Interview duration exceeded on one in ten occasions.

It is evident from the diagram that a small proportion of bottles will be nonconforming due to having weight below the lower specification limit, while a large proportion will be nonconforming due to having weight above the upper specification limit. These proportions are $F(486)$ and $1 - F(494)$ respectively, which you should verify to be 0.000233 and 0.308538 respectively. This gives a total proportion nonconforming of $0.000\,233 + 0.308\,538 = 0.308\,771$ which equates to $0.308\,771 \times 1\,000\,000 = 308\,771$ nonconforming bottles per million. You should confirm that this is the entry in Table 1.1 corresponding to a sigma quality level of 2

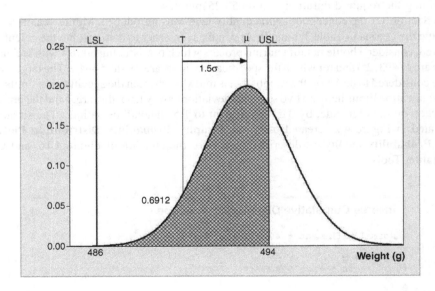

Figure 4.21   Bottle weight distribution with specification limits superimposed.

(apart from a small rounding discrepancy). Note that in Figure 4.21 has displayed probability 0.6912 for conformance corresponding to probability $1 - 0.6912 = 0.3088$ for nonconformance, confirming the above result on rounding.

Because the specification limits in the scenario discussed here are two standard deviations away from the target the process is said to have a sigma quality level of 2. The convention in Six Sigma is to quote the nonconformities per million opportunities when the process is operating with a mean that is 'off target' by 1.5 standard deviations. The use of a 1.5σ 'shift' in the calculation of sigma quality levels is controversial (Ryan, 2000, p. 522). As an exercise you should verify that the proportion is the same when the mean is off target, on the low side, by 1.5 standard deviations. A series of calculations of this type can be used to complete Table 1.1. A more comprehensive table is given in Appendix 1.

Figure 4.22 illustrates a 'Six Sigma' process, i.e. one for which the specification limits are six standard deviations away from the target. As illustrated by the solid curve, the process is operating 'on target'. Clearly with such a process the location could shift by an appreciable amount without any real impact on the proportion of product falling outside the specifications. For this process, a shift of 1.5 standard deviations from the target would lead to 3.4 nonconforming items per million. The dotted curve illustrates the distribution that would occur were the mean to shift upwards by 1.5 standard deviations.

Finally, consider again the $N(40, 8^2)$ distribution used to model the duration, X (minutes), of interviews. You should verify, using Minitab, that $P(X < 0) = 0.000\,000\,3$. Thus, according to the model, there is a very small probability that the duration of an interview could be negative, which is impossible. The probability density function of a normal distribution is defined for all values of the random variable being modelled, whether the values are feasible or not. Thus the model is 'wrong' in the sense that it allows the possibility of an impossible event. Yet the model could prove valuable to the team planning the survey and evaluating the performance of interviewers. Hence the comment quoted at the very beginning of the chapter is relevant.

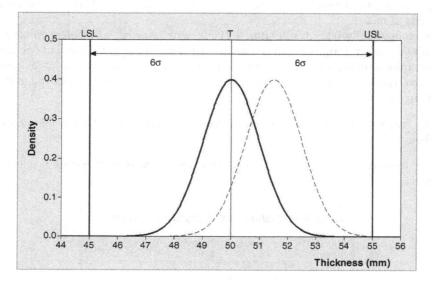

Figure 4.22   A 'Six Sigma' process operating on target.

Suppose that X is a random variable and that a second random variable, V, is defined as $V = kX$, where k is a positive constant. Two important results are:

$$\mu_V = k\mu_X, \quad \sigma_V = k\sigma_X.$$

Thus if k is thought of as a scale factor for converting values of X to values of V then that same scale factor must be used to convert the mean and standard deviation for X into those for V.

If X is normally distributed, then so is V.

Box 4.1   Multiple of a random variable.

4.3   Distribution of means and proportions

4.3.1   Two preliminary results

There are a number of results required for the development of control charts and of other statistical methods of value in quality improvement. These results will be introduced in this section, but theory and mathematical detail will be avoided. Two preliminary results are required.

The first preliminary result (Box 4.1) concerns *multiples of a random variable*. Informal justification of the numerical aspects may be obtained by converting the height data in inches, given in the Minitab pulse data set referred to earlier, to metres. This may readily be done using **Calc > Calculator...** to multiply the given heights in inches by the conversion factor 0.0254. The means and standard deviations of the two sets of height measurements are displayed in Panel 4.7. The reader is invited to verify that multiplication of the mean and standard deviation of the heights in inches by 0.0254 yields the mean and standard deviation of the heights in metres.

The second preliminary result (Box 4.2) concerns *sums of independent random variables*. If two random variables are independent then knowledge of the value of one does not yield any information about the value of the other. Independent random variables have zero correlation. However, the converse is not true. In words, the results summarized in Box 4.2 state that:

- the mean of the sum of a set of independent random variables is the sum of the their means;

- the variance of the sum of a set of independent random variables is the sum of their variances.

Descriptive Statistics: Height, Height (m)

```
Variable         Mean    StDev
Height         68.717    3.659
Height (m)     1.7454   0.0929
```

Panel 4.7   Descriptive statistics for height measurements.

Let X_1, X_2, X_3, ..., X_p be a set of independent random variables and let $T = X_1 + X_2 + X_3 + ... + X_p$, i.e. T is the sum of the set of random variables. Let $X_1, X_2, X_3, ..., X_p$ have means $\mu_1, \mu_2, \mu_3, ..., \mu_p$ and variances $\sigma_1^2, \sigma_2^2, \sigma_3^2, ..., \sigma_p^2$ respectively. Then the mean and variance of T are respectively

$$\mu_T = \mu_1 + \mu_2 + \mu_3 + ... + \mu_p,$$
$$\sigma_T^2 = \sigma_1^2 + \sigma_2^2 + \sigma_3^2 ... + \sigma_p^2.$$

If the Xs are normally distributed, then T is normally distributed.

Box 4.2   Sum of independent random variables.

Consider an assembly operation for automatic telling machines with three phases: set-up, build and test. Let the durations of each phase have means 15, 90 and 18 minutes, respectively. Suppose that the durations are independent of each other and that all three have uniform distributions with ranges of 6, 12 and 6 minutes respectively. This means that set-up is equally likely to take any time from 12 to 18 minutes, build any time from 84 to 96 minutes and test any time from 15 to 21 minutes. For a uniform distribution the probability density function is constant over the range of possible values and zero elsewhere. Statistical theory tells us that the variance of a uniform distribution is one-twelfth of the square of its range. Hence the variances for the durations of set-up, build and test will be 3, 12 and 3 minutes respectively.

According to the theory in Box 4.2, the mean and variance of the sum of a set of independent random variables are obtained by summing the individual means and the individual variances, respectively. Thus theory predicts that the total duration for the operation will have mean $15 + 90 + 18 = 123$ and variance $3 + 12 + 3 = 18$, which corresponds to a standard deviation of 4.24 minutes.

Minitab enables samples to be generated from uniform distributions so the theory can be checked by simulation of a series of assembly operations. Having named columns C1, C2, C3 and C4 respectively Set-up, Build, Test and Total, one can proceed to simulate data for 10 000 assembly operations as indicated in Figure 4.23. Use is required, three times, of **Calc > Random Data > Uniform....** Figure 4.23 shows the procedure about to be implemented for the third time in order to generate 10 000 values for the duration of the Test phase. Note the specification **Lower endpoint: 15** and **Upper endpoint: 21**. The Total duration can then be computed using **Calc > Row Statistics...**, with **Sum** selected as **Statistic**, Set-up, Build and Test entered in the **Input variables:** window and Total entered in the **Store result in:** window.

The descriptive statistics displayed in Panel 4.8 were obtained using **Stat > Basic Statistics > Display Descriptive Statistics....** Under **Statistics...** only **Mean, Standard deviation** and **Variance** were checked. The mean and standard deviation of the sample of 10 000 values of Total were 122.90 and 4.23 for the simulation carried out by the author. These are close to the values of 123.00 and 4.24 for the population mean and standard deviation given by the results on sums of random variables. The reader is invited to carry out the simulation for her/himself.

The histograms of the simulated data in Figure 4.24 illustrate the uniform distribution of the component times and also that the distribution of the Total duration has the appearance of a

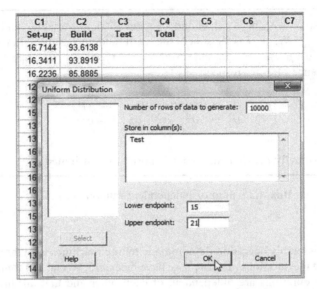

Figure 4.23 Generating data from a uniform distribution.

Descriptive Statistics: Set-up, Build, Test, Total			
Variable	Mean	StDev	Variance
Set-up	14.993	1.746	3.048
Build	89.913	3.444	11.859
Test	17.989	1.731	2.996
Total	122.90	4.23	17.86

Panel 4.8 Descriptive statistics for simulated data.

normal distribution – a fitted normal distribution curve is shown. This gives an indication of the importance of the normal distribution – the sum of a series of independent random variables with nonnormal distributions can often be adequately modelled by a normal distribution. If each of the independent random variables is normally distributed then the sum of the random variables is also normally distributed.

4.3.2 Distribution of the sample mean

The sample mean is widely used in control charting and other statistical methods of value in Six Sigma quality improvement. Box 4.3 summarizes the important results required for the sample mean. These results may be obtained by considering the mean of a random sample as a multiple of a sum of independent random variables.

If the reader finds the mathematics a bit daunting then perhaps some further Minitab simulation will aid understanding. A set of 1000 random samples from the $N(60, 2^2)$

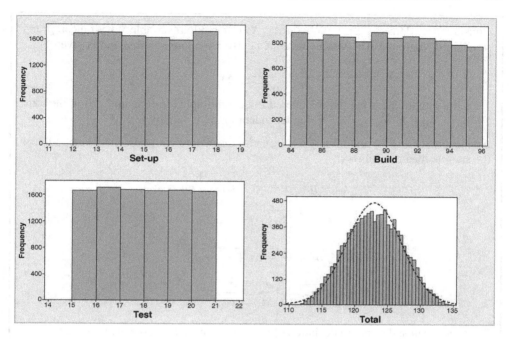

Figure 4.24   Histograms of simulated data.

distribution may be simulated using **Calc > Random Data > Normal…** to create four columns of 1000 values from the specified distribution. Each row of four values may then be considered to be a sample of $n = 4$ values from the $N(60, 2^2)$ distribution. **Calc > Row Statistics…**, with **Mean** as the selected **Statistic,** may then be used to compute the 1000 sample means and to store them in a column named Mean. The dialog involved in this last step is displayed in Figure 4.25.

The theory in Box 4.3 indicates that the mean of the population of sample means will be $\mu_{\bar{x}} = \mu = 60$ and that the standard deviation of the population of sample means will be $\sigma_{\bar{x}} = \sigma/\sqrt{n} = 2/\sqrt{4} = 1$. The theory also indicates that the distribution of the sample mean will be normal.

The histogram of the sample means, with fitted normal curve, shown in Figure 4.26 supports the theory. Note that the mean of the sample means of 60.03 and that the standard deviation of the sample means of 1.007, reported in the text box at the top right-hand corner of the display, are both close to the population values given by theory of 60.00 and 1.000, respectively. Again the reader is invited to carry out the simulation for her/himself.

Box 4.4 gives a very important result, the *central limit theorem*, concerning the distribution of the sample mean when the random variable of interest is *not* normally distributed. As an illustration of the central limit theorem, consider a weaving process that operates continuously and for which filament breaks occur at random at the rate of 2 per hour. This means that the number of breaks occurring per hour will have the Poisson distribution with parameter 2. Statistical theory shows that it follows that the time interval, in minutes, between filament breaks will have the exponential distribution with mean 30. (No details of the

Consider a random sample $X_1, X_2, X_3, \ldots, X_n$ of a random variable X with mean μ and standard deviation σ. The sample mean $\bar{X}$ is also a random variable and is given by:

$$\bar{X} = \frac{\sum_{i=1}^{n} X_i}{n} = \frac{1}{n}(X_1 + X_2 + X_3 + \ldots + X_n) = \frac{1}{n} T$$

Thus the sample mean can be considered to be a multiple (scale factor $1/n$) of the total $T = X_1 + X_2 + X_3 + \ldots + X_n$ of n independent random variables.

Application of the two preliminary results yields the mean mean and standard deviation of the sample mean as follows:

$$\mu_{\bar{X}} = \frac{1}{n}(\mu + \mu + \mu + \ldots + \mu) = \frac{1}{n} n\mu = \mu,$$

$$\sigma_{\bar{X}} = \frac{1}{n} \times \sigma_T = \frac{1}{n}\sqrt{\sigma_T^2}$$

$$= \frac{1}{n} \times \sqrt{\sigma^2 + \sigma^2 + \sigma^2 + \ldots + \sigma^2}$$

$$= \frac{1}{n} \times \sqrt{n\sigma^2} = \sqrt{\frac{\sigma^2}{n}} = \frac{\sigma}{\sqrt{n}}.$$

Thus the mean and standard deviation of the sample mean are respectively μ and $\sigma/\sqrt{n}$.

Many authors use the phrase 'standard error' in place of standard deviation in this context.

Box 4.3   Distribution of the sample mean.

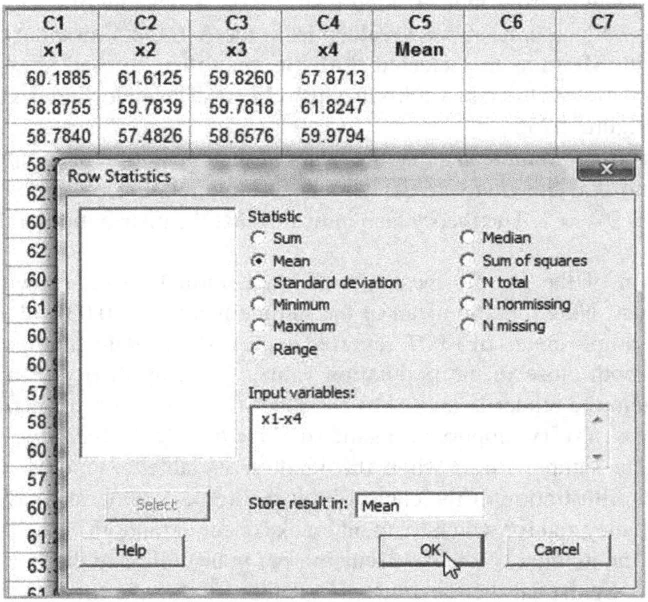

Figure 4.25   Calculation of the sample means.

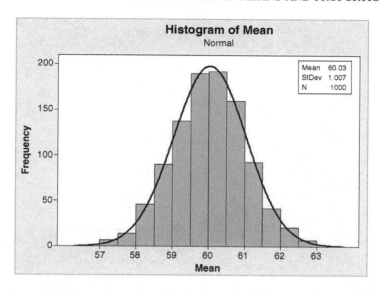

Figure 4.26 Histogram of sample means with fitted normal curve.

exponential distribution are provided in this book.) Scrutiny of log sheets maintained by the process operators yielded 1000 time intervals between filament breaks that are plotted in Figure 4.27 in the form of a density histogram with the probability density function of the exponential distribution with mean 30 superimposed. This type of distribution is positively skewed.

Minitab was used to simulate 1000 samples of $n = 4$ intervals between breaks and also 1000 samples of $n = 25$ intervals between breaks. The sample means were calculated and are displayed in the histograms in Figure 4.28 with fitted normal curves superimposed. With sample size $n = 4$ the distribution of the sample mean is positively skewed and the distribution is not normal. However, with sample size $n = 25$ the distribution of the sample mean is much more symmetrical and the normal distribution appears to provide and adequate model for the distribution of the sample mean.

The major importance of the central limit theorem is that it enables probability statements to be made about means of reasonably large samples regardless of whether or not the distribution of individual values is normal. As an example, consider a type of car tyre with life nonnormally distributed with mean 20 000 miles and standard deviation 1600 miles. Suppose that we wish to obtain the probability that a random sample of 64 of these tyres has mean life of 20 400 miles or greater. The solution is as follows.

Even if the random variable X is *not* normally distributed the sample mean will be *approximately* normally distributed with mean μ and standard deviation $\sigma/\sqrt{n}$. The larger the sample size n, the better the approximation. This result is known as the central limit theorem.

Box 4.4 The central limit theorem.

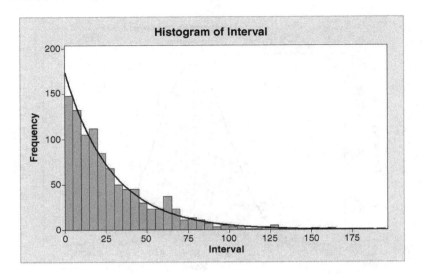

Figure 4.27   Histogram of intervals between breaks with exponential distribution.

Let X denote tyre life and let $\bar{X}$ denote the mean life of random samples of 64 tyres. By the central limit theorem $\bar{X}$ will be approximately normally distributed with mean $\mu_{\bar{X}} = \mu = 20\,000$ and standard deviation $\sigma_{\bar{X}} = \sigma/\sqrt{n} = 1600/\sqrt{64} = 200$. Use of **Calc** > **Probability Distributions** > **Normal...** gives the cumulative probability function value

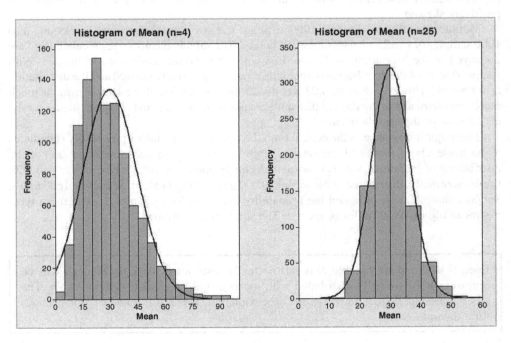

Figure 4.28   Histograms of means of samples of size $n = 4$ and of samples of size $n = 25$.

Let there be constant probability p that an item is nonconforming and let X denote the number of nonconforming items in random samples of n items. Thus X will have the binomial distribution with parameters n and p, i.e. X has the $B(n,p)$ distribution with mean $\mu_X = np$ and standard deviation $\sigma_X = \sqrt{npq}$, where $q = 1 - p$.

The proportion of nonconforming items is given by $V = X/n = kX$, where $k = 1/n$, so the theory in Box 4.1 gives

$$\mu_V = \frac{1}{n}np = p,$$

$$\sigma_V = \frac{1}{n}\sqrt{npq} = \sqrt{\frac{pq}{n}} = \sqrt{\frac{p(1-p)}{n}}.$$

The fact that the standard deviation of a proportion is $\sqrt{p(1-p)/n}$ is of fundamental importance in the creation of control charts for proportion of nonconforming items.

Box 4.5   Standard deviation of a proportion.

0.977 25 so the required probability is $1 - 0.977\,25 = 0.022\,75 \approx 1/44$. (Alternatively use can be made of **Graph > Probability Distribution Plot... > View Probability**.) The conclusion is that there is a 1 in 44 chance of obtaining a mean life of 20 400 or greater for a random sample of 64 of the tyres. Calculations of this nature lie at the heart of hypothesis testing which provides important tools for determining whether or not steps taken to improve a process have been effective. Hypothesis testing is introduced in Chapter 7.

4.3.3 Distribution of the sample proportion

For the development of control charts etc., a formula for the standard deviation of the proportion of nonconforming items in a sample is required. A proportion may be considered as a multiple of a random variable – details are presented in Box 4.5

Consider a scenario where samples of 10 items are taken at regular intervals from a process and checked. Conforming items are denoted by S and nonconforming items by D. The indicator random variable B is defined as having value 0 for a conforming item and value 1 for a nonconforming item. Table 4.9 gives results for one sample. For this sample the proportion of nonconforming items is $3/10 = 0.3$. This is also the mean of the sample of 10 values of the indicator random variable B for the sample. A proportion is a sample mean in disguise! Thus if the probability, p, of a nonconforming item remains constant as successive random samples are taken, the central limit theorem indicates that the series of proportions of nonconforming items

Table 4.9   Conformance record for a sample of 10 items.

Item No.	1	2	3	4	5	6	7	8	9	10
Status	S	D	S	S	D	S	S	S	D	S
Indicator B	0	1	0	0	1	0	0	0	1	0

will have an approximate normal distribution with mean p and standard deviation $\sqrt{p(1-p)/n}$.

In order to illustrate, **Calc > Random Data > Binomial...** was used to generate 1000 random samples from the binomial distribution with parameters $n = 100$ and $p = 0.2$. Thus the number of nonconforming items in random samples of 100 components from a process yielding constant probability 0.2 of a nonconforming item was being simulated for a total of 1000 samples. With **Calc > Calculator...** the simulated counts of nonconforming items were converted to proportions by dividing by 100. According to the above theory the proportions will be approximately normally distributed with mean $p = 0.2$ and standard deviation

$$\sqrt{\frac{p(1-p)}{n}} = \sqrt{\frac{0.2 \times 0.8}{100}} = 0.04.$$

The histogram of the proportions displayed in Figure 4.29 was created using **Graph > Histogram...**, the **With Fit** option being used to superimpose a fitted normal curve. The mean and standard deviation of the sample of 1000 proportions of 0.2037 and 0.04065 (displayed to the right of the histogram in Figure 4.29) are close to the population values of 0.20 and 0.04 respectively. The normal distribution also clearly provides an adequate model for the distribution of sample proportions.

The statistical models referred to in this chapter provide the foundations for much of what follows in this book. Readers who wish to gain a deeper and wider understanding of these statistical models would benefit from consulting the books by Montgomery and Runger (2010) and Hogg and Ledolter (1992).

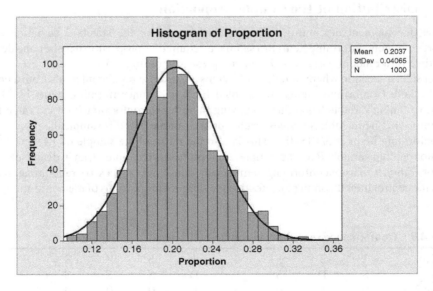

Figure 4.29 Histogram of proportions with fitted normal curve.

4.4 Multivariate normal distribution

The *multivariate normal distribution* provides a statistical model for some scenarios where two, three or more continuous random variables are of interest. For the case of two random variables the multivariate normal distribution is referred to as the *bivariate normal distribution*. A bivariate normal distribution for the two random variables X and Y is specified by the five parameters μ_x, μ_y, σ_x^2, σ_y^2 and σ_{xy}, i.e. by the means, variances and covariance. (The symbol σ_{xy} denotes the population covariance between random variables X and Y.) If X and Y have the bivariate normal distribution specified then X has the $N(\mu_x, \sigma_x^2)$ distribution and Y has the $N(\mu_y, \sigma_y^2)$ distribution. These are known as the marginal distributions of X and Y.

Suppose that a blow moulding process for plastic PET 500 ml bottles yields bottles with weight (g) and diameter (mm) having the bivariate normal distribution with means 25.0 and 72.0, variances 0.04 and 0.05 respectively and covariance -0.03. Thus the covariance matrix is $\begin{pmatrix} 0.04 & -0.03 \\ -0.03 & 0.05 \end{pmatrix}$.

In order to generate some data from this distribution via Minitab, first set up the means in column C1. Second, with commands enabled, use **Calc > Matrices > Read** to specify that the matrix comprises two rows and two columns and to name it Covariance. The default option **Read from keyboard** is accepted. The dialog box is shown in Figure 4.30. On clicking **OK**, the matrix can be entered following the data prompts as indicated in Panel 4.9. In presenting the data to Minitab one must adhere to the matrix pattern of two rows and two columns.

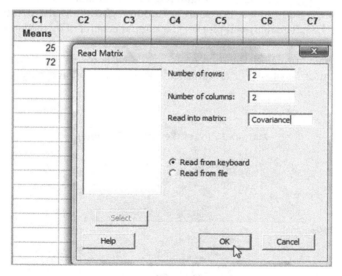

Figure 4.30 Reading data into a matrix.

```
MTB > Name m1 "Covariance"
MTB > Read 2 2 'Covariance'.
DATA> 0.04 -0.03
DATA> -0.03 0.05
2 rows read.
MTB >
```

Panel 4.9 Reading a matrix via the keyboard.

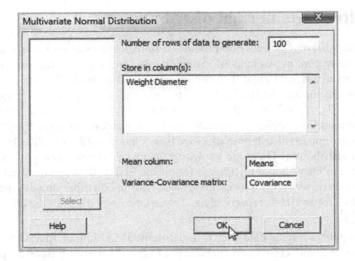

Figure 4.31   Generating data from a bivariate normal distribution.

Minitab now has the necessary information to enable simulated data to be generated for, say, a sample of $n = 100$ bottles. Use **Calc** > **Random Data** > **Multivariate** > **Normal** as shown in Figure 4.31. The marginal plot for the sample generated by the author is displayed in Figure 4.32.

The multivariate normal distribution provides the basis for an important type of control chart for the monitoring of two or more process variables simultaneously.

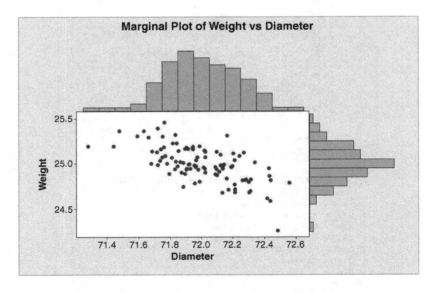

Figure 4.32   Marginal plot of simulated bivariate data.

4.5   Statistical models applied to acceptance sampling

4.5.1   Acceptance sampling by attributes

Acceptance sampling deals with the inspection and classification of a sample of units selected at random from a larger batch or lot and the ultimate decision about disposition of the batch – accept, reject or some other action. Acceptance sampling may be applied to incoming batches of parts from an external supplier that are to be used in a manufacturing process or to batches of incomplete product as they proceed from an internal 'supplier' to the next stage in a manufacturing process. In his chapter on acceptance sampling Montgomery (2009) states that acceptance sampling was a major component of quality improvement activities at the time of the Second World War but that more recently 'it has been typical to work with suppliers to improve their process performance through the use of [statistical process control] and designed experiments'. A brief introduction to acceptance sampling is included here as statistical models discussed earlier in this chapter are applied.

Consider the following single-sampling plan. A random sample of size $n = 400$ units from a batch of size $N = 40\,000$ units are inspected, and if the total number of nonconforming units is less than or equal to the acceptance number $c = 2$ then the batch is accepted; otherwise it is rejected. Suppose that the process that creates the units currently yields 0.5% nonconforming. We would therefore expect a batch of $N = 40\,000$ units to contain $M = 0.5\% \times 40\,000 = 200$ nonconforming units. The probability of acceptance of the batch is the probability that a random sample of $n = 400$ contains 2 or fewer nonconforming units. This probability may be calculated using **Calc > Probability Distributions > Hypergeometric...** and making the entries displayed in Figure 4.33.

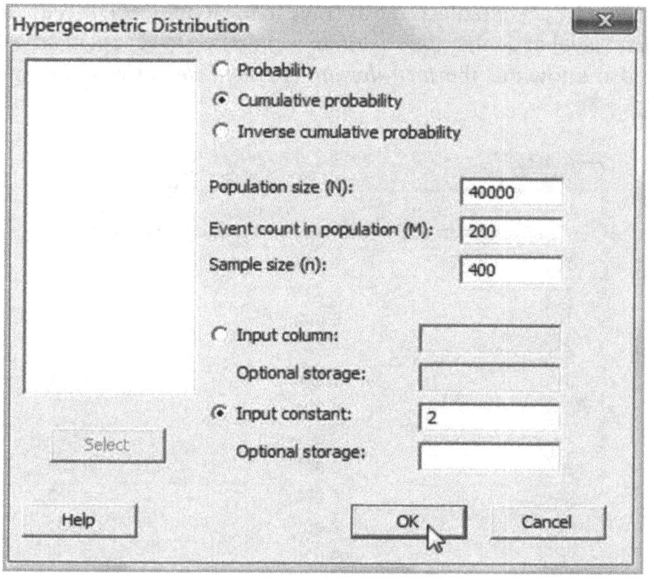

Figure 4.33   Calculation of probability of batch being accepted.

```
Cumulative Distribution Function

Hypergeometric with N = 40000, M = 200, and n = 400

x    P( X <= x )
2      0.676686
```

Panel 4.10 Probability of acceptance of batch with 0.5% nonconforming units.

The result in the Session window, displayed in Panel 4.10, indicates that the probability of obtaining 2 or fewer nonconforming items in the sample is 0.677. Thus the probability of a batch containing 0.5% defective units being accepted is 0.677. (The hypergeometric rather than the binomial distribution is used to do the calculation because the event probability is not constant. However, with large batch sizes and low proportions of nonconforming items the binomial may be used to approximate the hypergeometric. The interested reader will find that the binomial approximation yields 0.676 677 as compared with the 0.676 686 in Panel 4.10.) The reader is invited to verify that the probability of accepting a batch that contains 1% nonconforming units is 0.235 i.e. there is approximately a 1 in 4 chance that a batch containing 1% defective units would be accepted.

A series of calculations yields the table of acceptance probabilities displayed in Figure 4.34. If there are no defects in the lot then the probability of acceptance is 1.000. This means that a batch with no nonconforming items is certain to be accepted. (That is of course desirable! We are, of course, assuming that there are no inspection errors!) The operating characteristic (OC) curve for the sampling plan is a plot of acceptance probability versus proportion of nonconforming items (see Figure 4.35).

In order to design a single-sampling plan two concepts are widely used. The *acceptable quality level* (AQL) is the poorest level of quality for the manufacturing process that the customer would consider acceptable as the average in the long run. The *rejectable quality level* (RQL) is the poorest level of quality the customer is prepared to accept in an individual batch or lot. The RQL is also known as the *lot tolerance percent defective* (LTPD) and the *limiting quality level* (LQL).

Worksheet 1 ***	C1	C2
↓	Proportion nonconfroming	Probability of acceptance
1	0.000	1.000
2	0.001	0.992
3	0.002	0.954
4	0.003	0.881
5	0.004	0.784
6	0.005	0.677
7	0.006	0.569
8	0.007	0.469
9	0.008	0.378
10	0.009	0.300
11	0.010	0.235

Figure 4.34 Table of acceptance probabilities.

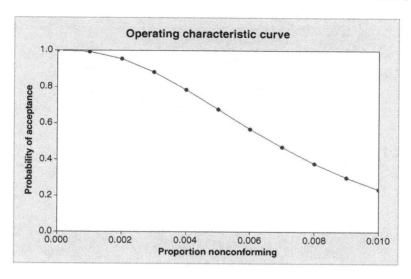

Figure 4.35   Operating characteristic curve.

For the purposes of illustration, suppose that in a particular scenario AQL is 2% and RQL is 8%. Suppose, too, that it considered desirable that there should be probability 0.95 of accepting a lot containing 2% defectives (the AQL) and that there should be probability 0.10 of accepting a lot containing 8% defectives (the RQL). Put another way, this means there would be a 0.05 probability $(1 - 0.95)$ of rejecting a lot containing 2% defectives and there would be a 0.90 $(1 - 0.10)$ probability of rejecting a lot containing 8% defectives. We can think of the probability 0.05 of rejecting a good batch (with proportion defective equal to the AQL of 2%) as being the *producer's risk* and the probability of and the probability of 0.10 of accepting a bad batch (with proportion defective equal to the RQL of 8%) as being the *consumer's risk*.

It is desirable, then, in this scenario, to have an OC curve which passes through the points (0.02, 0.95) and (0.08, 0.10). Nomograms and tables are available to determine appropriate values for the sample size n and the acceptance number c. Minitab facilitates the calculation using **Stat > Quality Tools > Acceptance Sampling by Attributes...** with the dialog displayed in Figure 4.36. Proportion defective was selected under **Units for quality levels:**, the alternative choices being Percent defective and Defectives per million. Note that it is not necessary to specify a lot (batch) size.

The key part of the Session window output is shown in Panel 4.11. The plan involves taking a random sample of 98 from the batch and acceptance of the batch if the number of nonconforming items found is less than or equal to 4. It was desired to have the OC curve pass through the points (0.02, 0.950) and (0.08, 0.100). The latter part of the Session window output indicates that the best that could be found was the OC curve that passed through (0.02, 0.953) and (0.08, 0.099). The OC curve is also displayed.

The supplied worksheet Unit_Reference_Codes.MTW contains a column named Unit Reference giving reference codes for a batch of 1200 units. In order to select a random sample of 98 units one could use **Calc > Random Data > Sample From Columns...** and enter **Number of rows to sample:** 98, **From columns:** 'Unit Reference', **Store samples in:** Sample. On doing this the author obtained the sequence UO2391, UQ2432, US2428, ...,

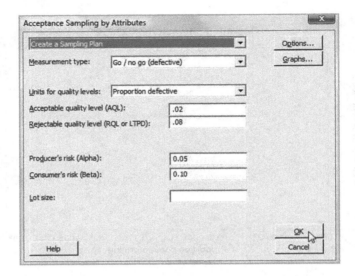

Figure 4.36   Designing an attributes single-sampling plan.

```
Generated Plan(s)

Sample Size          98
Acceptance Number     4

Accept lot if defective items in 98 sampled <= 4; Otherwise reject.
```

Panel 4.11   Specification of the required single-sampling plan.

US2446. The reader is invited to try this for her/himself – it is very unlikely that the same sequence will be obtained!

4.5.2   Acceptance sampling by variables

Rather than simply classifying units as either conforming or nonconforming, it will be possible in some circumstances to base the decision whether or not to accept a batch on the basis of a measurement on each unit in the sample taken. Consider glass bottles for which the specification limits for weight are 485 and 495 g and a scenario where AQL is 0.1% and RQL is 0.5% with producer and consumer risks of 0.05 and 0.10, respectively. Suppose that the standard deviation of bottle weight is unknown but that experience shows that bottle weight may be adequately modelled by a normal distribution. Use of **Stat** > **Quality Tools** > **Acceptance Sampling by Variables** > **Create/Compare...** with the dialog displayed in Figure 4.37 is required.

The key part of the Session window output is shown in Panel 4.12. The plan involves taking a random sample of 160 bottles from the batch, weighing them and calculating the sample mean and standard deviation. Data are provided in supplied worksheet Bottle_Weight_Sample.MTW. The reader is invited to verify that sample mean and standard deviation are respectively 490.94 and 1.18. The two Z-values required in Panel 4.12 may be calculated

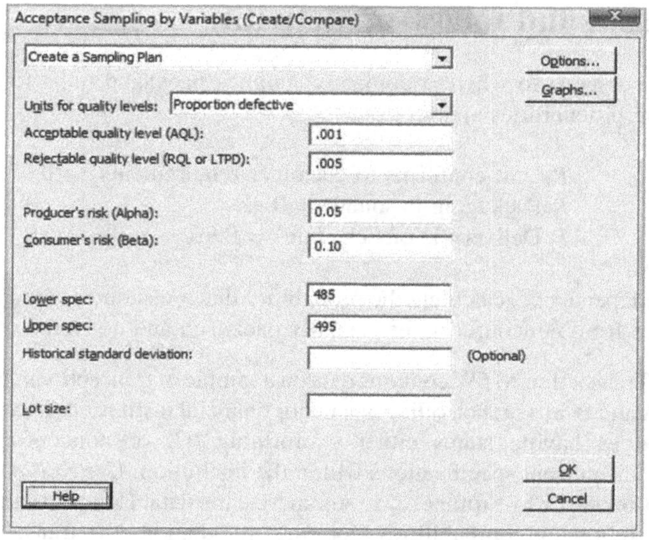

Figure 4.37 Designing a variables single-sampling plan.

```
Generated Plan(s)

Sample Size                          160
Critical Distance (k Value)          2.80110
Maximum Standard Deviation (MSD)     1.65685

Z.LSL = (mean - lower spec)/standard deviation
Z.USL = (upper spec - mean)/standard deviation
Accept lot if standard deviation <= MSD, Z.LSL >= k and Z.USL >= k; otherwise reject.
```

Panel 4.12 Specification of the required single-sampling plan.

as shown in Box 4.6. Since *both* Z-values *exceed* the critical distance (*k* value) of 2.80 *and* the sample standard deviation (1.18) is *less than* the maximum allowable standard deviation (MSD) of 1.66, the decision would be to accept the batch.

Further information on acceptance sampling for both attributes and variables, including Military Standard plans, may be found in Montgomery (2009) and in the references cited therein.

$$Z.LSL = \frac{Mean - Lower\ spec}{Standard\ deviation} = \frac{490.94 - 485}{1.18} = 5.03$$

$$Z.USL = \frac{Upper\ spec - Mean}{Standard\ deviation} = \frac{495 - 490.94}{1.18} = 3.44$$

Box 4.6 Calculation of the Z-values.

4.6 Exercises and follow-up activities

1. Consider a scenario where a company supplies packaged units to customers. The following probabilities apply:

   P(Unit conforms to customer requirements) $= 0.95$,
   P(Packaging is sound) $= 0.92$,
   P(Delivery is on schedule) $= 0.90$.

Assuming independence, calculate the probability that a customer who orders a unit will receive it free from nonconformities, soundly packaged and delivered on schedule.

2. The file Transaction.MTW contains data on a sample of transactions carried out by two teams, A and B, at a branch office of a major financial institution. Each transaction was classified as having status either conforming (C) or nonconforming (N-C) in terms of the current specifications within the institution. Use **Stat > Tables > Cross Tabulation and Chi-Square...** to summarize the data. Hence, write down estimates of the following probabilities for the population of transactions sampled: $P(A), P(B), P(C), P(\bar{C}), P(\bar{C}|A), P(\bar{C}|B)$. Do you think that status is independent of team?

3. An injection moulding process for the production of digital camera casings yields 10% defective. Denote by X the number of defective casings in random samples of 20 casings selected from the process output at regular intervals.

   (i) State the distribution of the random variable X and its parameters.

   (ii) Obtain $P(X \le 3)$, $P(X > 3)$, $P(X < 3)$.

   (iii) Use **Calc > Make Patterned Data > Simple Set of Numbers...** to set up the values of x from 0 to 20 in column C1 as indicated in Figure 4.38. (This facility can save much tedious typing!)

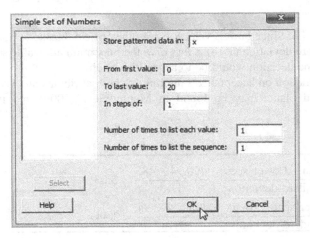

Figure 4.38 Setting up the sequence of values of x.

Table 4.10 Frequencies of goals per match.

Number of goals scored	Number of matches
0	7
1	17
2	13
3	14
4	7
5	5
6	0
7	1
8 or more	0

(iv) Tabulate and display the probability function $f(x)$ and tabulate the cumulative probability function $F(x)$.

(v) Calculate the mean and standard deviation of X.

4. Primary healthcare workers believe that anaphylactic shock reaction to immunization injections occurs with children in one case in two thousand. Immunization teams carry adrenalin packs in order to treat children suffering a shock reaction. Calculate, using Minitab, the probability that two adrenalin packs would be insufficient to treat the shock reactions occurring in a group of 1200 children.

5. Table 4.10 gives the numbers of goals scored in a the soccer matches played in the 2010 FIFA World Cup in South Africa.

 (a) Set up the data in Minitab in a column named Goals using the same method as was used for the V2 flying bomb data and obtain the mean number of goals per match.

 (b) Fit a Poisson distribution to the data. (You should find that you obtain expected frequencies 6.6, 15.0, 17.0, 12.9, 7.3, 3.3, 1.2, 0.4 and 0.2.)

 (c) In order to create the type of display in Figure 4.9 set up columns in Minitab as displayed in Figure 4.39. (Try using **Calc > Make Patterned Data > Text Values...** to create the column indicating the type of frequency.)

 (d) In order to create the chart use **Graph > Bar Chart...** and use the drop-down menu to select the option **Bars represent:** Values from a table.
 Select Cluster under **One column of values** and then complete the dialog boxes as shown in Figure 4.40. The use of the variable Type for attribute assignment under **Data View** means that bars representing observed frequencies in the chart have a different colour from those representing expected frequencies. Observe the good fit of the Poisson model to the observed data and note that this indicates that goals may be considered as random events in a time continuum.

 (e) Minitab actually provides under **Stat > Basic Statistics > Goodness-of-Fit Test for Poisson...** a formal method for assessing how well a Poisson distribution fits

↓	C1	C2	C3-T
	Goals	Frequency	Type
1	0	7.0	Observed
2	1	17.0	Observed
3	2	13.0	Observed
4	3	14.0	Observed
5	4	7.0	Observed
6	5	5.0	Observed
7	6	0.0	Observed
8	7	1.0	Observed
9	8	0.0	Observed
10	0	6.6	Expected
11	1	15.0	Expected
12	2	17.0	Expected
13	3	12.9	Expected
14	4	7.3	Expected
15	5	3.3	Expected
16	6	1.2	Expected
17	7	0.4	Expected
18	8	0.2	Expected

Figure 4.39   Data for bar chart creation.

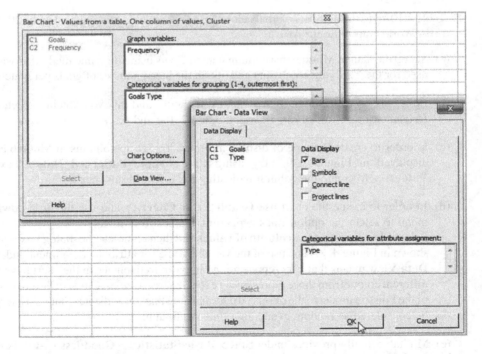

Figure 4.40   Creating a bar chart to compare observed and expected frequencies.

```
Goodness-of-Fit Test for Poisson Distribution

Data column: No. Goals

Poisson mean for No. Goals = 2.26563

No.                    Poisson                Contribution
Goals  Observed  Probability  Expected        to Chi-Sq
0            7    0.103765     6.6410           0.019410
1           17    0.235093    15.0459           0.253777
2           13    0.266316    17.0442           0.959612
3           14    0.201124    12.8720           0.098858
4            7    0.113918     7.2908           0.011595
>=5          6    0.079783     5.1061           0.156476


   N   N*  DF   Chi-Sq  P-Value
  64    0   4  1.49973    0.827
```

Panel 4.13   Assessing goodness-of-fit of a Poisson distribution.

the data. Enter **Variable:** Goals, leave the **Frequency:** window blank and under **Graphs...** check only **Bar chart of the observed and the expected values**. A similar bar chart to the one created earlier is produced, along with the Session window output in Panel 4.13.

In the procedure followed by Minitab numbers of goals per match with low expected frequencies are combined so that no expected frequency is less than 5. Informal support for the provision of a satisfactory model by a Poisson distribution comes from the relatively close match between the lengths of the bars representing observed frequencies and the lengths of the bars representing expected frequencies in the chart. Formal acceptance of the Poisson distribution as a satisfactory model is provided by the P-value well in excess of 0.05. Details of the chi-squared goodness-of-fit test used are not provided in this book.

6. Customers of a mortgage bank expect to have mortgage applications processed within 35 working days of submission. Given that the processing time at the bank can be modelled by the $N(28, 3^2)$ distribution, verify that 99% of applications would be processed within 35 working days. Following a successful Six Sigma project aimed at reducing processing time, it was found that, although variability in processing time was unchanged, the mean had dropped to 19 working days. The bank wishes to inform prospective customers that '99 times out 100 we can process your application in q working days'. Assuming that the normal distribution is still an adequate model, calculate q.

7. Use of the symbol Z for a random variable having the standard normal distribution, i.e. the $N(0, 1)$ distribution, is fairly widespread. Many statistical texts use the notation z_p for the value such that $P(Z \leq z_p) = 1 - p$. Thus for the case of $p = 0.1, 1 - p = 0.9$, the required value may be obtained using **Calc > Probability Distributions > Normal...** with the **Inverse cumulative probability** option.

Use Minitab to generate the worksheet in Figure 4.41. In order to round the values of z_p to two decimal places click on a cell in the column containing the values to be rounded. Next use **Editor > Format Column > Numeric...**, select **Fixed decimal** and change the number of decimal places to 2.

↓	C1	C2	C3
	p	1-p	zp
1	0.1000	0.9000	1.28
2	0.0500	0.9500	1.64
3	0.0250	0.9750	1.96
4	0.0100	0.9900	2.33
5	0.0050	0.9950	2.58
6	0.0025	0.9975	2.81
7	0.0010	0.9990	3.09
8	0.0005	0.9995	3.29

Figure 4.41   Worksheet to be created in Exercise 7.

8. The supplied worksheet Burst.MTW contains burst strength (psi) data for a sequence of bottles taken from mould number 67 at regular intervals during a production run.

   (i) Create a run chart of the data and note how it provides no indication that mould 67 behaved in other than a stable, predictable manner during the production run.

   (ii) Create a histogram of the data with a fitted normal distribution.

   (iii) Assess how well a normal distribution models burst strength by performing a normality test.

   (iv) Estimate the proportion of bottles from mould 67 failing to conform to the requirement that burst strength should be at least 250 psi. (Use the normal distribution with parameters estimated from the data, i.e. with mean 626.2 and standard deviation 117.5, to perform the calculation.)

   (v) Use the table in Appendix 1 to obtain the sigma quality level for mould 67.

9. In addition to columns and matrices, constants may be used in Minitab. With commands enabled the calculation, performed using **Calc > Calculator** and described towards the end of Section 4.1, may be performed using constants K1 and K2 as shown in Panel 4.14. Check the calculation involved in Exercise 4.2 using this method.

```
MTB > let k1=0.933189
MTB > let k2=k1**10
MTB > print k2

Data Display

K2     0.500837
```

Panel 4.14   Calculation using constants.

10. Follow the procedure described in Section 4.4 and simulate your own sample of weight and diameter data for 100 bottles. Obtain the correlation between weight and diameter for your sample and compare with the population value calculated from the covariance matrix used in the simulation.

11. Read the Minitab tutorial material at http://www.minitab.com/en-GB/training/ tutorials/accessing-the-power.aspx?id=1688&langType=2057 on acceptance sampling.

12. A single-sample acceptance attributes sampling plan is required to have a producer's risk of 0.06 for an acceptable quality level of 0.5% nonconforming, and a consumer's risk of 0.10 for a rejectable quality level of 5% nonconforming. Use Minitab to determine the appropriate plan.

13. It is desirable that the greatest torque required to loosen the cap on a type of food container should be 2.5 N m. For producer's risk of 0.05 for an acceptable quality level of 1% nonconforming, and a consumer's risk of 0.10 for a rejectable quality level of 8% nonconforming, determine the appropriate plan. Use Minitab to confirm that the sample size required is 27 and that the critical distance is 1.81. Data for a sample of 27 containers from a lot is provided in Torque.MTW. Verify that Z. USL $= 2.58$ and that therefore the lot would be accepted. Confirm this using **Stat > Quality Tools > Acceptance Sampling by Variables > Accept/ Reject Lot....**

5

Control charts

The fact that the criterion which we happen to use has a fine ancestry in highbrow statistical theorems does not justify its use. Such justification must come from empirical evidence that it works. (Shewhart, 1931, p. 18)

Overview

Control charts or process behaviour charts have been used for nearly 90 years to monitor process performance. Although originally developed for use in manufacturing industry they are now widely applied to processes involving the provision of services in fields such as finance and healthcare.

This chapter deals with a wide variety of control charts and with their creation, interpretation and maintenance via Minitab. Variables charts enable the monitoring of continuous random variables (measurements), while attribute charts monitor discrete random variables (counts). The consequences of tampering with processes are illustrated. Reference is made to auto-correlated data and feedback adjustment. Time-weighted control charts will be introduced, as will multivariate charts for the simultaneous monitoring of two or more variables.

The term 'control charts' suggests that these tools have a role only in the control phase of Six Sigma projects. However, as Figure 1.4 indicates with reference to the transient ischaemic attack and stroke clinic project, they may also be employed during the measure, analyse and improve phases. Indeed, the team involved with the project resolved that the control charts should continue to be maintained as a control measure once the project had formally ended.

5.1 Shewhart charts for measurement data

5.1.1 I and MR charts for individual measurements

'The general idea of a control chart was sketched out in a memorandum that Walter Shewhart of Bell Labs wrote on May 16, 1924' (Ryan, 2000, p. 22) Reference has already been made in

---

Table 5.1   Starch temperature data.

Observation	1	2	3	4	5	6	7	8	9	10
Time	08:00	08:15	08:30	08:45	09:00	09:15	09:30	09:45	10:00	10:15
Temperature	27.2	27.6	26.8	27.2	27.1	26.6	27.6	27.7	27.5	26.6
Observation	11	12	13	14	15	16	17	18	19	20
Time	10:30	10:45	11:00	11:15	11:30	11:45	12:00	12:15	12:30	12:45
Temperature	27.2	26.7	25.9	27.1	27.6	27.5	28.3	26.5	29.0	27.2

Chapter 2 to variation due to common causes and due to special causes. Caulcutt (2004, p. 37) referred to Shewhart's thinking as follows:

> He suggested that a process is acted upon by countless factors, many of which have little effect on the measured performance. Nonetheless, these minor factors, or 'common causes', are important because they are jointly responsible for the random variation in performance. If, in addition a process is acted upon by a major factor, or a 'special cause', the process will change and this change may be revealed by a violation of the control chart rules.

Liquid starch is used in the packaging industry in the manufacture of corrugated paper. Starch temperature is monitored at a manufacturing plant, which operates continuously, by recording temperature (°C) at intervals of 15 minutes. A set of 20 consecutive observations of temperature on 2 August 2010 while the Blue shift team (one of three) was running the process is given in Table 5.1, along with the time of observation.

One of the new features in Release 16 of Minitab is the **Assistant** menu. Selection of **Assistant** > **Control Charts...** yields the flow chart displayed in Figure 5.1. The first question to consider is: what is the data type? Temperature is a continuous random variable – one may think of measuring temperature using a mercury thermometer and the endpoint of the column

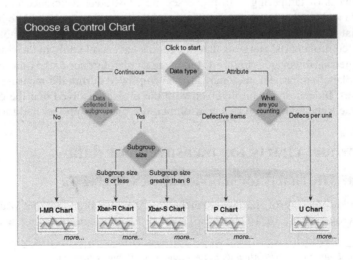

Figure 5.1   Flow chart for chart selection.

being located at any position on the continuous scale marked on the glass body of the instrument. The next question to address is whether or not the data are collected in subgroups. The answer is negative as a single temperature measurement is made every 15 minutes. This leads to the choice **I-MR Chart** – *individual* values and the *moving range* of these values are plotted in what is actually a pair of charts.

Clicking on **more. . .** underneath the **I-MR Chart** icon yields guidelines on collecting the data and using the chart. Clicking on the icon itself yields a simplified menu for creation of the chart with the assumption that the individual values to be charted have already been set up in a column of a worksheet. The author has opted to introduce the reader immediately to the full menus for the creation of individual value and moving range charts.

Columns called Date, Shift, Time and Temperature were set up in a Minitab worksheet. (In order to enter the times at which the temperatures were recorded, i.e. 8 : 00 to 12 : 45 in intervals of 15 minutes, one may use **Calc** > **Make Patterned Data** > **Simple Set of Date/ Time Values. . .** with **Patterned Sequence** specified as **From first date/time:** 08 : 00 **To last date/time:** 12 : 45 **In steps of:** 15 with **Step unit:** Minute.) The temperature data must first be entered into a column, along with any other relevant data in other columns. Use of **Stat** > **Control Charts** > **Variables Charts for Individuals** > **Individuals. . .** yields the dialog box in Figure 5.2.

Temperature is entered in **Variables:** to be charted. Under **I Chart Options. . .**, clicking on the **Estimate** tab, clicking on the down arrow to select **Use the following subgroups when estimating parameters** and inserting 1 : 20 in the window ensures that all 20 measurements will be used in the calculation of the chart limits. Defaults were accepted otherwise. In addition, **Stamp** was checked under **Scale. . .** and Time selected under **Stamp columns:** This yields the basic individuals control chart for the starch temperature displayed in Figure 5.3.

Prior to discussion of the chart the reader is invited to enter Date, Shift, Time and Temperature into four columns of a worksheet and to recreate the chart. Note that up to three Stamp columns may be selected, so as more data come to hand one could select Shift and Date, in addition to Time, in order to aid chart interpretation. On creating the chart the reader may find that the $\bar{X}$, LCL and UCL reference lines are labelled outwith the chart area. In order to have the chart appear as in Figure 5.3 select **Tools** > **Options. . .**, double-click on **Control**

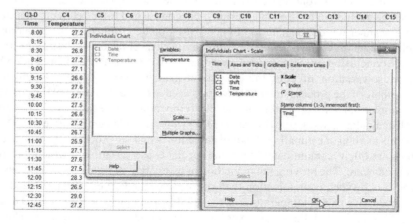

Figure 5.2   Creation of an individuals chart for temperature.

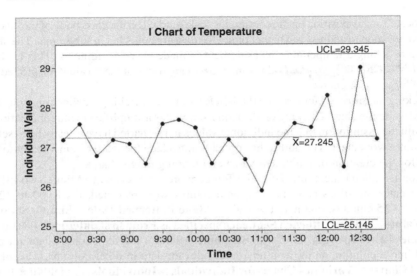

Figure 5.3   Individuals chart for temperature.

Charts and Quality Tools and then click on **Other** and check **Display control limit/center line labels for all stages**. The author recommends use of this setting as, when stages are used as in creating the chart in Figure 1.4, it is useful to have an indication of process performance levels given on the chart for all the stages.

The 'naked run chart' has been 'clothed' through the addition of a centre line (CL) corresponding to the mean, $\bar{X} = 27.245$, of the 20 observations of temperature together with the LCL and UCL reference lines. These are the upper and lower control limits (or upper and lower chart limits), respectively. They are 'three sigma limits' placed at 'three sigma' below and above the centre line respectively. Sigma in this context refers to an estimate of the process standard deviation obtained from the data. Since all 20 points lie between the 'tramlines' formed by the LCL and UCL it is conventional to conclude that the process is exhibiting only common cause variation. (Signals of evidence of special cause variation other than the occurrence of a point beyond the chart limits will be considered later in the chapter.) The process can be deemed to be in a state of statistical control and to be behaving in a stable and predictable manner within the natural limits of variation determined by the upper and lower chart limits. (None of the P-values on the Minitab run chart of the data is less than 0.05, which supports the conclusion from the control chart that there is no evidence of anything other than common cause variation affecting the process for maintenance of the starch temperature.)

The estimate of the process standard deviation used in the computation of the chart limits is *not* the sample standard deviation of the 20 Temperature observations. The estimate is obtained by calculating the 19 moving range (MR) values as indicated in Table 5.2. The reason for use of this method of estimating standard deviation is that the process data used to compute chart limits are often 'contaminated' by some special cause variation of which the creator of the chart is unaware. The moving range method of estimation of process standard deviation is influenced less by such contamination than the sample standard deviation method.

Each successive pair of temperatures is regarded as a sample of $n = 2$ values. The first pair has range $27.6 - 27.2 = 0.4$, the second pair has range $27.6 - 26.8 = 0.8$ and so on. The mean of the 19 moving ranges is 0.7895. Values of the factor, d_2, which may be used to convert a mean

Table 5.2 Calculation of moving ranges.

Observation	1	2	3	4	5	6	7	8	9	10
Temperature	27.2	27.6	26.8	27.2	27.1	26.6	27.6	27.7	27.5	26.6
MR	*	0.4	0.8	0.4	0.1	0.5	1	0.1	0.2	0.9
Observation	11	12	13	14	15	16	17	18	19	20
Temperature	27.2	26.7	25.9	27.1	27.6	27.5	28.3	26.5	29.0	27.2
MR	0.6	0.5	0.8	1.2	0.5	0.1	0.8	1.8	2.5	1.8

range for a set of samples from a normal distribution into a standard deviation estimate, can be found in Appendix 2 or obtained via Help. Use **Help > Methods and Formulas > Quality and process improvement > Control charts > Variable Charts for Individuals > Methods for estimating standard deviation**. Clicking on the table link at the end of the Average moving range heading reveals that, for sample size $n = 2$, the value of d_2 is 1.128. (Reference to **Estimate** under **I Chart Options ...**, in the dialog involved in the creation of Figure 5.2, reveals that the default **Method for estimating standard deviation** with **Subgroup size = 1** is to use **Average moving range** with **Length of moving range:** 2. The user may, if desired, specify that the standard deviation be estimated using moving ranges of length greater than 2 and may specify use of median moving range rather than average moving range.) The calculation of the chart limits is displayed in Box 5.1.

The values obtained in Box 5.1 agree with those displayed on the Minitab control chart in Figure 5.3. Since $3/1.128 = 2.66$ the calculations may be streamlined by use of the formulae:

$$LCL = \overline{X} - 2.66 \times \overline{MR},$$
$$UCL = \overline{X} + 2.66 \times \overline{MR}.$$

An estimate of process standard deviation (process sigma) is given by

$$\frac{\text{Mean moving range}}{d_2} = \frac{\overline{MR}}{d_2} = \frac{0.7895}{1.128} = 0.6999.$$

The lower chart limit is

$$LCL = \overline{X} - 3 \times \text{Estimated sigma}$$
$$= 27.245 - 3 \times 0.6999 = 27.245 - 2.0997 = 25.145,$$

and the upper chart limit is

$$UCL = \overline{X} + 3 \times \text{Estimated sigma}$$
$$= 27.245 + 3 \times 0.6999 = 27.245 + 2.0997 = 29.345.$$

Box 5.1 Calculation of chart limits.

These formulae would be required should the reader wish to create an individual chart using pencil and paper or via a spreadsheet package. The formulae, together with those for other control charts covered later in this chapter, are given in Appendix 3. The above formulae apply only when the moving range used is of length 2.

Having found no evidence of any special cause variation on the control chart, it could be adopted for process monitoring – with the limits calculated from the first 20 observations being employed. (It is not desirable, in general, to update the chart limits as new data become available.) Thus when the next observation of temperature becomes available all that is required is for the value to be plotted on the chart. In order to do this via Minitab, right-click on the active chart and from the menu click on **Update Graph Automatically**. On typing the next temperature value of 26.2 into the Temperature column in the worksheet the control chart will be automatically updated. Employment earlier of the option **Use the following subgroups when estimating parameters:** 1 : 20 under **Estimate** ensures that the chart limits remain those calculated from the initial 20 observations.

The reader will have observed that, under **Estimate**, one may select the default option **Omit the following subgroups when estimating parameters:**. The author prefers generally to specify the data to be used in the calculations rather than the data to be omitted. Had the chart initial set of observations yielded a chart with, say, the 17th point outside the chart limits and there was a known special cause associated with that observation, then revised chart limits could be obtained with the 17th observation omitted from the calculations. This could be achieved with **Use the following subgroups when estimating parameters:** 1 : 16 18 : 20 under **Estimate**.

The additional point lies within the chart limits so there is no signal of a possible special cause. On plotting the next four values 26.5, 25.6, 26.3 and 24.1 the chart will be as shown in Figure 5.4. The reader is invited to enter the data as described above and to create the chart in Figure 5.4 for her/himself. The reader should note that if a centre line label or chart limit label is obscured it may readily be moved to a better location by left-clicking, keeping the mouse button depressed and dragging.

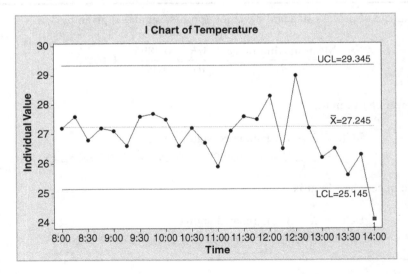

Figure 5.4   Individuals chart for temperature with additional data points.

The 25th point on the chart lies below the lower chart limit, so this provides evidence that a special cause may be affecting the process. Note that the plotting symbol for this 'out of control' point is a (red) square annotated with a 1. The reason for this is that there are a number of tests for special causes of which the first on the list provided by Minitab is the occurrence of a point outside the chart limits, i.e. of a point lying at a greater distance than three standard deviations from the centre line.

The maintenance engineer was subsequently called in and found a defective heating element, which he replaced. Thus the special cause of variation in the process was removed. One could then proceed to continue to monitor temperature using the chart with the limits established using the first 20 observations. In the case of major changes to the process it might be advisable to start the whole charting process again, i.e. to gather a series of initial temperature readings and to plot an initial chart. If there are no points outside the limits on this new chart then it can be adopted for further routine monitoring. If there are points outside the limits then a search should be made for potential special causes.

The moving ranges may also be plotted in a second control chart. For the 25 temperatures the chart of the 24 moving ranges is shown in Figure 5.5. It was created using **Stat > Control Charts > Variables Charts for Individuals > Moving Range...**, clicking on **MR Options...** and then on the **Estimate** tab, entering **Variables:** Temperature and specifying **Use the following subgroups when estimating parameters:** 1 : 20. The upper limit was calculated from the initial set of 19 moving ranges as follows:

$$\text{UCL} = 3.267 \times \overline{MR} = 3.267 \times 0.7895 = 2.579.$$

No reference will be made in this book to underlying theory concerning the distribution of ranges of samples from a normal distribution and to the derivation of the above upper limit formula. The 'centre' line on the chart is plotted at the level of the mean moving range. The lower chart limit is effectively 0 when ranges of pairs of measurements are used to estimate

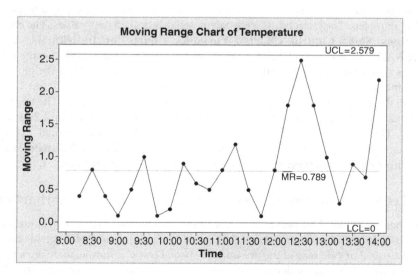

Figure 5.5 Moving range chart with limits based on first 20 temperatures.

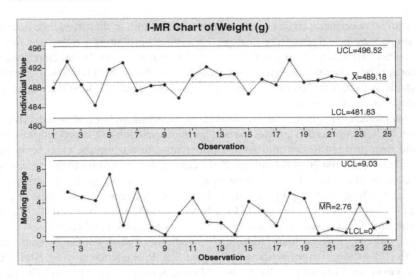

Figure 5.6   Individual and moving range charts for weight.

process standard deviation. All the points plotted lie within the chart limits. Note that, since the moving range can never take a negative value, a point below the (effective) lower chart limit is impossible.

The individuals chart or X chart can signal changes in the process location. Ideally the moving range chart would only signal changes in the process variability. However, changes in the process location, which are not accompanied by any change in process variability, can also yield points above the upper limit on the moving range chart. Montgomery (2009, p. 264) urges caution in the use of Shewhart control charts for individual measurements.

Note that Minitab offers the facility to create the charts separately or as a pair. The I-MR pair of charts – individuals and moving range – for the 25 bottle weights given in Table 2.1 is shown in Figure 5.6. The chart limits were calculated using all 25 data values. The charts were created using **Stat** > **Control Charts** > **Variables Charts for Individuals** > **I-MR...**, selecting **I-MR Options...**, clicking on the **Estimate** tab and specifying **Use the following subgroups when estimating parameters:** 1 : 25. The data are available in Weight1A.MTW.

No points fall outside the three-sigma control chart limits on either chart, so extended charts with those limits could subsequently be used to monitor the process. The updated chart with the addition of the data for the further Weights recorded in Table 2.2 is shown in Figure 5.7. The extended data set is available in Weight1B.MTW and the reader should note that, in creating the chart in Figure 5.7, it is necessary to specify **Use the following subgroups when estimating parameters:** 1 : 25 under **Estimate** via **I-MR Options....**

5.1.2 Tests for evidence of special cause variation on Shewhart charts

The run chart in Figure 2.14 provided evidence of special cause variation, yet no points fall outside the limits on either control chart in Figure 5.7. In addition to a point outside chart limits providing a signal of evidence of the presence of special cause variation, there are a number of other tests used to provide evidence of the presence of special causes of variation. The tests available in Minitab for the Shewhart charts provided are accessed via the **Tests** tab under

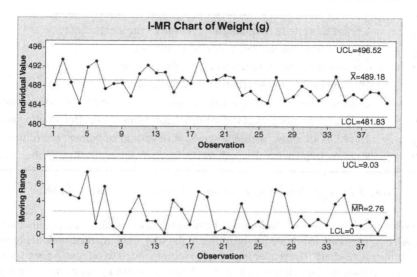

Figure 5.7 Updated individual and moving range charts for weight.

<Chart Type> **Options....** (<Chart Type> represents **I Chart** or **MR** or **I-MR** etc., depending on the particular chart or charts being used.) The tests available for the individuals chart are displayed in Figure 5.8.

The default test checks for the occurrence of a point more than three standard deviations from the centre line, i.e. for a point outside the chart limits. This test is referred to as Test 1 in Minitab. Note that it is listed first in the dialog box displayed in Figure 5.8. The default versions of all eight tests used in Minitab are listed in Box 5.2. The user can select which of the tests

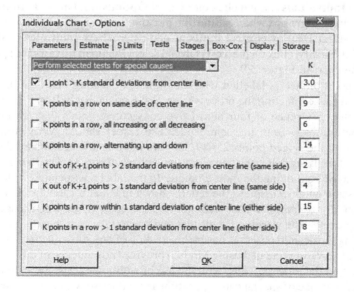

Figure 5.8 Tests for evidence of special causes.

Test 1	1 point more than 3 standard deviations from center line
Test 2	9 points in a row on same side of center line
Test 3	6 points in a row, all increasing or all decreasing
Test 4	14 points in a row, alternating up and down
Test 5	2 out of 3 points > 2 standard deviations from center line (same side)
Test 6	4 out of 5 points > 1 standard deviation from center line (same side)
Test 7	15 points in a row within 1 standard deviation of center line (either side)
Test 8	8 points in a row > 1 standard deviation from center line (either side)

Box 5.2   Tests for evidence of special causes available in Minitab.

he/she wishes to apply or use the drop-down menu to select the application of either all tests or no tests. All eight tests are available for the individuals chart, but only the first four are available for the moving range chart. The reader is strongly recommended to refer to Appendix 4, where an example of evidence of special cause variation being provided by each one of the eight tests is displayed. (If desired, the reader may define the tests differently from the defaults listed in Box 5.2. For example, many practioners use 8 rather than 9 points in a row in Test 2. The change may be made locally by changing the value of K from 9 to 8 in the second box in the column on the right of the dialog box in Figure 5.8, but it will revert to the default value of 9 in any new Minitab project. If desired the change may be made global by using **Tools** > **Options. . .**, double-clicking on **Control Charts and Quality Tools**, clicking on **Tests** and changing 9 to 8 in the window for Test 2.)

Another option provided for control charts is the positioning of horizontal reference lines/control limits at any number of standard deviations from the centre line the user desires. For the weight data in Weight1B.MTW select **Stat** > **Control Charts** > **Variables Charts for Individuals** > **Individuals. . .** and click on **I Chart Options. . . .** On the **Estimate** tab, select **Use the following subgroups when estimating parameters** and enter 1 : 25 beneath it. On the **S Limits** tab, select **Display control limits at** and enter **These multiples of the standard deviation:** 1 2 3. Finally, click on the **Tests** tab and select **Perform all tests for special causes**. This yields the chart in Figure 5.9.

The 29th point plotted is labelled with the digit 6, indicating that there is a signal of a possible special cause affecting the process from Test 6 on the list in Box 5.2. Test 6 involves checking for the occurrence of four out of five consecutive points that are more than one standard deviation away from the centre line. The reader should verify from scrutiny of Figure 5.9 that, of the ringed points 25–29, four are more than one standard deviation away from the centre line, lying below the one-sigma lower limit of 486.73. The figure also shows the message displayed on moving the mouse pointer to the label 6 associated with the 29th data point. The Session window displays the report, shown in Panel 5.1, on the chart just created.

Thus when 29 observations have been plotted there is evidence that a special cause is affecting the process. The warning means that one has to take care that the limits have been calculated using the desired observations in the creation of any subsequent chart. (A run chart of the first 29 observations created using Minitab provides no evidence of any special cause of variation affecting the process. Thus, with the use of additional tests, the individuals control chart provides evidence of special cause variation and it does so with fewer observations than the run chart in this case.)

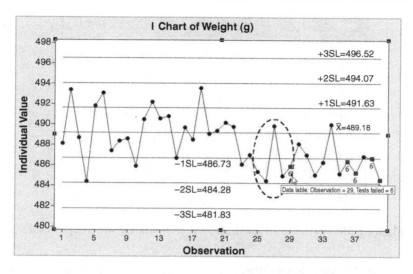

Figure 5.9   Alternative individuals chart of weight data.

Test Results for I Chart of Weight (g)

```
TEST 6. 4 out of 5 points more than 1 standard deviation from center line (on
     one side of CL).
Test Failed at points: 29, 36, 37, 39, 40

* WARNING * If graph is updated with new data, the results above may no
          * longer be correct.
```

Panel 5.1   Session window report on chart in Figure 5.9.

In the next section charts for samples or subgroups of measurements are introduced, together with the facility in Minitab to chart data from different stages in the history of a process on the same diagram. This facility may be used with all of the Shewhart charts considered in this chapter.

5.1.3   Xbar and R charts for samples (subgroups) of measurements

In many situations processes are monitored using samples or subgroups of product. The third column in the Minitab worksheet Camshaft.MTW (available in the Minitab Sample Data folder supplied with the software) gives the length (mm) of a series of 20 samples (or subgroups) of size $n = 5$ camshafts taken from supplier 2. Reference to the flow chart from the **Assistant** menu displayed in Figure 5.1 leads to the widely used procedure of computing the sample means (Xbar) and the sample ranges (R) and plotting both series of values in sequence with appropriate centre lines and control limits added. The tests available in Minitab for the Xbar chart of means and the R chart of ranges match those for the individuals and moving range charts, respectively.

To create the charts using Minitab use **Stat > Control Charts > Variables Charts for Subgroups > Xbar-R. . . .** In this case, where all the measurements are in a single column, the default **All observations for a chart are in one column:** option is used. Clicking on the

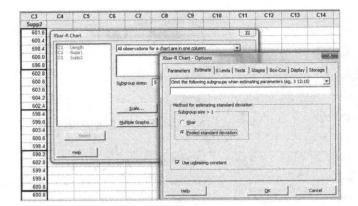

Figure 5.10   Creation of mean and range charts for subgroups of size 5.

Estimate tab under **Xbar-R Options...** reveals that the default **Method for estimating standard deviation** with **Subgroup size > 1** is to use **Pooled standard deviation**. Note that **Use unbiasing constant** is also checked by default.

One could opt to use **Rbar**, the mean of all the sample ranges, in order to estimate the standard deviation of camshaft length. Historically this was a widely used method, but statistical theory shows that the use of pooled standard deviation yields better estimates than use of mean sample ranges. Readers with an interest in the technical details may find it helpful to consult the paper by Mahmoud *et al.* (2010). Throughout this chapter the Xbar-R charts presented have all been created using the default option to estimate standard deviation using pooled standard deviation and the unbiasing constant.

The completed dialog is displayed in Figure 5.10, the option to perform all tests for special causes having been selected. Bold rectangles have been added to the image to indicate the first four samples/subgroups of camshaft length. The charts in Figure 5.11 were obtained. There are three signals on the Xbar chart indicating potential special cause variation affecting supplier 2's process – there is evidence that the process is not in a state of statistical control, i.e. that it is not behaving in a stable, predictable manner.

Let us suppose that discussion with those responsible for the process led to identification of assignable causes of variation for subgroup 2 (machine fault) and subgroup 14 (operator error) but not for subgroup 9. When signals of potential special cause variation lead to identification of actual special causes it is normal to recalculate the chart limits with the corresponding subgroups, 2 and 14 in this case, omitted from the calculations and to scrutinize the revised charts. This may be achieved using **Xbar-R Options...**, clicking on the **Estimate tab** and specifying 2 14 under **Omit the following subgroups when estimating parameters**, but the author prefers to employ **Estimate** specifying 1 3 : 13 15 : 20 under **Use the following subgroups when estimating parameters**. The **Labels** facility was used to add footnotes indicating the actions taken. The resulting chart pair is shown in Figure 5.12. (If the special causes identified have not been eliminated then one should be cautious about recalculating limits. The Xbar chart with the existing limits has successfully detected these causes and so has the potential to do so again should they recur. Otherwise there is a risk of 'tightening' the Xbar chart limits to such an extent that the chart starts to yield too many false alarm signals of special cause variation.)

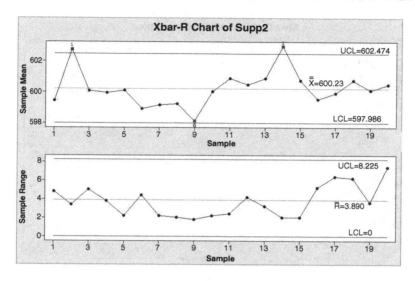

Figure 5.11   Mean and range charts for supplier 2.

The means for subgroups 2 and 14 fall outside the new control limits on the Xbar chart, but these subgroups can now effectively be ignored. Subgroup 9 no longer gives a signal on the Xbar chart, but subgroup15 now does. However, this is a spurious signal since the point corresponding to subgroup 14 has been counted as one of the four from five consecutive points more than one standard deviation from the centre line (same side). There are no signals on the R chart, so the decision might well now be taken to begin monitoring the process by taking further subgroups of five camshafts at regular intervals and plotting the means and ranges on charts using the limits displayed on the charts in Figure 5.12.

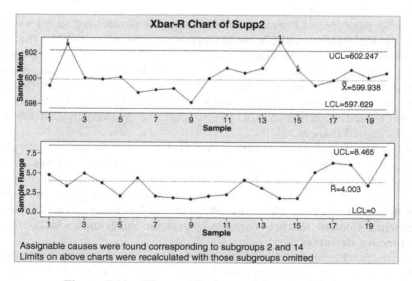

Figure 5.12   Xbar and R charts with revised limits.

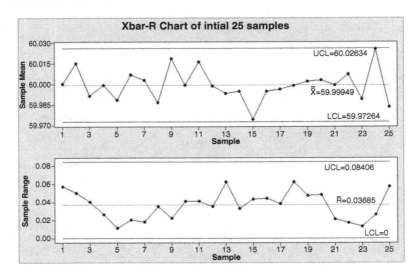

Figure 5.13   Initial Xbar and R charts for rods process.

Some simulated data will now be used to demonstrate how different types of process changes are indicated by signals on Xbar and R charts. Imagine a process which cuts extruded plastic into rods and that, under ideal conditions, the rods have length (mm) which is normally distributed with mean 60.00 and standard deviation 0.02, i.e. $N(60.00, 0.02^2)$. **Calc > Random Data > Normal...** was used to generate four columns of 25 values from the $N(60.00, 0.02^2)$ distribution. The values were rounded to three decimal places and stored in the supplied worksheet Rods.MTW. Each row of four values from the columns may be regarded as a sample/subgroup of size $n = 4$ from the normal distribution specified. Xbar and R charts of the initial data are shown in Figure 5.13.

In creating these charts **Observations are in one row of columns:** x1-x4 was specified and, from the **Estimate** tab under **Xbar-R Options...**, one may select **Use the following subgroups when estimating parameters:** and specify 1 : 25. With all available tests applied there are no signals of potential special cause variation on the charts, so the decision could be taken to use Xbar and R charts with the limits shown to monitor the process. (Suppose that you forgot to select Perform all tests for special causes on the **Tests** tab. The Edit Last Dialog icon ⊞ may be used, or alternatively Ctrl + E, to access the most recently used dialog box and to make any desired changes.)

Before proceeding to look at further simulated data for the process, details of the calculation of the limits for the Xbar and R charts are presented in Boxes 5.3 and 5.4. Readers may skip the details in these boxes as the charts may be employed effectively without familiarity with technical details. If required, the formulae and constants involved are available from the **Help** menu via **Help > Methods and Formulas > Quality and process improvement > Control charts > Variable Charts for Subgroups > Methods for estimating standard deviation**.

A further 15 subgroups of four length measurements were generated using the same distribution as for the initial 25 subgroups that were charted in Figure 5.12. Then a further 20 subgroups were generated using different distributions to illustrate four different scenarios

The means, $\bar{x}$, of the 25 samples of $n = 4$ lengths have mean $\bar{\bar{x}} = 59.999\,49$, the double bar notation indicating that the value is the mean of a set of means. (Since the normal distribution used to generate the simulated data had mean $\mu = 60.00$ it is not surprising that the mean of the 25 means is close to 60.00 and thus provides a good estimate of the process mean μ.)

The pooled standard deviation is the square toot of the mean of the 25 sample variances in this case, where the samples all have the same size, and is $0.017\,837\,1$. This has to be divided by the unbiasing constant $c_4(76) = 0.996\,672$, available in the linked table in **Methods and Formulas**, to yield the estimate $0.017\,896\,7$, of the process standard deviation (process sigma). (This estimate is close to the standard deviation $\sigma = 0.02$ of the distribution used to generate the simulated data.)

The theory of the distribution of the sample mean from Section 3.4 yields three-sigma Xbar chart limits of

$$\mu_{\bar{X}} \pm 3\sigma_{\bar{X}} = \mu \pm 3\frac{\sigma}{\sqrt{n}}.$$

The chart limits are

$$\bar{\bar{x}} \pm 3 \times \frac{\text{Estimated sigma}}{\sqrt{n}} = 59.999\,49 \pm 3\frac{0.017\,896\,7}{\sqrt{4}}$$

$$= 59.999\,49 \pm 0.026\,845\,0 = (59.972\,64,\ 60.026\,34).$$

These limits are in agreement with those on the Xbar chart in Figure 5.13.

Box 5.3   Calculation of Xbar chart limits.

The three-sigma R chart limits are

$$\mu_R \pm 3\sigma_R = d_2\sigma \pm 3d_3\sigma,$$

where the constants d_2 and d_3 may be read from the linked table in **Methods and Formulas**. The centre line is placed at the estimated μ_R given by chart limits are

$$d_2 \times \text{Estimated sigma} = 2.059 \times 0.017\,896\,7 = 0.036\,85).$$

The chart limits are

$$d_2 \times \text{Estimated sigma} \pm 3d_3 \times \text{Estimated sigma}$$

$$= 2.059 \times 0.017\,896\,7 \pm 3 \times 0.8794 \times 0.017\,896\,7$$

$$= (-0.010\,36, 0.084\,06).$$

Since range cannot be negative, the lower chart limit is effectively 0. These limits and the centre line are in agreement with those on the R chart in Figure 5.13. Sample size has to be at least 7 for nonzero lower limits to occur on an R chart.

Box 5.4   Calculation of R chart limits.

in terms of process changes. Thus one can think of the process change, i.e. that a special cause of variation took effect, occurring at some time between the taking of the 40th and 41st samples. The data sets are supplied as RodsScenario1.MTW, RodsScenario2.MTW, RodsScenario3.MTW and RodsScenario4.MTW. In all four cases:

- limits on the charts (see Figures 5.14–5.16) are those calculated from data for the first 25 samples;

- all tests for evidence of special cause variation were performed.

In scenario 1 (process mean increased, process standard deviation unchanged), subgroup 42 gives rise to the first signal of the process change on the Xbar chart in Figure 5.14. In practice, on plotting the data for this sample, action would be taken to seek a special cause of variation affecting the process. Thus it took just two subgroups to signal the process change. However, if the samples were taken at 15-minute intervals this could correspond to up to half an hour of production of less satisfactory rods from the point of view of the customer. The number of samples required to signal a process change is referred to as the *run length*. (It must be emphasized that chart limits, or control limits as they are referred to by some, are *not* specification limits.)

In scenario 2 (process mean decreased, process standard deviation unchanged), it may be seen from Figure 5.15 that the first signal of the process change was from the 44th sample. Thus it took four subgroups to flag the change in this scenario. With a smaller change in the process mean in this scenario than in the first, it is not surprising that the run length is greater. (Note that, as in scenario 1, there is a spurious signal on the R chart. Recall that, in simulating the data for both scenarios 1 and 2, no change was made to the process standard deviation, i.e. to process variability.)

In scenario 3 (process standard deviation increased, process mean unchanged) it took just one sample for the R chart to signal the likely occurrence of a special cause affecting process variability. With increased variability the limits on the Xbar chart in Figure 5.16 are too close together from sample 41 onwards. Thus an increase in process variability typically yields signals on the Xbar chart as well as on the R chart. Thus it is advisable to examine the R chart first when employing Xbar and R charts for process monitoring purposes and to interpret the charts as a pair.

A major increase in variability, such as the one illustrated in this scenario, could have a major impact on process capability. Thus the process owners would most likely wish to take action quickly after the data for subgroup 41 were plotted to eliminate any special cause found to be affecting the process. The changes in location in the first two scenarios could also impact on process capability in a detrimental manner. The final scenario illustrated refers to a situation where process variability is reduced. Reduction of variability is fundamental to achieving quality improvement.

In scenario 4 (process standard deviation decreased, process mean unchanged) Figure 5.17 indicates that it took 13 subgroups for a signal to appear on the R chart. Note that the reduction in process variability does not give rise to signals on the Xbar chart. The reason for this is that from sample 40 onwards the Xbar chart limits are too far apart. Were we dealing with real as opposed to simulated data then perhaps a deliberate change was made to the process after subgroup 40 was taken, possibly as the result of a Six Sigma project undertaken to identify ways to reduce process variability. In any case it would clearly be desirable to maintain the reduced variability.

For the purposes of illustration let us suppose that, between the times at which samples 40 and 41 were taken, a new feed control system was fitted to the machine which cut the

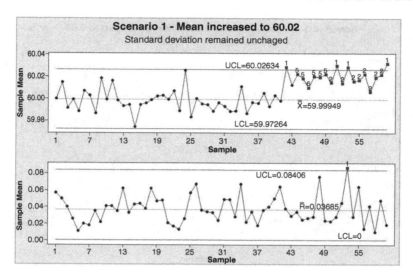

Figure 5.14   Xbar and R charts for an increase in process mean (scenario 1).

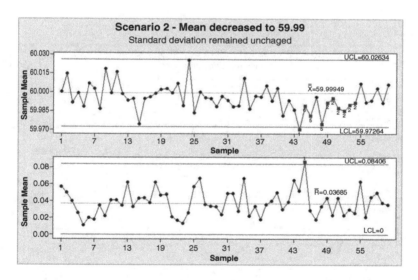

Figure 5.15   Xbar and R charts for a decrease in process mean (scenario 2).

extruded plastic into rods. In addition to the four columns containing the subgroups of four length measurements across their rows, a fifth column consisting of 40 values of 1 followed by 20 values of 2, indicating the two phases of operation monitored, has been added to the worksheet. Some of the dialog involved may be viewed in Figure 5.18. Points to note are:

- **Observations are in one row of columns:** x1-x4 has been specified

- Under **Options...** and **Estimate** one has to select **Use the following subgroups when estimating parameters:** and specify **both** 1 : 25 **and** 41 : 60.

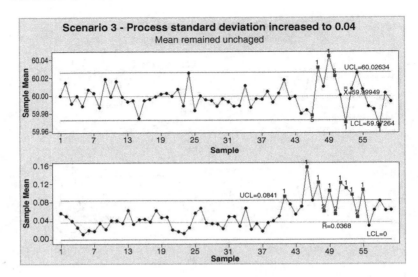

Figure 5.16   Xbar and R charts for an increase in process standard deviation (scenario 3).

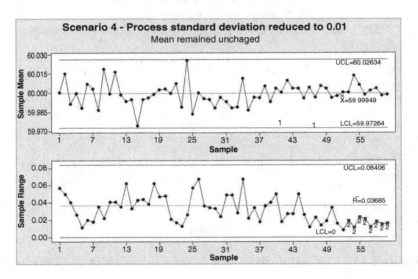

Figure 5.17   Xbar and R charts for a decrease in process standard deviation (scenario 4).

- Under **Options...** and **Stages** one has to specify selection of **Phase** in order to **Define stages (historical groups) with this variable:**. The default option to use both **When to start a new stage** and **With each new value** may be used. (Alternatively, **When to start a new stage** and **With the first occurrence of these values:** 1 2 could be employed.)

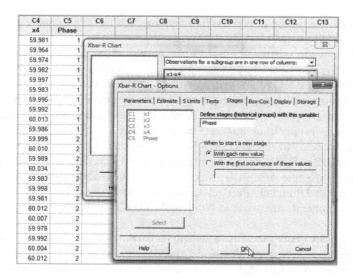

Figure 5.18 Defining stages in the creation of Xbar and R charts.

The charts are displayed in Figure 5.19. In order to have the limits and centre lines labelled for all stages use <Chart> **Options** > **Display** > **Other** and check **Display control limits/center line labels for all stages**.

The 20 subgroups plotted for the second stage may be regarded as initial data for a new Xbar-R chart pair. There is a signal from the fourth of the second-phase subgroups on the R chart, so many experts would recommend reviewing the situation after another few subgroups have been obtained before adopting the new charts for further routine monitoring. Montgomery (2009, p. 297) states that when used in this manner 'the control chart becomes a

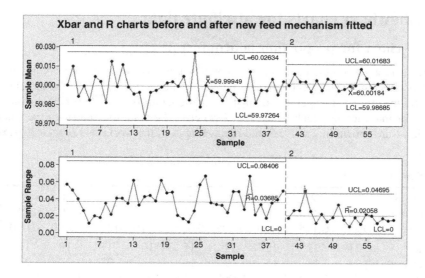

Figure 5.19 Xbar and R charts with two stages.

logbook in which the timing of process interventions and their subsequent effect on process performance is easily seen'.

As an alternative to mean and range (Xbar and R) charts, mean and standard deviation (Xbar and S) charts may be used when data for process monitoring are collected in subgroups. The flow chart in Figure 5.1 recommends the use of Xbar and S charts whenever sample size is 8 or more. An example of the use of an Xbar and S chart pair is provided as an exercise. Other topics on Shewhart control charts for measurement data, such as the use of individuals charts to check that a process is operating 'on target' and triple charts (I-MR-R/S charts) are referred to in follow-up exercises. No reference will be made to zone charts in this book.

Tests 1, 2 (with 8 points in a row), 5 and 6 are referred to as the Western Electric Company (WECO) rules, and some practitioners prefer to use these four tests rather than all eight available in Minitab. Ultimately the decision on which tests to use lies with the process team. The following comments are made in NIST/SEMATECH (2005, Section 6.3.2):

> While the WECO rules increase a Shewhart chart's sensitivity to trends or drifts in the mean, there is a severe downside to adding the WECO rules to an ordinary Shewhart control chart that the user should understand. When following the standard Shewhart 'out of control' rule (i.e., signal if and only if you see a point beyond the plus or minus 3 sigma control limits) you will have 'false alarms' every 371 points on the average . . . . Adding the WECO rules increases the frequency of false alarms to about once in every 91.75 points, on the average . . . . The user has to decide whether this price is worth paying (some users add the WECO rules, but take them 'less seriously' in terms of the effort put into troubleshooting activities when out of control signals occur).

Readers wishing to construct control charts for measurement data without using Minitab may find the factors in Appendix 2 and the formulae in Appendix 3 of value.

5.2   Shewhart charts for attribute data

5.2.1   P chart for proportion nonconforming

Consider a large e-commerce company at which there is concern over complaints from customers concerning inaccurate invoices being e-mailed to them. During the measure phase of a Six Sigma project aimed at improving the situation, random samples of 200 invoices were checked for inaccuracies, each week, for 20 weeks. The data, together with the calculated proportions, are shown in Table 5.3 and are available in the worksheet Inaccurate1.MTW.

Table 5.3   Invoice data.

Week no.	1	2	3	4	5	6	7	8	9	10
No. inaccurate	23	23	20	21	17	22	24	20	18	17
Proportion	0.115	0.115	0.100	0.105	0.085	0.110	0.120	0.100	0.090	0.085
Week no.	11	12	13	14	15	16	17	18	19	20
No. inaccurate	24	17	15	19	19	22	27	23	23	18
Proportion	0.120	0.085	0.075	0.095	0.095	0.110	0.135	0.115	0.115	0.090

The upper chart limit is given by

$$UCL = \bar{p} + 3\sqrt{\frac{\bar{p}(1-\bar{p})}{n}}$$

$$= 0.103 + 3\sqrt{\frac{0.103 \times 0.897}{200}} = 0.103 + 0.0645 = 0.1675.$$

The lower chart limit is

$$LCL = \bar{p} - 3\sqrt{\frac{\bar{p}(1-\bar{p})}{n}}$$

$$= 0.103 - 3\sqrt{\frac{0.103 \times 0.897}{200}} = 0.103 - 0.0645 = 0.0385.$$

Box 5.5 Calculation of P chart limits.

In this scenario counts are being made of the number of items (invoices) that are defective (contain one or more inaccuracies), so according to the flow chart in Figure 5.1 the appropriate control chart in this situation is a P chart or chart for proportion defective. Some refer to items as being nonconforming rather than defective. The mean of the 20 proportions is $\bar{p} = 0.103$ and the chart centre line is plotted at this value. This is taken as an estimate of the population proportion, p, and the calculation of the three-sigma limits for the chart are made using the formulae, given in Appendix 3, incorporating the standard deviation of a proportion stated in Chapter 4. The calculations are displayed in Box 5.5.

To create the chart with Minitab, the number of inaccurate invoices is entered into a column labelled No. Inaccurate and use made of **Stat > Control Charts > Attributes Charts > P . . . .** No. Inaccurate is entered under **Variables:** as the variable to be charted and the sample/subgroup size is specified using **Subgroup sizes:** 200. The default versions of all four available tests were implemented under **P Chart Options. . . > Tests**. This may be achieved by checking each of the four tests or by selecting **Perform all tests for special causes** from the menu under **Tests**. The four available tests are the same as the first four of the eight available with both charts for individuals (X) and for means (Xbar). Examples of patterns yielding signals of possible special cause variation from the tests are given in Appendix 4. The chart is displayed in Figure 5.20. The reader should note that Minitab did not require a column of proportions in order to create the chart.

There are no signals of any potential special causes affecting the process so it can be deemed to be in a state of statistical control, operating in a stable and predictable manner with approximately 10% of invoices having inaccuracies. It can therefore be agreed to use the chart with the centre line and limits shown for future process monitoring.

Before proceeding to look at further data from the process the reader is invited to re-create the P chart using **Assistant > Control Charts** and clicking on **more. . .** under **P chart**.

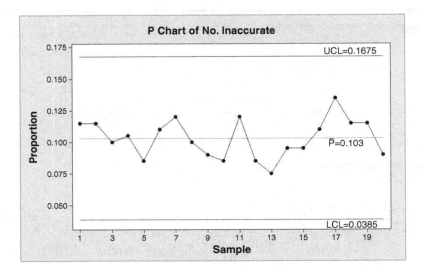

Figure 5.20   P chart of invoice data.

Click on **Attribute data** and read the two descriptions, then under **Attribute data** click on **Next**. Having read the two descriptions, under **Defective items** click on **Next** to obtain the screen shown in Figure 5.21. Clicking on the $+$ icons yields further details. Finally the reader is invited to click on the create chart icon and reproduce the chart in Figure 5.20. (Hint: You will need to select **Estimate from the data** under **Control limits and center line**.) This leads to the creation of three items in the Graphs folder – a Stability Report that includes the P chart, a Summary Report and a Report Card. The clean bill of health from the Report Card supports the decision to 'roll out' the chart in Figure 5.20 for further monitoring.

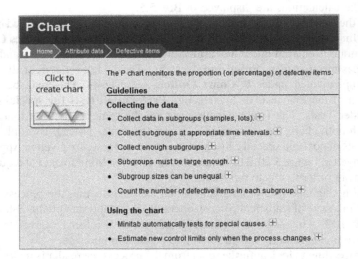

Figure 5.21   P chart guidelines from the **Assistant** menu.

Table 5.4 Further invoice data.

Week No.	21	22	23	24	25	26	27	28	29	30
No. Inaccurate	10	16	17	15	18	22	13	10	11	11
Proportion	0.05	0.08	0.085	0.075	0.09	0.11	0.065	0.05	0.055	0.055
Week No.	31	32	33	34	35	36	37	38	39	40
No. Inaccurate	8	10	16	17	15	18	22	13	10	11
Proportion	0.04	0.05	0.08	0.085	0.075	0.09	0.11	0.065	0.05	0.055

At the end of the 20-week period during which the above data were gathered, changes planned by the project team were introduced. Data for the next 20 weeks are given in Table 5.4 and the data for all 40 weeks are available in the worksheet Inaccurate2.MTW.

The extended control chart for proportions was created using **Stat > Control Charts > Attributes Charts > P...** and is shown in Figure 5.22. Note that the limits are those based on the data for the first 20 weeks so this was indicated using **Estimate** under **P Chart Options....** (The P chart dialog available via the **Assistant** menu does not permit specification of the samples to be used in the calculation of the limits.) The samples to be used in the calculation of the limits may be specified in two ways: either **Estimate > Omit the following subgroups...** and enter 21 : 40 or **Estimate > Use the following subgroups...** and enter 1 : 20. The author prefers always to use the latter method.

The first signal appears from the sample taken during week 35. Reference to the list of tests in Appendix 4 indicates that a signal arising from Test 2 results from the occurrence of nine consecutive points on the same side of the centre line. These points have been ringed in the plot, and on moving the mouse pointer to the label 2 beside the ninth point the sample number and the test failed are displayed. Thus the chart provides evidence that the process changes have led to improvement, in the form of a reduction in the proportion of inaccurate invoices. Thus it is appropriate to introduce an additional column named Phase in the worksheet to indicate the two phases of operation of the process, before and after changes.

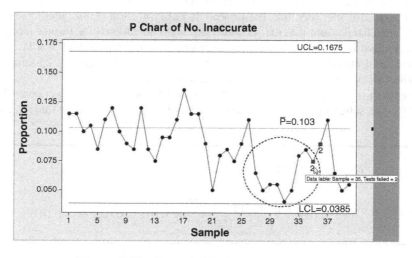

Figure 5.22 Extended P chart of invoice data.

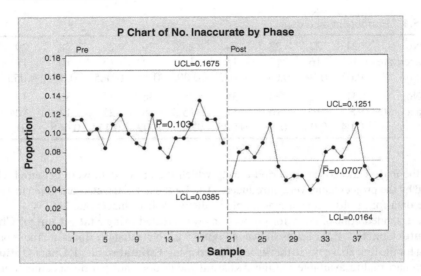

Figure 5.23   P charts for the two phases.

The first 20 cells in the column named Phase could contain the text Pre and the remaining 20 the text Post. This may be achieved using **Calc** > **Make Patterned Data** > **Text Values...** with **Store patterned data in:** Phase, **Text values:** Pre Post, **Number of times to list each value:** 20, and **Number of times to list the sequence:** 1. In order to create the revised chart displayed in Figure 5.23 all the data values were used in the computation of limits. Again this may be specified in two ways: either **Estimate** > **Omit the following subgroups...** and leave the window blank, or **Estimate** > **Use the following subgroups...** and enter 1:40. In addition, under **P Chart Options...** > **Stages** one has to enter Phase under **Define stages (historical groups) with this variable:**.

Figure 5.23 indicates that the changes have reduced the proportion of inaccurate invoices to around 7%. It also indicates that the process is behaving in a stable, predictable manner following the changes and that the second chart could be adopted for further monitoring.

Clearly there is room for further improvement. Let us suppose that at a later date the proportion has dropped to around 2%, with mean proportion for a series of 25 samples being 0.018. The calculation of the Lower Chart Limit is shown in Box 5.6.

A negative proportion is impossible so there is, strictly speaking, no lower control limit on the P chart with subgroup size 200. However, Minitab inserts a horizontal line at zero on the

$$
\begin{aligned}
\text{LCL} &= \bar{p} - 3\sqrt{\frac{\bar{p}(1-\bar{p})}{n}} \\
&= 0.018 - 3\sqrt{\frac{0.018 \times 0.982}{200}} = 0.018 - 0.028 = -0.010
\end{aligned}
$$

Box 5.6   Calculation yielding a negative lower chart limit.

$$n > \frac{9(1 - \bar{p})}{\bar{p}}$$

Box 5.7 Criterion to ensure a nonzero lower limit on a P chart.

P chart in such cases, labelled LCL $= 0$. With no lower limit, the possibility of evidence of a further drop in the proportion of nonconforming invoices being signalled by a point falling below the lower limit is not available. To reinstate this option the sample size can be increased. Some mathematical manipulation shows that, to ensure a lower limit exists on a P chart with three-sigma limits, the inequality in Box 5.7 must be satisfied. For $\bar{p} = 0.018$, the formula gives $n > 491$. Thus, once the monthly proportion of nonconforming invoices had dropped to around 2%, monthly samples of, say, 500 invoices would provide the opportunity to detect further improvement through a signal from a point on the chart falling below the lower limit.

In some situations it is not possible to have constant sample size but it is still possible to create a P chart. The chart limits are no longer horizontal parallel lines but have a stepped appearance, the limits being closer together for larger samples and wider apart for smaller samples. As an example, consider the data in Table 5.5 giving monthly admissions of stroke patients to a major hospital together with the numbers of those patients treated in the acute stroke unit.

The data were set up in three columns in the supplied Minitab worksheet ASU.MTW. The first contains the month in which the data were collected in the date format Jan-02, Feb-02 etc. and was set up using **Calc > Make Patterned Data > Simple Set of Date/Time Values...** with **Patterned Sequence** specified as **From first date/time:** Jan-02 **To last date/time:** Dec-03 **In steps of:** 1 with **Step unit:** Month and defaults otherwise. The second column contained the monthly counts of patients admitted with a diagnosis of stroke, and the third the number of those patients who receive treatment in the acute stroke unit.

Part of the dialog involved in creating a P chart for the proportion of patients receiving treatment in the acute stroke unit is shown in Figure 5.24. **Subgroup size:** is specified by selecting the column named Strokes. Use of the **Scale...** facility enables the horizontal axis of the chart to be 'stamped' with the months when the data were collected. Limits were based on the first 15 observations as changes were made to the process of managing stroke patients at the hospital at the end of March 2003, corresponding to observation 15. The chart is displayed in Figure 5.25. The centre line is placed at 0.6352 and labelled $\bar{P} = 0.6352$. However, this value is not the mean of the first 15 proportions; it is the total number of stroke patients receiving ASU care for the first 15 months (477) divided by the total number of stroke patients admitted in the

Table 5.5 Monthly stroke admissions.

2002	Jan	Feb	Mar	Apr	May	Jun	Jul	Aug	Sep	Oct	Nov	Dec
Strokes	47	58	35	49	58	56	50	45	51	53	61	47
ASU	31	34	29	28	30	35	31	37	26	33	37	32
2003	Jan	Feb	Mar	Apr	May	Jun	Jul	Aug	Sep	Oct	Nov	Dec
Strokes	51	38	52	43	43	44	49	42	38	58	39	36
ASU	32	32	30	34	31	28	32	37	32	41	26	31

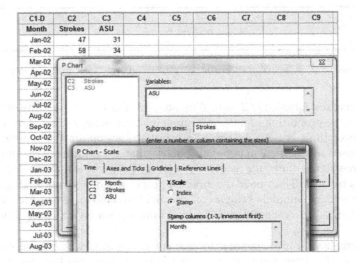

Figure 5.24   Creation of a P chart with variable sample size.

first 15 months (751). In the case of constant sample size the result of this calculation is the same as the result of taking the mean of the corresponding proportions. The UCL and LCL values displayed apply to the final sample that had size 36. The reader is invited to use the formulae in Box 5.5 to confirm the LCL of 0.3945 and UCL of 0.8758 displayed.

From the signals on the chart it would appear that the process changes have led to a greater proportion of stroke patients receiving acute stroke unit care. (Of course the theory underlying the P chart is based on the binomial distribution for which the probability of care in the acute stroke unit would remain constant from patient to patient. In reality this is unlikely to be the case. Wheeler and Poling (1998, pp. 182–184) and Henderson *et al.* (2008) refer to this issue.

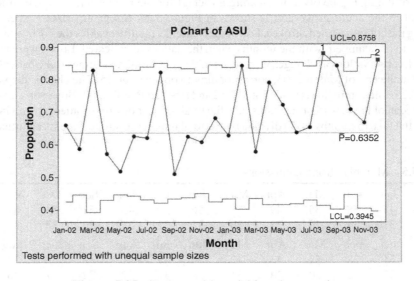

Figure 5.25   P chart with variable subgroup size.

However, the chart does display the data in an informative way and methods presented in Chapter 7 can be used to test formally whether or not the proportion of patients receiving acute stroke unit care has increased.)

The assumptions underlying valid use of a P chart are as follows:

1. Samples of n (not necessarily a constant) items provide the areas of opportunity.

2. Each item may be classified to either possess, or not possess, an attribute. Usually the attribute is nonconformance with specifications.

3. The probability, p, that an item possesses the attribute of interest is constant.

4. The status of any item with regard to possession of the attribute is independent of that of any other item.

5.2.2 NP chart for number nonconforming

The number defective or NP chart is exactly equivalent to the P chart, the only difference being that the number defective is plotted instead of the proportion defective. The NP charts for the number of inaccurate invoices data in Tables 5.3 and 5.4 in the previous section are displayed in Figure 5.26. Note that this is simply a scaled version of the chart in Figure 5.23 – e.g. the upper chart limit in the post-change phase in Figure 5.26 is 200 times the upper chart limit in the post-change phase on the P chart in Figure 5.23.

The author believes that, since the NP chart plots the number of defective items rather than the proportion of defective items, it is less directly informative than the P chart. One advantage of the NP chart over a P chart is that it is much simpler to update a pencil and paper version of an NP chart as no calculation is required – the count of defective items in the sample is plotted directly on to the chart. The underlying assumptions for valid use of an NP chart are the same as for the P chart.

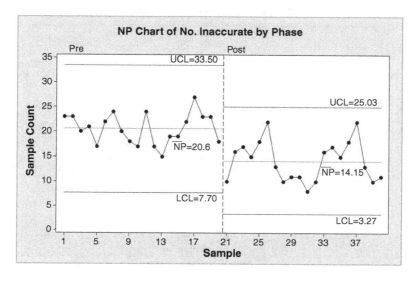

Figure 5.26   NP charts for the two phases.

Table 5.6   Counts of nonconformities in ATMs.

ATM	1	2	3	4	5	6	7	8	9	10	11	12	13	14	15
No. of nonconformities	5	4	7	9	4	6	5	8	9	11	5	10	6	6	5
ATM	16	17	18	19	20	21	22	23	24	25	26	27	28	29	30
No. of nonconformities	4	7	10	6	9	8	8	4	8	8	4	4	7	3	12

5.2.3   C chart for count of nonconformities

The C chart is used to plot the count of defects/nonconformities in equal 'areas of opportunity' for these to occur. These 'areas of opportunity' may be in time, space or segments of product. The number of yarn breakages per hour on a monofilament spinning frame, the number of nonconformities (imperfect solder joints, missing or damaged components etc.) on a printed circuit board taken at regular intervals from a process are respectively time and space examples.

Table 5.6 gives counts of the number of nonconformities detected during verification of automatic telling machines of a particular type. The data are in time sequence and are available in ATM.MTW. A C chart of the data is shown in Figure 5.27 created using **Stat > Control Charts > Attributes Charts > C...** with the counts from the above table previously entered into a column named No. Nonconformities. The chart limits calculations are shown in Box 5.8. Tests 1 to 4 inclusive (see Appendix 4) are available for the C chart in Minitab. These were all applied in the creation of the chart in Figure 5.27. There is no evidence from the chart of any special cause affecting the process. Thus the chart could be employed for further monitoring of the process.

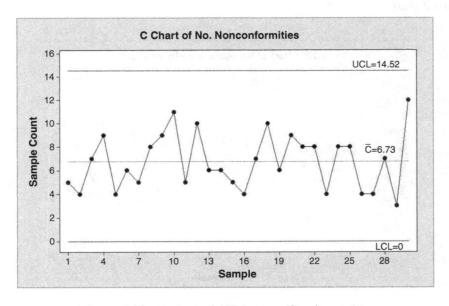

Figure 5.27   C chart of ATM nonconformity counts.

The upper chart limit is given by
$$\text{UCL} = \bar{p} + 3\sqrt{\bar{p}}$$
$$= 6.733 + 3 \times 2.595 = 14.52.$$

The lower chart limit is
$$\text{LCL} = \bar{c} - 3\sqrt{\bar{c}}$$
$$= 6.733 - 3 \times 2.595 = -1.51.$$

As a count of nonconformities can never be negative, Minitab sets the LCL to 0.

Box 5.8   Calculation of limits for a C chart.

The assumptions underlying valid use of a C chart are as follows:

1. The counts are of nonconformities or events.

2. The nonconformities or events occur in a defined region of space or period of time or segment of product referred to as the area of opportunity.

3. The nonconformities or events occur independently of each other, and the probability of occurrence of a nonconformity or event is proportional to the size of the area of opportunity.

5.2.4   U chart for nonconformities per unit

The U chart may be employed when counts of nonconformities are made over a number of units of product. The worksheet Faults.MTW contains the number of faults detected in each of 30 consecutive hourly samples of 40 retractable single-use syringes from a pilot manufacturing process. A U chart of the data is shown in Figure 5.28 created using **Stat > Control**

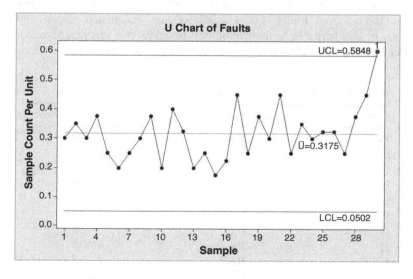

Figure 5.28   U chart of faults per syringe.

Charts > **Attributes Charts** > **U. . . .** Tests 1 to 4 inclusive (see Appendix 4) are available for the U chart in Minitab. These were all applied in the creation of the chart in Figure 5.28. There is evidence from the chart of a special cause affecting the process since the last point is above the upper chart limit. The sample size for a U chart can be variable and Minitab enables U charts with variable sample size to be created in the same way as in the case of the P chart. A column that indicates the sample sizes has to be specified in the **Subgroup sizes:** window in the dialog, as was done in the dialog displayed in Figure 5.24 for the chart in Figure 5.25.

Reference to the Assistant flow chart for Shewhart control chart selection that is displayed in Figure 5.1 reveals no reference to the C chart. However, a C chart is the special case of the U chart with constant size 1 for all subgroups. The reader is invited to re-create the C chart in Figure 5.27 using the U chart facility.

Readers wishing to construct control charts for attribute data without using Minitab may find the formulae in Appendix 3 of value. This completes the material on Shewhart control charts in this book. Montgomery (2009, pp. 330–344) provides guidelines for the implementation of these charts. After a brief discussion of funnel plots, we turn our attention to time-weighted control charts.

5.2.5 Funnel plots

Although not strictly control charts, funnel plots will be included here because the underlying statistical modelling is identical to that for the P chart. In order to introduce the funnel plot, consider the situation where a customer has records of counts of nonconforming units for a number of suppliers as shown in Table 5.7. (Having the data in time sequence is crucial for the correct use of control charts, but is not so for funnel plots; indeed the data for a funnel plot typically applies to the same time period.)

The funnel plot, with three sigma limits, is shown in Figure 5.29. The plot gets its name from the funnel-like shape of the curves defining the limits. The proportions for each supplier are plotted against the number of units tested. The centre line corresponds to the mean proportion nonconforming across all ten suppliers of $\bar{p} = 228/2000 = 0.114$, approximately 11%. The three-sigma limits for a supplier are calculated using the formula $\bar{p} \pm 3\sqrt{\bar{p}(1-\bar{p})/n}$, where n is the number of units tested for that supplier. Should limit values greater than 1 or negative values be obtained then they should be set to 1 or 0, respectively.

Table 5.7 Records of nonconforming units for ten suppliers.

Supplier	Units tested	No. nonconforming
A	200	19
B	150	41
C	60	8
D	400	48
E	80	11
F	250	13
G	160	19
H	200	18
I	360	38
J	140	13

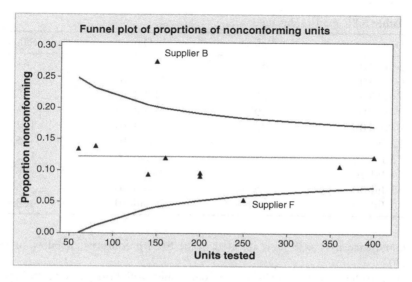

Figure 5.29   Funnel plot of proportions of nonconforming units.

Suppliers B and F plot outside the limits. As supplier B falls above the upper limit the chart provides evidence that supplier B produces nonconforming units at a significantly higher rate (27.3%) than the overall rate of 11.4%. Similarly, there is evidence that supplier F performs significantly better (3.6%) than the overall rate of 11.4%.

Spiegelhalter (2002) discusses the use of funnel plots for institutional comparisons in healthcare and also calculation of the limits using the binomial probability distribution rather than the normal approximation method used above. League tables are often produced when comparing performance across institutions, but some argue that identifying the institutions that stand out from the crowd via a funnel plot analysis and then investigating the performance of these can lead to insights that lead to quality improvement. Minitab does not have a facility for the direct creation of funnel plots. A follow-up exercise is provided and details of how a funnel plot may be created using Minitab are provided in the notes on the exercise on the book's website.

5.3   Time-weighted control charts

5.3.1   Moving averages and their applications

Time-weighted control charts plot information derived not only from the most recent sample obtained but also from the most recent and earlier samples. Two types of time-weighted control charts will be discussed: the exponentially weighted moving average (EWMA) chart and the cumulative sum (CUSUM) chart.

Before considering the EWMA chart, moving averages will be introduced.

Consider daily sales of pizzas at Halcro Snacks, which operates a fast-food kiosk, on Mondays to Fridays inclusive each week, in a business park. The data (Table 5.8) are available in Pizzas.xls.

Having set up the data in two columns of a worksheet, **Graph** > **Time Series Plot...** > **Simple** may be used to display the data. The **Time/Scale...** button may be used to **Stamp**

Table 5.8 Daily sales data.

Day	Sales	Day	Sales
5-Jul-10	54	19-Jul-10	63
6-Jul-10	93	20-Jul-10	110
7-Jul-10	55	21-Jul-10	70
8-Jul-10	59	22-Jul-10	83
9-Jul-10	143	23-Jul-10	177
12-Jul-10	86	26-Jul-10	65
13-Jul-10	95	27-Jul-10	112
14-Jul-10	58	28-Jul-10	66
15-Jul-10	75	29-Jul-10	56
16-Jul-10	146	30-Jul-10	163

the horizontal time axis with Day, selected in the **Stamp columns:** window, as shown in Figure 5.30.

A cyclical pattern is evident in the level of daily sales, with Friday having the highest level each week. A moving average of length 5 is a natural way to summarize the data in view of the five-day operation. (In this context 'average' implies the mean.) Each consecutive set of five daily sales data includes data for a Monday, Tuesday, Wednesday, Thursday and Friday. By the end of the first five days of trading, total daily sales were

$$54 + 93 + 55 + 59 + 143 = 404,$$

so the first moving average of length 5 is

$$54/5 + 93/5 + 55/5 + 59/5 + 143/5 = 404/5 = 80.8.$$

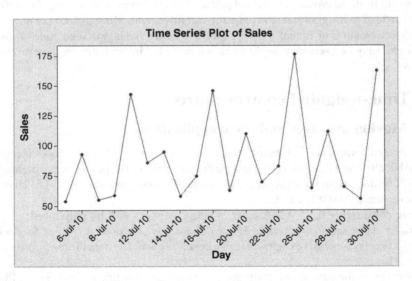

Figure 5.30 Time series plot of sales data.

Day	1	2	3	4	5	6	7	8	9	10	11	12	13	14	15	16	17	18	19	20
Weight	0.2	0.2	0.2	0.2	0.2															
Sales	54	93	55	59	143	86	95	58	75	146	63	110	70	83	177	65	112	66	56	163
MA (Length 5)	*	*	*	*	80.8															

Day	1	2	3	4	5	6	7	8	9	10	11	12	13	14	15	16	17	18	19	20
Weight		0.2	0.2	0.2	0.2	0.2														
Sales	54	93	55	59	143	86	95	58	75	146	63	110	70	83	177	65	112	66	56	163
MA (Length 5)	*	*	*	*	80.8	87.2														

etc.

Day	1	2	3	4	5	6	7	8	9	10	11	12	13	14	15	16	17	18	19	20
Weight																0.2	0.2	0.2	0.2	0.2
Sales	54	93	55	59	143	86	95	58	75	146	63	110	70	83	177	65	112	66	56	163
MA (Length 5)	*	*	*	*	80.8	87.2	87.6	88.2	91.4	92	87.4	90.4	92.8	94.4	100.6	101	101.4	100.6	95.2	92.4

Figure 5.31   Schematic for calculation of moving averages.

An equivalent way to obtain this is to multiply each one of five consecutive daily sales counts by 0.2 and sum the five products. The set of five factors 0.2, 0.2, 0.2, 0.2 and 0.2 are known as **weights**. The set of weights for calculation of a moving average must sum to 1. (With quarterly data, for example, one could employ a moving average of length 4 with weights 0.25, 0.25, 0.25 and 0.25.) The first two and the final moving average calculations are displayed in schematic form in Figure 5.31.

The first moving average becomes available on day 5 and the final one on day 20. They can be readily calculated and plotted in Minitab. Use **Stat > Time Series > Moving Average...**, select Sales as the **Variable:** and specify the **MA length:** as 5. Under **Graphs...** check **Plot smoothed vs. actual**. Under **Storage...** check **Moving averages**, accept defaults otherwise and click **OK, OK**. The plot in Figure 5.32 is obtained.

In Figure 5.31 the first moving average value of 80.8 appears below 5 in the row giving the day number. This moving average is plotted (default in Minitab) against the 5th day. The moving average provides a smoothing technique for data in the form of time series. Scrutiny of the moving average plot enables any underlying trend or other major pattern in the series to be

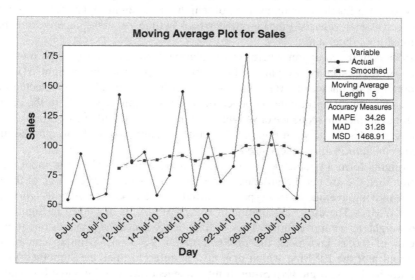

Figure 5.32   Time series plot of sales with moving average of length 5.

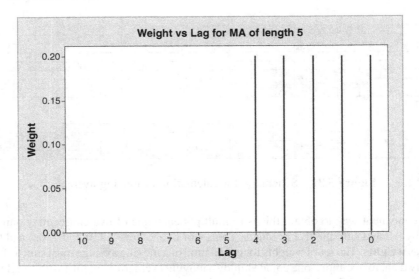

Figure 5.33 Weights for moving average of length 5.

identified. In this case the plot of the moving average plot indicate that sales appear to follow an initial upward trend, with a suggestion of a downward trend latterly. The accuracy measures in the panel to the right of the plot will be explained later in the section.

As far as the calculation of the moving average values is concerned, the reader is invited to visualize the shaded template in Figure 5.31 being moved from one group of five sales figures to the next. With each group, the sales figures are multiplied by the adjacent weights and the products summed to give the moving average value. The current observation may be considered to have lag 0, the immediately prior observation lag 1 and so on. A plot of the weights versus lag for the moving average of length 5 is given in Figure 5.33. The scale has been reversed as a lag value of 1 means one step back in time, i.e. in the negative direction on a conventional horizontal axis.

Consider now the situation where, in calculating a moving average, the most recently observed value is assigned weight α, the observation prior to that weight $\alpha(1-\alpha)$, the observation prior to that weight $\alpha(1-\alpha)^2$ and so on, where the number α is selected such that $0 < \alpha \leq 1$. The sum of an infinite sequence of weights of this type is 1. In the case where $\alpha = 0.4$, for example, the sequence of weights would be $0.400, 0.240, 0.144, 0.086, \ldots$. A plot of weight versus lag is shown in Figure 5.34. Minitab refers to α as the *weight*, but the term *smoothing constant* is also widely used.

The weights form a geometric series and decrease exponentially. The more recent the observation then the greater the influence, or weight, it has in the calculation of the moving average. A moving average of this type is referred to as an *exponentially weighted moving average* (EWMA). The direct calculation of some EWMAs is displayed in Figure 5.35.

Each weight, as we move further back in time, is 0.6 times the previous one in this case. This factor of 0.6 is known as the *discount factor* $\theta = 1 - \alpha$. Minitab can be used to calculate and plot the EWMA, which may also be referred to as the smoothed value. Use **Stat > Time Series > Single Exp Smoothing...**, select **Sales** as the **Variable:** and specify the **Weight to Use in Smoothing** by Use: 0.4 (the smoothing parameter, denoted by Alpha in

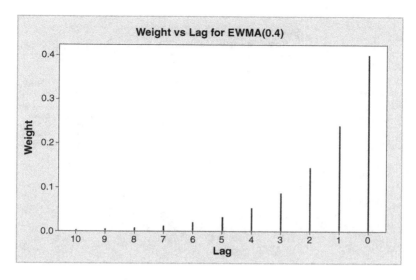

Figure 5.34   Weights for an exponentially weighted moving average.

Minitab). Values for the smoothing parameter usually lie between 0 and 1. Under **Graphs...** select **Plot smoothed vs. actual**. Under **Storage...** check **Smoothed data**, under **Time...** enter **Stamp:** Day and click **OK, OK**. The plot in Figure 5.36 is obtained. Again the plot of the moving average indicates that Sales appear to follow an initial upward trend with a suggestion of a downward trend latterly.

The calculation, in practice, of EWMAs (the smoothed values) in Minitab is outlined in Box 5.9. These first two smoothed values may be confirmed from the column of smoothed data created in the worksheet and named SMOO1 by the software. Note, too, that the final three smoothed values agree with those obtained in the schematic in Figure 5.37 (Under **Options...** the user may change the number of observations used to calculate a smoothed value for time 0 from the default number of six.)

Day	...	11	12	13	14	15	16	17	18	19	20
Weight	...	0.004	0.007	0.011	0.019	0.031	0.052	0.086	0.144	0.240	0.400
Sales	...	63	110	70	83	177	65	112	66	56	163
EWMA(0.4)											110.7

Day	...	11	12	13	14	15	16	17	18	19	20
Weight	...	0.007	0.011	0.019	0.031	0.052	0.086	0.144	0.240	0.400	
Sales	...	63	110	70	83	177	65	112	66	56	163
EWMA(0.4)										75.8	110.7

Day	...	11	12	13	14	15	16	17	18	19	20
Weight	...	0.011	0.019	0.031	0.052	0.086	0.144	0.240	0.400		
Sales	...	63	110	70	83	177	65	112	66	56	163
EWMA(0.4)									88.9	75.8	110.7

etc.

Figure 5.35   Schematic for calculation of an EWMA.

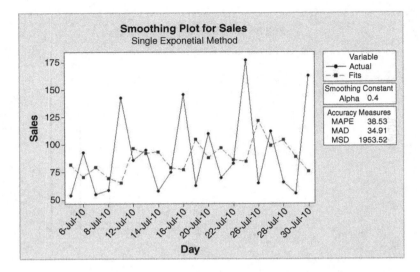

Figure 5.36   Plot of EWMA and original data.

The cyclical nature of the sales data made the choice of length 5 for the simple moving average a natural one. An arbitrary choice of 0.4 was made for the smoothing constant in order to introduce the concept of the exponentially weighted moving average. So far we have considered the moving average as a means of smoothing time series data. The moving average evaluated at the current point in time may be used as a forecast of the value at the next point in

For single exponential smoothing it may be shown that the smoothed value at time t is given by

$$\alpha \times (\text{Data value at time } t) + (1 - \alpha)(\text{Smoothed value at time } t - 1).$$

The observed data values are considered to be at times 1, 2, 3, ..., so to 'kick-start' the calculations a smoothed value for time $t = 0$ is required. The default in Minitab is to take the mean of the first six data values to be the smoothed value at time $t = 0$. From Table 4.8 it may be verified that this value is 81.667. Now the smoothed value at time 1 is

$$\alpha \times (\text{Data value at time } 1) + (1 - \alpha)(\text{Smoothed value at time } 0)$$
$$= 0.4 \times 54 + 0.6 \times 81.667$$
$$= 70.600,$$

and the smoothed value at time 2 is

$$\alpha \times (\text{Data value at time } 2) + (1 - \alpha)(\text{Smoothed value at time } 1)$$
$$= 0.4 \times 93 + 0.6 \times 70.600$$
$$= 79.560,$$

etc.

Box 5.9   Calculation of exponentially weighted moving averages.

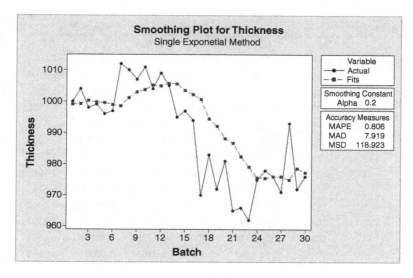

Figure 5.37   Forecasts of Thickness.

time, i.e. as a 'one-period-ahead forecast'. By way of illustration, consider the data in Table 5.9 and worksheet Thickness.MTW, giving thickness (in angstroms, Å) of a nitride layer measured at a fixed location on the first wafer in each one of 30 successive batches of wafers from a microelectronics fabrication process. In terms of quality, the ability to forecast is of value – if the forecast thickness for the next batch does not lie within specifications for thickness, then there is the possibility of taking some action to adjust the process in some way to ensure that thickness for the next batch is satisfactory.

With the data set up in two columns, Batch and Thickness, use **Stat** > **Time Series** > **Single Exp Smoothing. . .** , select Thickness in **Variable:** and specify the **Weight to Use in Smoothing** as **Use:** 0.2 (the default value for the smoothing parameter in Minitab). A title can be created under **Options. . . .** Under **Time. . .** select **Stamp** and specify Batch. Under **Graphs**. check **Plot predicted vs. actual**. Under **Storage. . .** check **Fits (one-period-ahead forecasts)** and **Residuals**. The plot in Figure 5.37 is obtained.

The lowest panel to the right of the plot gives three measures of accuracy for the forecasts. The first few rows of the worksheet, which includes the stored fits (one-period-ahead forecasts) and residuals and four additional columns, calculated by the author in order to indicate how the measures of accuracy are computed, are displayed in Figure 5.38.

The error (or deviation or residual – yes, all three names appear in this context in Minitab!) is the actual value observed minus the forecast (fitted value) made of that value, that is to say, RESIDUAL = DATA − FIT; this is a simple rearrangement of the formula DATA = FIT + RESIDUAL. Thus, for example, for batch 11 the error is $1004 - 1005.2114 = -1.2114$. This

Table 5.9   Thickness data for nitride layer.

Batch	1	2	3	4	5	6	7	8	9	10	11	12	13	14	15
Thickness	1000	1004	998	999	996	997	1012	1010	1007	1011	1004	1009	1005	995	997

Batch	16	17	18	19	20	21	22	23	24	25	26	27	28	29	30
Thickness	994	970	983	972	981	965	966	962	975	978	976	971	993	972	976

▦ THICKNESS.MTW ***

↓	C1	C2	C3	C4	C5	C6	C7	C8
	Batch	Thickness	FITS1	RESI1	% Error	APE	AD	SD
1	1	1000	999.0000	1.0000	0.10000	0.10000	1.0000	1.000
2	2	1004	999.2000	4.8000	0.47809	0.47809	4.8000	23.040
3	3	998	1000.1600	-2.1600	-0.21643	0.21643	2.1600	4.666
4	4	999	999.7280	-0.7280	-0.07287	0.07287	0.7280	0.530
5	5	996	999.5824	-3.5824	-0.35968	0.35968	3.5824	12.834
6	6	997	998.8659	-1.8659	-0.18715	0.18715	1.8659	3.482
7	7	1012	998.4927	13.5073	1.33471	1.33471	13.5073	182.446
8	8	1010	1001.1942	8.8058	0.87186	0.87186	8.8058	77.542
9	9	1007	1002.9554	4.0446	0.40165	0.40165	4.0446	16.359
10	10	1011	1003.7643	7.2357	0.71570	0.71570	7.2357	52.356
11	11	1004	1005.2114	-1.2114	-0.12066	0.12066	1.2114	1.468
12	12	1009	1004.9691	4.0309	0.39949	0.39949	4.0309	16.248

Figure 5.38   Columns C5–C8 indicate computation of accuracy measures.

error as a percentage of the observed thickness for batch 11 of 1004 gives the percentage error as $-0.120\,66\%$. APE in Figure 5.38 is the absolute percentage error, AD the absolute deviation (error) and SD the squared deviation (error). Mean APE (MAPE), mean AD (MAD) and mean SD (MSD) may all be used as measures of forecast accuracy and are displayed to the right of the plot in Figure 5.37. Interested readers are invited to calculate the entries displayed in columns C5–C8 and to verify the values for MAPE, MAD and MSD displayed in Figure 5.37.

Forecasts were also generated using values 0.4, 0.6 and 0.8 for the smoothing constant alpha. The accuracy measures obtained with the four values for the smoothing constant are displayed in Table 5.10. Of the four smoothing parameters tested, an alpha of 0.6 performs best in that it gives the lowest values for MAPE, MAD and MSD. Minitab also provides a procedure for the selection of an optimal value for the smoothing constant based on the fitting of an autoregressive integrated moving average (ARIMA) time series model to the data. (These models are not considered in this book.) This procedure was implemented by checking **Optimal ARIMA** as **Weight to Use in Smoothing**. For the thickness data this yields an alpha of 0.5855. In addition, **Generate forecasts** was checked with **Number of forecasts:** set to 1 and **Starting from origin:** 30 specified. The corresponding plots are shown in Figure 5.39.

Comparison of this plot with the one in Figure 5.37 reveals the superior performance of the smoothing parameter 0.5855 over that of the smoothing parameter 0.2. The reader should note

Table 5.10   Accuracy measures for a series of values of the smoothing constant.

Smoothing constant α	MAPE	MAD	MSD
0.2	0.806	7.92	118.9
0.4	0.684	6.73	82.8
0.6	0.650	6.40	76.5
0.8	0.662	6.51	82.0

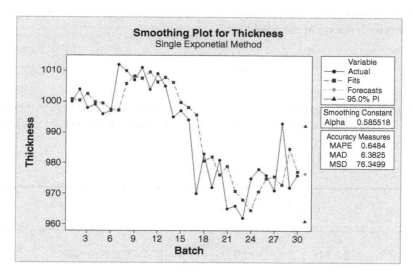

Figure 5.39   Forecasts with optimal alpha.

that the accuracy measures displayed in the panel to the right of the plot are marginally lower than those listed in Table 5.10 for smoothing constant 0.6. The diamond shaped symbol on the right of the plot represents the forecast thickness of 976.5 for the 31st batch. The triangular symbols define lower and upper 95% prediction limits of 960.9 and 992.1 for the forecast. These limits are such that 95 times out of 100 in the long term the actual observation will lie in the interval. The forecast and the prediction limits are displayed in the Session window.

In addition to their application for smoothing and forecasting, moving averages may be plotted in control charts. As moving averages are linear combinations of random variables the results on linear combinations of random variables presented in Chapter 4 may be used to obtain formulae for standard deviations of moving averages. The mathematics will not be presented. Details may be found in Montgomery (2009, pp. 419–430). Only the EWMA control chart will be considered in this book.

5.3.2   Exponentially weighted moving average control charts

The EWMA control chart performs well at detecting small changes in a process mean. For example, for a normally distributed random variable, the average run length (ARL) for detection of a one standard deviation shift in the process mean using an individuals chart is approximately 44. For an EWMA chart with smoothing constant 0.4 the ARL for detection of such a shift is approximately 11. However, the EWMA chart does not detect relatively large shifts in a process mean as quickly as a Shewhart chart. Hunter (1989, pp. 13–19) demonstrated that the EWMA chart with smoothing constant 0.4 performs similarly to the Shewhart chart employing Minitab tests 1, 2 (with eight points in a row employed rather than the default nine), 5 and 6, i.e. the Western Electric Rules. A discussion can be found in Montgomery (2009, pp. 422–424).

It can be shown that the standard deviation of the EWMA is given by the expression in Box 5.10. For the temperature data of Table 5.1 the estimates of process mean and standard deviation, based on the first 20 data values, were 27.245 and 0.699 9, respectively.

The standard deviation of the EWMA is

$$\sigma \sqrt{\frac{\alpha}{2-\alpha} \left[1 - (1-\alpha)^{2i}\right]},$$

where i is the sample number and α is the smoothing parameter. In the long term the value of $(1-\alpha)^{2i}$ becomes negligible since $0 < \alpha < 1$ and the standard deviation is effectively

$$\sigma \sqrt{\frac{\alpha}{2-\alpha}}.$$

Box 5.10   Standard deviation of EWMA.

$$27.245 \pm 3 \times 0.699\,9 \sqrt{\frac{0.4}{2-0.4}} = 27.245 \pm 1.049\,85$$

$$UCL = 28.295$$
$$LCL = 26.195$$

Box 5.11   Calculation of long-term EWMA chart limits.

The long-term three-sigma limits for an EWMA control chart with smoothing parameter 0.4, for example, could therefore be calculated as shown in Box 5.11.

When all 25 data values (available in Temperature.MTW) have been entered into a column, to create the chart in Minitab use is made of **Stat > Control Charts > Time-Weighted Charts > EWMA. . . .** Under **Subgroup sizes:** 1 was specified. The **Weight of EWMA:** is the smoothing constant and 0.4 was selected. Under **EWMA Options. . .** and **Estimate**, use of samples 1 : 20 and the **Rbar** method were indicated so that the estimates of process mean and standard deviation would be the same as those used in the setting up of the individuals chart of the data earlier. The chart in Figure 5.40 was obtained. As with the individuals chart in Figure 5.3 the 25th point is below the lower chart limit, thus providing evidence of a potential special cause affecting the process. (The reader should note the narrower limits for the first few samples.)

Consider again the sequence of weights α, $\alpha(1-\alpha)$, $\alpha(1-\alpha)^2$, $\alpha(1-\alpha)^3$, . . . used in exponential smoothing. With $\alpha = 1$ these weights become 1, 0, 0, 0 . . . so that the EWMA would simply be the most recent observation. Just for fun, the reader is invited to create the EWMA chart of temperature as in Figure 5.40 but with smoothing constant 1 and to check that it is identical to the Shewhart chart in Figure 5.4.

Further simulated data for the rod cutting process described in Section 5.1 will be used to illustrate the use of the EWMA for data collected in subgroups. In the scenario illustrated in Figure 5.41 the process mean was changed from 60.000 to 60.008 after sample 40. The procedure computes the sample means and applies the EWMA methodology to the sample means. The subgroups comprised the rows of columns x1, x2, x3 and x4. Observe that the EWMA chart, with limits smoothing constant 0.2, signalled this process change at sample number 54. (Limits were based on the first 25 samples and the default pooled standard deviation

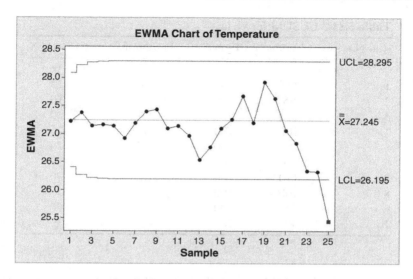

Figure 5.40   EWMA control chart of temperature.

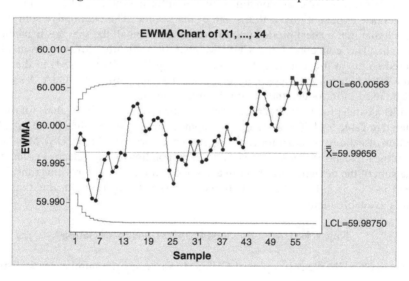

Figure 5.41   EWMA control chart for rod length.

with unbiasing constant method of estimation of process standard deviation was used.) The data are available in RodsScenario5.MTW. The reader is invited to verify that an Xbar chart also signals the process change at sample number 59 via Test 2. This example illustrates the ability of the EWMA chart to detect a small shift in the process mean earlier than a Shewhart chart.

5.3.3 Cumulative sum control charts

The most recently plotted point in a Shewhart control chart contains only information from the most recent sample. The latest plotted point in a control chart of moving averages of length 5

Table 5.11 CUSUM for yield.

Run No. i	Yield	Target	Deviation	CUSUM S_i
0	*	*	*	0
1	115	100	15	15
2	95	100	− 5	10
3	110	100	10	20
4	105	100	5	25
5	95	100	− 5	20
6	100	100	0	20
7	110	100	10	30
8	100	100	0	30
9	90	100	− 10	20
10	85	100	− 15	5

contains equally weighted information from the five most recent samples. The latest plotted point in an EWMA control chart contains weighted information from all the samples, but the weights decrease exponentially with the age of the sample. The cumulative sum (CUSUM) chart is such that in the latest plotted point information from all the samples is included with equal weight. The concept of the CUSUM chart is radically different from all charts encountered so far in this chapter. It was proposed by Ewan Page (1954) in the UK thirty years after Walter Shewhart proposed the charts that bear his name in the USA. A simple set of data will be used to introduce the basic principle of the chart.

Consider a batch process with target for yield of 100 units. The yield values for runs 1 to 10 are tabulated in Table 5.11. The first step in computing the CUSUM values is the calculation of the deviation of yield from target for each run. It is necessary to define a CUSUM value of zero corresponding to run no. 0. The CUSUM value corresponding to, say, run no. 3 is denoted by S_3 and is the sum of the deviation for that run and the deviations for all previous runs, i.e. in this case $S_3 = 15 + (− 5) + 10 = 20$. Calculation can be speeded up on observing that once data for a run is available then:

$$\text{New CUSUM} = \text{Previous CUSUM} + \text{New deviation.}$$

The basic CUSUM chart is a plot of CUSUM (S_i) versus run number (i) as displayed in Figure 5.42.

Table 5.12 Mean yields for some sets of consecutive runs.

Set	Runs	Mean yield	Segment joining	Slope of segment
1	4 to 9 inclusive	100	P3 to P9	0
2	1 to 5 inclusive	104	P0 to P5	4
3	5 to 9 inclusive	99	P4 to P9	−1

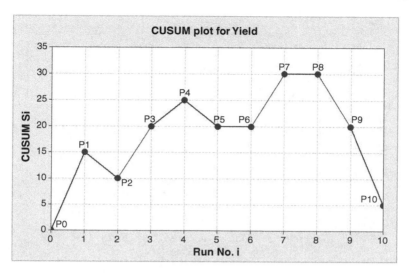

Figure 5.42   CUSUM plot for yield.

The reader is invited to confirm the mean yield values given in the third column of Table 5.12. The set of runs 4 to 9 inclusive has mean 100. The slope of the line segment joining the points P3 and P9 in the plot may be calculated as

$$\frac{S_9 - S_3}{9 - 3} = \frac{20 - 20}{6} = 0.$$

The reader is invited to check the other two slopes given in Table 5.12. Observe that the slope indicates by how much the mean yield for a consecutive set of runs differs from the target yield of 100. Thus the fundamental property of the CUSUM plot is that the slope indicates the process mean performance over the corresponding time period.

In order to illustrate this, a set of 90 yields was simulated using Minitab. The first 40 were from the $N(100, 10^2)$ distribution, the next 20 were from the $N(95, 10^2)$ distribution and the final 30 from the $N(103, 10^2)$ distribution. The target was again taken to be 100. The CUSUM plot is shown in Figure 5.43. The data are available in the worksheet Yields.MTW.

Reference lines have been added to indicate the three phases in terms of the distribution of yield. In the first phase the 'horizontal' appearance of the CUSUM plot corresponds to the process operating 'on target'. In the second phase the downward trend in the plot corresponds to a process operating 'below target'. In the final phase the upward trend corresponds to a process operating 'above target'. Formal detection of signals of possible special cause variation may be carried out using a V-mask. To illustrate, let us suppose that we decide to set up a CUSUM chart with a V-mask when the first 25 yields are available from the simulation referred to above. The data are available in Yields25.MTW. In Minitab we would need to use **Stat > Control Charts > Time-Weighted Charts > CUSUM....** With **All observations for a chart are in one column:** selected, Yield is specified as the variable to be charted. **Subgroup sizes:** 1 and **Target:** 100 are entered. Using **CUSUM Options...**, under **Estimate**, 1 : 25 may be entered via **Use the following subgroups when estimating parameters**. Under **Plan/Type**, for **Type of CUSUM**, choose **Two-sided (V-mask)** and for **Center on subgroup:** enter 25. Accept the default **CUSUM plan** and all other defaults. The

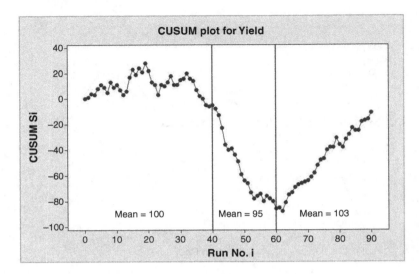

Figure 5.43   CUSUM for simulated Yield data.

resulting chart is displayed in Figure 5.44. (The author considers it potentially misleading that, having specified **Target:** as 100 during the creation of the chart, Target $= 0$ appears as a legend to the right of the plot, so he clicks on it and deletes it.)

The midpoint of the line segment, which forms the blunt end of the V-mask, is placed on the point specified using **Center on subgroup:**. If all the previously plotted points are 'embraced' by the arms of the mask then the process may be deemed to be in a state of statistical control, exhibiting no signal of any potential special cause variation. The mask can then be adopted for further monitoring of the process and can be thought of as being moved to each new point as it is plotted.

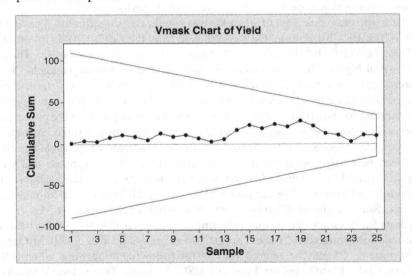

Figure 5.44   CUSUM chart with V-mask for first 25 simulated Yields.

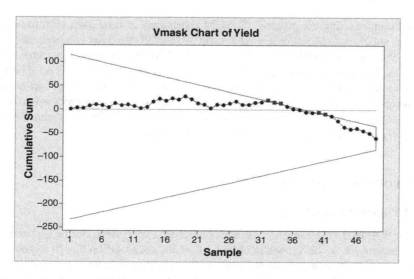

Figure 5.45   CUSUM chart with signal of potential special cause variation.

On plotting the point for the 49th run the chart appears as shown in Figure 5.45. The fact that at least one point is not embraced by the arms of the V-mask is the signal of potential special cause variation from this type of CUSUM chart. As with the EWMA chart, the CUSUM chart is very sensitive to small changes in the process mean.

If one selects **One-sided (LCL,UCL)** as the **Type of CUSUM,** with all other choices as before then the chart in Figure 5.46 is the result. This version of the CUSUM chart consists of two one-sided CUSUM plots. Here the point below the LCL for run 49 signals potential special

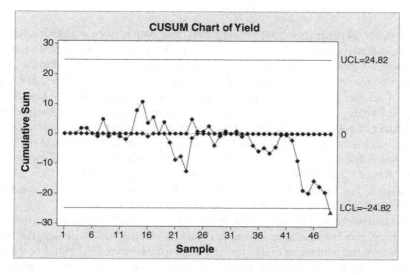

Figure 5.46   CUSUM chart with signal of potential special cause variation.

cause variation affecting the process. CUSUM charts may also be used with measurements recorded in subgroups. Montgomery (2009, pp. 400–419) provides comprehensive details. Caulcutt (1995, pp. 108–109) refers to the use of CUSUM charts for 'post mortem' analysis of process data. This approach may even be used with data for which there is no specific target value. By creating a CUSUM chart of the first type considered, with target set equal to the overall mean for the data series, one can often gain useful insights into process performance – marked changes in slope indicate the likelihood of changes in the process mean. A paper by Henderson *et al.* (2010) provides examples and data sets.

5.4   Process adjustment

5.4.1   Process tampering

The late Dr W. Edwards Deming often carried out funnel experiments during his presentations. These experiments were developed in order to illustrate the assertion that 'if anyone adjusts a stable process to try to compensate for a result that is undesirable, or for a result that is extra good, the output will be worse than if he had left the process alone' (Deming, 1986, pp. 327–331). In the experiments a target point is set up on a sheet of foam placed on a table and marbles are dropped, one by one, through a funnel onto the foam. Initially the funnel is aimed directly at the target. Following each drop the point of impact of the marble is recorded and one of a series of four rules is applied to determine the next point of aim.

1. Leave aim unchanged.

2. Adjust aim from previous aim position to 'compensate' for the deviation from target of the last bead dropped.

3. Adjust aim to opposite side of target from point of rest of last bead.

4. Adjust aim to point of rest of last bead.

Results from the experiments may be simulated, and output from simulations of the four scenarios, generated using a Minitab macro written by Terry Zeimer in 1991 (http://www.minitab.com/en-GB/support/macros/default.aspx?q=deming+funnel&collection=LTD), is displayed in Figure 5.47. Further information on the macro will be provided in Chapter 11. (With one of the four plots active, **Editor** > **Layout Tool...** was used to display all four plots in a single Minitab graph.) The points represent 100 impact points under each scenario.

   The scaling of both axes is the same in each plot and the target is at the central point of the grid in both. The top right-hand plot (Rule 2) exhibits greater variation about the target than does the top left-hand plot (Rule 1). Rule 1 is optimum, Rule 2 yields stability but increased variability, Rule 3 leads to instability with oscillation and Rule 4 to what is known as random walk behaviour. The fundamental point of the experiment is to demonstrate that tampering with a stable process leads to increased variability in performance.

   In order to demonstrate the effect of Rule 2 type 'tampering' on a process, consider again the rod-cutting process operating 'on target' and producing rods with lengths which are normally distributed with mean 60.00 mm and standard deviation 0.02 mm. For the purposes of illustration consider an individuals chart, with three-sigma limits, for the length of a single rod selected at random from the process output at regular intervals. The chart limits are at 59.94

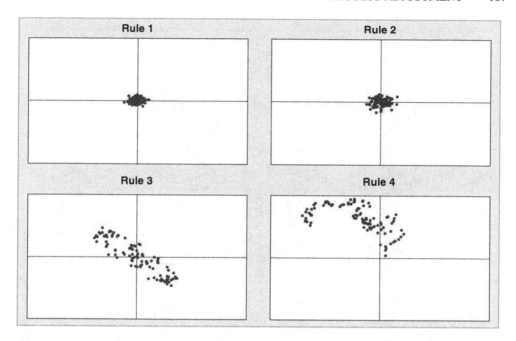

Figure 5.47   Simulations of the Deming funnel experiments.

and 60.06. If an operator of the process applied Rule 2 type tampering then he would respond to an observed length of 60.03 mm by *reducing* the process aim by 0.03 mm *from the previous aim* and to an observed length of 59.98 by *increasing* the process aim by 0.02 mm *from the previous aim*. Were an operator of the process to apply Rule 3 type tampering then he would respond to an observed length of 60.03 mm by *changing* the process aim to 0.03 mm *below target*, i.e. to 59.97, and to an observed length of 59.98 by *changing* the process aim to 0.02 mm *above target*, i.e. to 60.02. Under Rule 4 type tampering an operator would respond to an observed length of 60.03 mm by *changing* the process aim to 60.03 mm and to an observed length of 59.98 mm by *changing* the process aim to 59.98 mm. Individuals control charts for 100 simulated values under all four rules are shown in Figure 5.48. In each case the scaling on the vertical axis is the same to facilitate visual comparison of performance.

The increased variability under Rule 2 is again apparent, and some of the plotted points lie outside the historical chart limits in the case of Rule 2. The standard deviation of the set of 100 lengths obtained under Rule 1 is 0.0187, which is close to the specified standard deviation of 0.02. The standard deviation of the 100 lengths obtained under Rule 2 is 0.0263. Theoretically it can be shown that, under Rule 2, the variability, as measured by standard deviation, is increased by a factor of $\sqrt{2}$, i.e. by approximately 40%. Thus the tampering leads to increased variability, with a consequent reduction in the capability of the process to meet customer specifications.

Under Rule 1 the process displays what Shewhart referred to as controlled variation and what is widely referred to as common cause variation. A point outside the control limits on the Shewhart individuals chart would be taken as a signal of the possible occurrence of a special cause of variation. Such a signal would lead those involved in running the process to search for

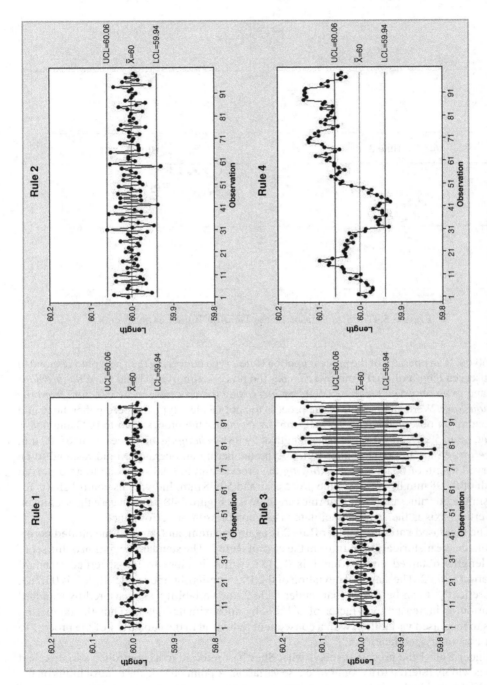

Figure 5.48 Individuals charts for the four funnel rules.

any special causes, e.g. a damaged cutting tool or a new and poorly trained operator. Having identified such special causes, effort would typically be made to eliminate them. Of course, some special causes may correspond to evidence of improved process performance, in which case it would be desirable to retain rather than eliminate.

5.4.2 Autocorrelated data and process feedback adjustment

George Box and Alberto Luceño applaud the Six Sigma strategy for quality improvement for the recognition that 'even when best efforts are made using standard quality control methods, the process mean can be expected to drift' (Box and Luceño, 2000, pp. 297–298). When successive process observations are not independent the data is said to be *auto-correlated*. The autocorrelation structure in the data enables a forecast of the next observation to be made from the available data. With the availability of a compensating factor whose effect on the process is known, appropriate adjustment to the level of the factor can be made in order to correct the predicted deviation from target. The procedure may be referred to as *feedback adjustment*.

Independence may be checked informally by examining a scatterplot of X_i versus X_{i-1}, i.e. of each observation plotted against the previous observation. The scatterplot in Figure 5.49 is for the rod cutting process operating without tampering under Rule 1. Clearly there is no lag 1 autocorrelation, i.e. no correlation between X_i and X_{i-1}.

The type of scatterplot in Figure 5.49 may be thought of as the 'fingerprint' of a typical process for which successive observations are independent. A formal analysis may be carried out by constructing a correlogram or autocorrelation function, consisting of a line graph of the autocorrelations at lag k plotted against k. The lag 1 correlation is the correlation between X_i and X_{i-1}, -0.134 in this case, and the lag 2 autocorrelation is the correlation between X_i and X_{i-2}, 0.168 in this case, etc. Correlograms generated using Minitab include significance limits indicating any autocorrelations which differ significantly from 0. The plot in Figure 5.50 was created using **Stat > Time Series > Autocorrelation**. None of the autocorrelation line

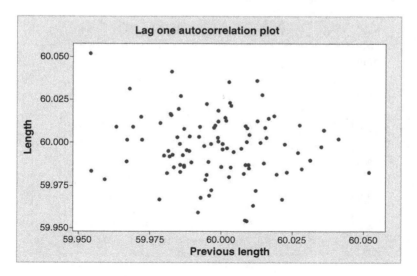

Figure 5.49 Lag 1 autocorrelation plot for data under Rule 1.

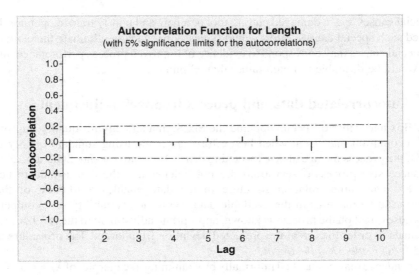

Figure 5.50 Correlogram for lengths obtained under Rule 1.

segments protrude beyond the limits, so there is no evidence of dependence in the time series of rod lengths in this case.

Consider a low-pressure chemical vapour deposition (LPCVD) process used in the fabrication of microelectronic circuits. A nitride layer is to be built up to a target thickness of 1000 Å on successive batches of silicon wafers. Let X_i represent a measurement of the thickness of the layer on a test wafer selected from the ith batch. Data from a simulated realization of such a process are plotted in Figure 5.51.

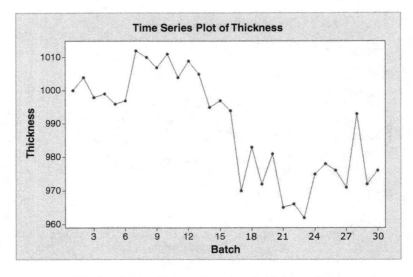

Figure 5.51 Time series plot of thickness data.

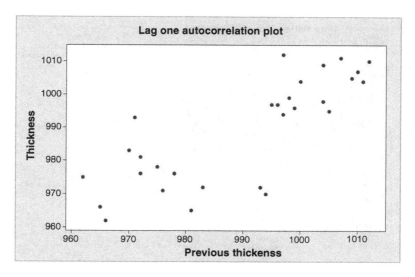

Figure 5.52   Lag 1 autocorrelation plot for thickness.

Here the thickness appears to shift and drift with time. A scatterplot of X_i versus X_{i-1} is shown in Figure 5.52. Unlike the scatterplot in Figure 5.49, the one in Figure 5.52 exhibits positive autocorrelation at lag 1. This scatterplot may be thought of as a typical fingerprint of a process for which successive observations are not independent.

The autocorrelation function (correlogram) is shown in Figure 5.53. Montgomery (2009, p. 446) comments that for such variables, even with moderately low levels of autocorrelation, conventional control charts will 'give misleading results in the form of too many false alarms'.

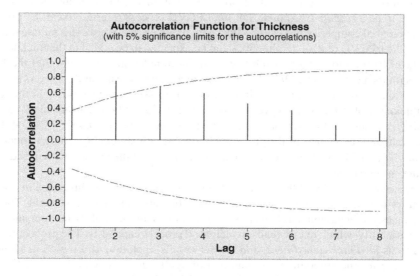

Figure 5.53   Correlogram for thickness.

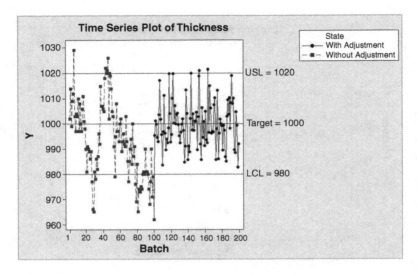

Figure 5.54 Run chart for batches both with and without adjustment.

Here the line segments representing the autocorrelation at both lag 1 and lag 2 protrude through the upper limit. Thus there is evidence that successive observations are not independent for this process. The autocorrelation structure in the data enables a forecast of the next observation to be made from the data. With the availability of both a forecasting procedure and a compensating factor whose effect on the process is known, appropriate adjustment to the level of the factor can be made in order to correct the predicted deviation from target. The exponentially weighted moving average is a forecasting tool used in industry in this context.

The assumption is made that the effect of any change in the level of the compensating factor will be complete by the time the next observation is made, i.e. that the process may be considered to be a *responsive system*. For the LPCVD process referred to earlier, processing time is a potential compensating factor with a gain of 30, i.e. for every extra minute the wafers remain in the LPCVD reactor another 30 Å can be expected to be added to the thickness of the nitride layer. If the most recent batch had spent 34 minutes in the reactor and the forecast thickness for the next batch was 940 Å, then, provided there were no random errors involved, a control action or adjustment of $+2$ minutes to the processing time would yield the required target thickness of 1000 Å. However, if the forecast was obtained from an EWMA with smoothing constant 0.4 then the actual adjustment made would be $0.4 \times (+2) = +0.8$ minutes. A simulated realization of the process without the adjustment procedure in operation is shown in Figure 5.54, together with a simulated realization of the process with the adjustment procedure in operation.

Comparison of the plot of thickness for the second 100 batches (adjustment in operation) with that for the first 100 batches (no adjustment in operation) reveals the benefit. The earlier 'wandering mean' behaviour has been replaced with much more stable behaviour and reduced process variability, which in turn leads to increased process capability. The procedure may be referred to as *feedback adjustment*, and further details, including discussion of choice of a suitable value for the smoothing constant and of applications, may be found in Henderson (2001), Montgomery (2009, p. 529) and Box and Luceño (1997).

5.5   Multivariate control charts

When monitoring the location of a single measured quality characteristic that remains stable and predictable, using a Shewhart chart with three-sigma limits and no other tests for evidence of special cause variation, the frequency of false alarm signals is 1 in 371 in the long term. In other words, one sample in 371 would give rise to a point plotting outside the chart limits although no special cause was affecting the process. If the locations of six independent measured quality characteristics which all remained stable and predictable were monitored using Shewhart charts the false alarm rate would be 1 in 62 in the long term, i.e. there would be six times as may false alarms to deal with. Typically a multivariate set of quality characteristics will not be independent.

For dependent bivariate random variables having a bivariate normal distribution, the equivalent of a point lying between the three-sigma limits in the univariate case is a point lying within a control ellipse in the scatterplot. Evidence of special cause variation could be overlooked through monitoring of the quality characteristics separately. Although the creation of control ellipses for bivariate data is quite feasible, the time sequence of the observations cannot be readily indicated. With three or more variables, representation of control ellipsoids and hyper-ellipsoids is impractical.

Hotelling's T^2 statistic may be plotted in a control chart in order to monitor a group of measured dependent quality characteristics. The theory underlying the chart assumes that the variables have a multivariate normal distribution. In order to construct the chart, estimates have to be made from the data of the means of the variables and of their covariance matrix. Data on short-circuit current (x) and fill factor (y) for photovoltaic cells, where a single cell was sampled at regular intervals from production, are in PV.MTW. The T^2 chart may be thought of as the multivariate equivalent of the Xbar chart, and Minitab also provides a generalized variance chart, which may be thought of as the multivariate equivalent of the R chart or S chart.

In order to create the charts, use **Stat** > **Control Charts** > **Multivariate Charts** > **Tsquared-Generalized Variance....** The two variables to be charted are selected and **Subgroup sizes:** 1 specified. Estimation was based on the first 30 samples. The charts are shown in Figure 5.55.

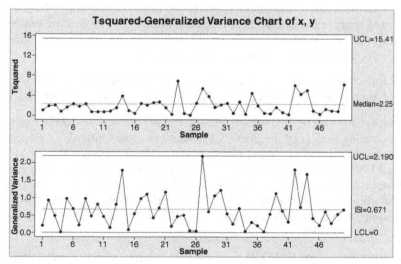

Figure 5.55 T^2 and generalized variance charts.

There are no signals of potential special cause variation affecting the process. Montgomery (2009, p. 499) gives a detailed account of multivariate control charts.

5.6   Exercises and follow-up activities

1. For the data you collected for Exercise 1 in Chapter 2, use Minitab to create an appropriate control chart.

2. Weekly checks of water quality are made at a chemical plant that manufactures products for use in the manufacture of microelectronic circuits. Values of the water quality index (WQI) are provided in Water.MTW. Create an individuals and moving range pair of control charts and verify that the WQI appears to be stable and predictable. Verify also that it is reasonable to consider WQI to be normally distributed by creating a normal probability plot of the data. Obtain estimates of the mean and standard deviation of the distribution from information on the control charts and compare with the estimates given on the normal probability plot.

3. In a continuous process for the manufacture of glass the soda level in the molten glass in the furnace is monitored daily. A series of 20 consecutive daily values are given in the worksheet Soda.MTW.

    (i) Create an individuals chart of the data, specifying that estimation is to be carried out using subgroups 1 to 20 and with all the available tests implemented. Observe that the process is in a state of statistical control and that therefore the decision may be taken to monitor soda level using the individuals chart created.

    (ii) Right-click on the chart and select **Update Graph Automatically**. Use **Window > Tile** to ensure that the control chart and the worksheet with the data may be viewed simultaneously.

    (iii) Add the next five data values 12.96, 12.88, 12.89, 13.09 and 12.79 to the worksheet and observe the data points being added to the plot as they are entered into the worksheet. You should observe that Test 6 signals possible special cause variation affecting soda level on plotting the final value.

    (iv) Recreate the chart using **S Limits** under **I Chart Options. . .** with 1 2 3 inserted. Observe from the revised chart how Test 6 has given rise to the signal.

4. Control charts may be used to ascertain whether or not there is evidence that a process is not operating on target through use of **Parameters** under **Options. . . .** Suppose that the target level for soda in the previous exercise was 13.00. With the 20 daily soda levels in Soda.MTW create an individuals chart using **Parameters** to set the mean at 13.00. (Do not enter a value in the **Standard deviation:** window.) Note how the chart yields a signal from the eleventh sample that the soda level is off target. The use of a control chart in setting or checking process aim is described in detail by Wheeler and Chambers (1992, pp. 194–204).

5. The worksheet PilotOD.MTW gives data on samples of four output shafts taken at regular intervals from a production process. (Data reproduced by permission of the

Statistics and Actuarial Science Department, University of Waterloo, Canada, from Steiner *et al.*, 1997, p. 6). The values represent deviations from nominal (micrometers) for the diameter of the pilot.

(i) Create Xbar and R charts of the data and verify that, when only Test 1 is used, the 15th sample signals a possible special cause affecting the process.

(ii) Verify that when all tests are used the 4th and 15th samples both provide signals.

(iii) Given that there was a problem identified with the process during the period when the 15th sample was taken, create the charts with the 15th sample omitted from the calculations of the limits and comment on process behaviour.

6. Refer again to the camshaft length data discussed in Section 5.1.3 and create Xbar and R charts for supplier 2 as in Figure 5.12, but before doing so create a column named Subgroup containing the numbers 1 to 20 each repeated 5 times using **Calc > Make Patterned Data > Simple Set of Numbers. . . .**   The following entries are required:

Store patterned data in: Subgroup

From first value: 1

To last value: 20

In steps of: 1

Number of times to list each value: 5

Number of times to list the sequence: 1

In addition, use the **Data Options. . .** facility to exclude the points corresponding to subgroups 2 and 14 from the plots. The dialog required is shown in Figure 5.56.

A text box was added to the charts indicating the nature of the special cause identified for subgroup 2. In Figure 5.57 the mouse pointer is shown positioned over the text tool icon in the process of creating text indicating the nature of the special cause – operator error – corresponding to the omitted subgroup 14. Once text has been entered double-clicking on it yields a menu that may be used, for example, to change font size.

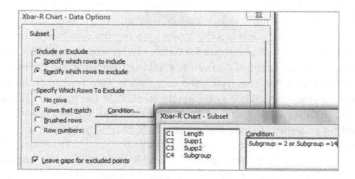

Figure 5.56   Specify subgroups for exclusion from plot.

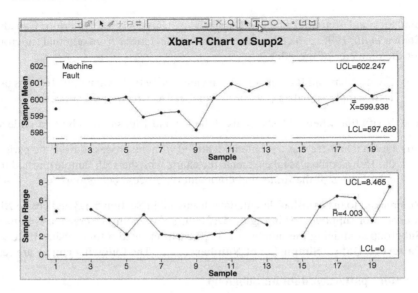

Figure 5.57   Xbar and R charts with gaps for excluded subgroups.

Although Minitab's Assistant flow chart suggests that the subgroup size should be at least 9 (see Figure 5.1) for Xbar and S charts to be used, analyse the data for supplier 2 using them and demonstrate that the same conclusions would be reached as via the Xbar and R charts.

7. The file Etch.MTW contains data on a dry etch process which etches silicon dioxide off silicon wafers during a batch microelectronic fabrication process (Lynch and Markle, 1997, pp. 81–83). The data are ©Society for Industrial and Applied Mathematics and ©American Statistical Association and are reproduced by permission of both organizations. During each batch run 18 wafers were processed and etch rate (angstroms per minute) was measured at nine positions on each wafer in a selection of six wafers from the batch. We will assume that the set of 54 measurements from each run constitutes a rational subgroup for the creation of Xbar and S (mean and standard deviation) charts of the 27 subgroups. In order to set up the data for charting use **Data** > **Stack** > **Rows...** as indicated in Figure 5.58. There were three phases involved. The first nine runs were carried out when the multi-wafer etch tool was only use intermittently, and the second nine runs were carried out when the tool was in regular use. Before the final nine runs were made the mass flow controller for CHF_3 was recalibrated. Note how it is necessary to expand the Phase column during the stacking operation into a new column named Stage. Note too that in specifying the name Etch Rate for the column in which the stacked data is to be stored it is necessary to enclose the name in single quotes.

Each consecutive group of 54 values in the Etch Rate column constitutes a subgroup for charting purposes. Create the Xbar and S charts by Stage and comment on the changes in process performance. Montgomery (2009, p. 251) advises that Xbar and S charts should be used in preference to Xbar and R charts when either the subgroup size is greater than 10 or the sample size is variable. Compare Xbar and R charts of the data with Xbar and S charts.

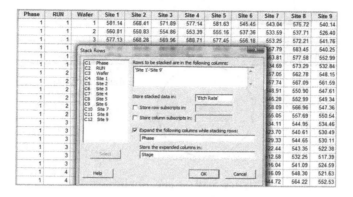

Figure 5.58   Stacking rows.

8. Bisgaard and Kulachi (2000) refer to a problem with off-centre bottle labels that had 'bothered management for some time'. The excessive variation in the position of the labels detracted from the appearance of an expensive product and there was concern that this was affecting the company's share of the decorating market.

The line foreman believed the labels were off centre because the there was a lot of variation in bottle diameter. He said that the Quality Control Department had attempted a capability study but had 'got nowhere'. The Maintenance Department claimed that the specifications were too tight and that the labels varied as well as the bottles. In an attempt to gain some insight into the problem, the deviations of label heights from target for 60 consecutive bottles were measured. The data for this example, available in the file Labels.xls, are from 'Finding assignable causes' by Bisgaard and Kulachi and are reproduced with permission from *Quality Engineering* (© 2000 American Society for Quality).

(i) Treat the data as 12 consecutive subgroups of size 5 and create Xbar and R charts and comment.

(ii) Treat the data as 60 consecutive individual measurements and create an individuals chart. Note the repetitive pattern.

(iii) The schematic diagram in Figure 5.59 indicates how labels were applied to bottles by a rotating drum with six label applicators spaced around its surface. Given that the first bottle in the data set had its label applied by the first applicator, set up a column named Applicator containing the sequence of numbers 1 to 6 repeated 10 times using **Calc > Make Patterned Data > Simple Set of Numbers....** The following entries are required:

Store patterned data in: Applicator
From first value: 1
To last value: 6
In steps of: 1
Number of times to list each value: 1
Number of times to list the sequence: 10

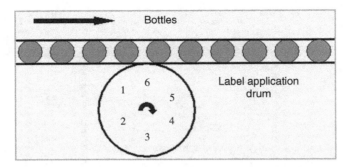

Figure 5.59   Label applicator.

(iv) Unstack the data to obtain columns of Deviations for each applicator and verify, by creating individuals charts for each, that the individual applicators appear to be performing in a stable, predictable manner.

(v) Create a boxplot of Deviation by Applicator and comment.

According to the authors, the moral of this story is that for process investigations, data should be plotted not only in time order but also in any other way that makes sense and preferably as individuals and not just as averages.

9. Montgomery (2009, pp. 292–298) gives an example on the manufacture of cans for frozen orange juice. The cans were spun from cardboard stock and a metal base panel attached. Every 30 minutes during production a sample of 50 cartons was inspected and the number of nonconforming cans recorded. Data for the first 30 samples are given in Cans1.xls and are reproduced by permission of John Wiley & Sons Inc., New York.

(i) Create a P chart of the data and verify that samples 15 and 23 signal the occurrence of possible special cause variation.

Process records indicated that there was a problem with a new batch of raw material at the time that sample 15 was taken and that an inexperienced operator had been involved in running the process at the time sample 23 was taken. As assignable causes could be found for these two 'out of control' points it was decided to recalculate the limits with those samples excluded.

(ii) Create the revised P chart. The author suggests that under **P Chart Options...** > **Estimate** you specify the sample to be used by employing **Use the following subgroups ...** and entering 1 : 14 16 : 22 24 : 30.

You should find that sample 21 now signals possible special cause variation. No assignable cause could be determined so it was decided to use the current chart with centre line at 0.215 and lower and upper limits of 0.041 and 0.389 for further process monitoring. At the same time, as the proportion of non-onforming cans was running at over 20%, it was decided to have adjustments made to the machine which produced the cans. A further 24 samples were taken and the extended data set is provided in Cans2.xls.

(iii) Update the P chart to show the additional data and note how it indicates that the adjustments were beneficial. (Note that under **P Chart Options...** > **Estimate** you will need to specify the sample to be used by employing **Use the following subgroups** ... and entering 1 : 14 16 : 22 24 : 30).

(iv) Create a Phase column with value 1 in the first 30 rows and value 2 in the next 24 rows and use it to create P charts for the two phases. The author suggests that under **P Chart Options...** > **Estimate** you specify the sample to be used by employing **Use the following subgroups** ... and entering 1 : 14 16 : 22 24 : 30 31 : 54. You should find that the chart for Phase 2 has centre line at 0.111 and lower and upper chart limits of 0 and 0.244, respectively.

Further data from the second phase of operation of the process are provided in Cans3. xls. Chart all the available data and comment on process performance.

Repeat the exercise using individuals charts of the actual proportions of nonconforming cans in the samples.

10. A department within a major organization prepares a large number of documents each week, with the numbers being similar from week to week. Table 5.13 gives the number of errors detected each week during final checks for a series of 15 weeks.

   (i) Create a C chart of the data.

   (ii) Given that a senior member of staff responsible for document preparation was on sick leave during week 4, explain why the chart with revised upper limit of 14.83, obtained on omitting the data for Week 4, could be 'rolled out' for routine monitoring.
   Additional data are given in Table 5.14.

   (iii) Plot the additional data, with the revised limits used in (ii), explain how the chart provides evidence of process improvement and state what action you would recommend.

11. The worksheet PCB1.MTW gives counts of nonconformities on samples of 10 printed circuit boards taken from successive batches of a particular type of board built in work cell A at a factory.

   (i) Create a U chart of the data and use the formulae in Appendix 3 to check the centre line and chart limits.

Table 5.13   Data for weeks 1–15.

Week	1	2	3	4	5	6	7	8	9	10	11	12	13	14	15
No. of errors	7	8	3	22	1	3	10	3	13	9	13	10	4	2	7

Table 5.14   Data for weeks 16–30.

Week	16	17	18	19	20	21	22	23	24	25	26	27	28	29	30
No. of errors	3	4	6	7	6	4	3	6	3	4	4	2	5	9	2

(ii) Since the sample size is constant here a C chart may be used. Create a C chart of the data and note that it is a scaled version of the U chart.

The advantage of the U chart, in terms of assessing process performance, is that it displays nonconformities *per unit*.

In work cell B a different type of board is manufactured and the sample size used for the monitoring of nonconformities varies. The worksheet PCB2. MTW gives counts of nonconformities on a series of samples of boards.

(iii) Create a U chart of these data and comment on process performance.

12. In Section 5.2.1 an example on the use of a P chart to monitor the proportion of stroke patients receiving acute stroke unit care was given. It was also noted that the assumption of a binomial distribution is unlikely to be valid. An alternative approach to the use of a P chart in this case is to compute the proportion of patients receiving acute stroke unit care for each month and to create an individuals chart of these proportions.

(i) Retrieve the data from the worksheet ASU.MTW, calculate the proportions and create an individuals chart of the proportions with limits based on the first 15 samples and all available tests implemented.

There are various points to note. First, the centre line on the individuals chart is at 0.646 4 as opposed to 0.635 2 in Figure 5.25. This is because the P chart procedure calculates the centre line as the total number receiving acute stroke unit care in the first 15 months (477) divided by the total number of stroke patients in the first 15 months (751). Second, the UCL is 1.024 7, an impossible value for a proportion! Third, unlike the P chart in Figure 5.25 there are no signals indicating an improved proportion of patients receiving acute stroke unit care. However, note that the last nine points are very close to being on the upper side of the centre line.

(ii) Re-create the chart using **S Limits** under **I Chart Options...** to **Place bounds on control limits**, check the two boxes and enter 0 and 1 respectively since the variable to be charted is a proportion.

13. In the manufacture of aerosol cans height is a critical dimension and is measured at three locations equally spaced round the can. During a production run a can was selected every 10 minutes and three height measurements obtained for a sequence of 40 cans. The data are available in Aerosols.MTW.

(i) Treat each row of the three columns of heights as a subgroup/sample of three heights and create Xbar and R charts.

Note that there are many signals on the Xbar chart. However, this is an incorrect approach. The problem is that the underlying assumption of independence is violated. The three heights in each subgroup/sample are from the same can. Had the samples/subgroups comprised a single height measurement from each of three different cans then use of Xbar and R charts would have been valid.

The correct approach is to use:

- an individuals chart of the means of the sets of three height measurements;
- a moving range chart for these means;
- a range (or standard deviation) chart for the sets of three height measurements.

(ii) Use **Calc > Row Statistics...** to create a column of means for the sets of three measurements and display the means in individuals and moving range charts.

Scrutiny of these two charts and the earlier R chart reveals no signals of possible special cause variation. The use of the ranges of the sets of three heights in (i) gave an estimate of standard deviation that is too small because it only measured variation *within* cans. This gave rise to limits on the Xbar chart that were too close together, hence the signals noted earlier.

(iii) Use **Stat > Control Charts > Variables Charts for Subgroups > I-MR-R/S (Between/Within)...** with subgroups specified across the three height columns to create the triple chart display of the data and verify that the charts obtained are the three discussed earlier.

Wheeler and Chambers (1992, pp. 221–226) discuss these under the heading three-way control charts.

14. Set up the funnel plot data in Table 5.7 in Minitab and create the funnel plot in Figure 5.29.

6

Process capability analysis

In general, process capability indices have been quite controversial. (Ryan, 2000, p. 186)

Overview

Capability indices are widely used in assessing how well processes perform in relation to customer requirements. The most widely used indices will be defined and links with the concept of sigma quality level established. Minitab facilities for capability analysis of both measurement and attribute data will be introduced.

6.1 Process capability

6.1.1 Process capability analysis with measurement data

Imagine that four processes produce bottles of the same type for a customer who specifies that weight should lie between 485 and 495 g, with a target of 490 g. Imagine, too, that all four processes are behaving in a stable and predictable manner as indicated by control charting of data from regular samples of bottles from the processes. Let us suppose that the distribution of weight is normal in all four cases, with the parameters in Table 6.1. The four distributions of weight are displayed in Figure 6.1, together with reference lines showing lower specification limit (LSL), upper specification limit (USL) and Target (T). How well are these processes performing in relation to the customer requirements?

In the long term the fall-out, in terms of nonconforming bottles, would be as shown in the penultimate column of Table 6.1. The fall-out is given as number of parts bottles) per million (ppm) that would fail to meet the customer specifications. The table in Appendix 1 indicates that these fall-outs correspond to sigma quality levels of 4.64, 3.50, 2.81 and 3.72 respectively for lines 1–4. Scrutiny of the distributions (the voices of the processes) with reference to the specification limits (the voice of the customer) reveals the following points:

Six Sigma Quality Improvement with Minitab, Second Edition. G. Robin Henderson.
© 2011 John Wiley & Sons, Ltd. Published 2011 by John Wiley & Sons, Ltd.

Table 6.1 Parameters for the distributions of weight with fall-out and sigma quality level (SQL).

Process	Mean	Standard deviation	Fall-out (ppm)	SQL
Line 1	490	1.5	858	4.64
Line 2	492	1.5	22 752	3.50
Line 3	490	3.0	95 581	2.81
Line 4	487	0.9	13 134	3.72

- Line 1 is performing as well as it can with the process mean 'on target'.

- Line 2 could perform as well as line 1 if the mean could be adjusted down from 492 to the target of 490. Adjustment of process location can often be a relatively easy thing to achieve.

- Line 3 is performing as well as it can with the process mean on target but it is inferior to lines 1 and 2 because of its greater variability. Reduction of variability would be required to improve the performance of line 3, and this can often be a relatively difficult thing to achieve.

- Line 4, although currently performing less well than line 1, has the potential to give the lowest fall-out of all four processes if the mean can be adjusted upward from 487 to 490.

Sigma quality levels are intended to encapsulate process performance in a single number. However, one must beware the danger of judging a process purely on the basis of its sigma

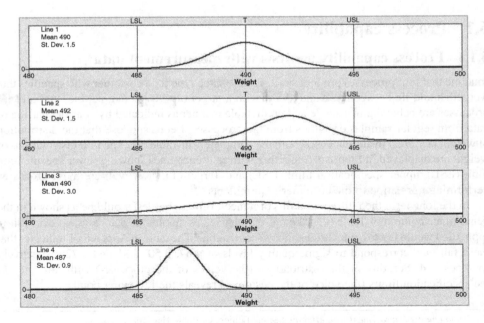

Figure 6.1 Distributions of weight for the four processes.

The process capability index C_p is given by

$$C_p = \frac{\text{Customer tolerance range}}{\text{Natural tolerance range}} = \frac{\text{USL} - \text{LSL}}{6\sigma}.$$

For line 1 this gives

$$C_p = \frac{495 - 485}{6 \times 1.5} = \frac{10}{9} = 1.11.$$

Box 6.1   Calculation of C_p for line 1.

quality level. In the case of the above four processes, reliance solely on sigma quality level could blind one to the high potential performance of line 4.

Process capability indices are designed to do the same as sigma quality levels, to encapsulate in single numbers process performance with respect to customer requirements. They can be said to measure the extent to which the 'voice of the process' is aligned with the 'voice of the customer'. The fundamental fact which underpins the indices is that 99.73% of observations from a normal distribution lie between $\mu - 3\sigma$ and $\mu + 3\sigma$, i.e. in a range of three standard deviations on either side of the mean. These values are often referred to as the *natural tolerance limits* for the process. Note that the proportion 0.27% of observations will lie outside the natural tolerance range in the case of a normal distribution. The *customer tolerance range* is the range of values that the customer will tolerate, i.e. from the lower specification limit to the upper specification limit. The process capability index C_p is defined as the ratio of the customer tolerance range to the natural tolerance range. Its calculation for line 1 is displayed in Box 6.1.

The reader is invited to perform the calculations for the other three lines and to confirm the entries in Table 6.2 for the four processes. Note that line 4 'tops the league' in terms having the highest C_p value, lines 1 and 2 have the same intermediate value and line 3 has the lowest. The index C_p measures the *potential* capability of a process. Thus, although lines 1 and 2 have the same potential capability, their actual capability in terms of fall-out and SQL values differs because line 2 is not operating on target. Thus a disadvantage of the index C_p is that it does not take process location into account. The process capability index C_{pk} does take process location into account. Its calculation for line 2 is displayed in Box 6.2.

The reader is invited to perform the calculations for the other three lines and to confirm the entries in Table 6.3 for the four processes.

Table 6.2   SQL and C_p values for the four lines.

Process	Mean	Standard deviation	SQL	C_p
Line 1	490	1.5	4.64	1.11
Line 2	492	1.5	3.50	1.11
Line 3	490	3.0	2.81	0.56
Line 4	487	0.9	3.72	1.85

The process capability index C_{pk} is given by

$$C_{pk} = \min\left[C_{pl}, C_{pu}\right] = \min\left[\frac{\mu - \text{LSL}}{3\sigma}, \frac{\text{USL} - \mu}{3\sigma}\right].$$

For line 2 this gives

$$C_{pk} = \min\left[\frac{492 - 485}{3 \times 1.5}, \frac{495 - 492}{3 \times 1.5}\right] = \min\left[\frac{7}{4.5}, \frac{3}{4.5}\right] = \min[1.56, 0.67] = 0.67.$$

Box 6.2   Calculation of C_{pk} for line 2.

The index C_{pk} measures the *actual* capability of a process. In the type of scenario discussed here, $C_p = C_{pk}$ when the process mean coincides with the target value mid-way between the LSL and USL, i.e. when the process is centred. (Another benefit of the C_{pk} index is that it may be calculated in situations where there is only one specification limit, e.g. a customer requirement could be that a tensile strength has to be at least 25 N/mm^2 or that cycle time must be no greater than 40 minutes.)

Table 6.4 gives values of C_p and C_{pk} for C_p ranging from 0.5 to 2.0, first with the process centred and second with the process off centre by a 1.5 standard deviation shift. Also given are the corresponding fall-out counts of nonconforming product in ppm and the sigma quality levels. In particular, note that a Six Sigma process corresponds to a C_p value of 2.0 and a C_{pk} value no less than 1.5.

In the above discussion of the four lines it was assumed that we had perfect knowledge of the process behaviour. We now turn to the assessment of process capability in a situation where the capability indices have to be estimated from process data.

In Exercise 3 in Chapter 2, reference was made to bottle weight data stored in the worksheet Bottles.MTW. The data were collected as subgroups of size 4 and Xbar and R charts are shown in Figure 6.2. The Minitab default and recommended pooled standard deviation method for estimating the process standard deviation was used, yielding 2.039 15.

All available tests for evidence of special cause variation were applied. No signals were obtained from the charts so it appears that the process was behaving in a stable, predictable manner. Thus it is reasonable to stack the data into a single column and consider it as a sample of 100 observations from the distribution of bottle weight.

The normal probability plot in Figure 6.3 indicates that a normal distribution provides an adequate model for the data. Figure 6.4 shows a histogram of the 100 observations of weight

Table 6.3 C_p and C_{pk} values for the four lines.

Process	Mean	Standard deviation	C_p	C_{pk}
Line 1	490	1.5	1.11	1.11
Line 2	492	1.5	1.11	0.67
Line 3	490	3.0	0.56	0.56
Line 4	487	0.9	1.85	0.74

Table 6.4 C_p and C_{pk}, with fall-out rates and sigma quality levels.

Process centred			Process off-centre by 1.5σ			
C_p	C_{pk}	Fall-out (ppm)	C_p	C_{pk}	Fall-out (ppm)	Sigma quality level
0.50	0.50	133 614	0.50	0.00	501 350	1.5
0.60	0.60	71 861	0.60	0.10	382 572	1.8
0.70	0.70	35 729	0.70	0.20	274 412	2.1
0.80	0.80	16 395	0.80	0.30	184 108	2.4
0.90	0.90	6 934	0.90	0.40	115 083	2.7
1.00	1.00	2 700	1.00	0.50	66 811	3.0
1.10	1.10	967	1.10	0.60	35 931	3.3
1.20	1.20	318	1.20	0.70	17 865	3.6
1.30	1.30	96	1.30	0.80	8 198	3.9
1.40	1.40	27	1.40	0.90	3 467	4.2
1.50	1.50	6.8	1.50	1.00	1 350	4.5
1.60	1.60	1.6	1.60	1.10	483	4.8
1.70	1.70	0.34	1.70	1.20	159	5.1
1.80	1.80	0.067	1.80	1.30	48	5.4
1.90	1.90	0.012	1.90	1.40	13	5.7
2.00	2.00	0.002	2.00	1.50	3.4	6.0

with a fitted normal curve superimposed and reference lines indicating the lower specification limit of 485 and upper specification limit of 495. The target value for weight will be considered to be $T = 490$, the mid-point of the specification range. The diagram gives a visual representation of process capability and indicates that some of the bottles measured failed to meet

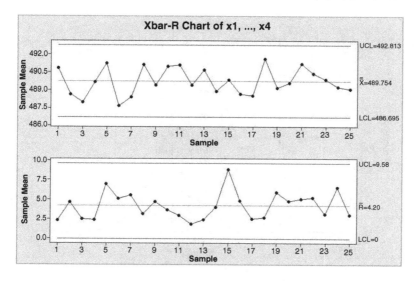

Figure 6.2 Xbar and R charts of bottle weight data.

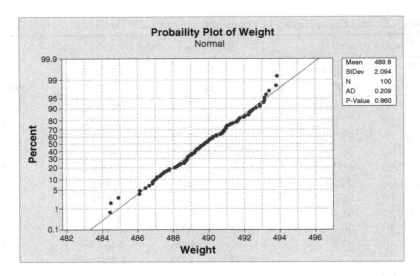

Figure 6.3 Normal probability plot of the stacked data.

customer requirements through being too light. One way in which we can proceed to estimate process capability is as detailed in Box 6.3.

Many authors deem a process with a C_p value less than 1 to be incapable. The fact that C_{pk} is less than C_p indicates that the process is not centred. The estimates of process mean and standard deviation may be used to predict fall-out for the process, as it is currently operating, at 14 913 ppm, with 9866 ppm predicted to be below the LSL and 5047 ppm predicted to be above the USL.

In order to perform the capability analysis using Minitab one can use **Stat** > **Quality Tools** > **Capability Analysis** > **Normal...**. Under **Options...** the **Target** was specified as

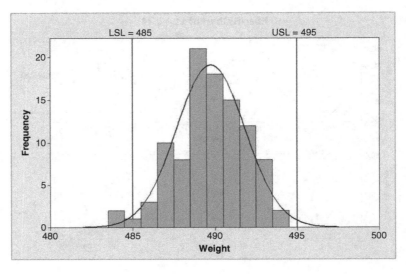

Figure 6.4 Histogram of weight with fitted normal curve and specification limits.

From the Xbar chart, $\bar{\bar{X}} = 489.754$, so we can take an estimate of the process mean to be $\mu = 489.754$. The pooled standard deviation estimate of the process standard deviation is $\hat{\sigma} = 2.039$. We then obtain

$$C_{\text{p}} = \frac{\text{USL} - \text{LSL}}{6\hat{\sigma}} = \frac{495 - 485}{6 \times 2.039} = \frac{10}{12.234} = 0.82,$$

$$C_{\text{pk}} = \min\left[C_{\text{pl}}, C_{\text{pu}}\right] = \min\left[\frac{\mu - \text{LSL}}{3\hat{\sigma}}, \frac{\text{USL} - \mu}{3\hat{\sigma}}\right]$$

$$= \min\left[\frac{489.754 - 485}{3 \times 2.039}, \frac{495 - 489.754}{3 \times 2.039}\right]$$

$$= \min\left[\frac{4.754}{6.117}, \frac{5.246}{6.117}\right] = \min[0.78, 0.86] = 0.78.$$

Box 6.3   Estimation of C_{p} and C_{pk} for bottle weight.

490. Under **Estimate...** the default **Pooled standard deviation** was chosen as the means of estimating process standard deviation in order to be able to compare Minitab output directly with the results calculated in Box 6.3 and the option **Use unbiasing constants to calculate overall standard deviation** was checked. Defaults were accepted otherwise. The dialog box is shown in Figure 6.5, with the arrangement of the data indicated and specification limits

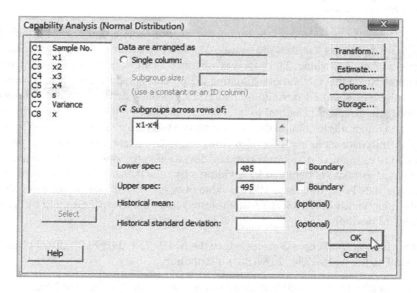

Figure 6.5   Dialog for process capability analysis via Minitab.

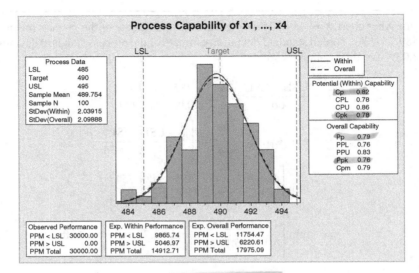

Figure 6.6 Process capability output.

entered. The output is displayed in Figure 6.6. (Process Capability™ is a trademarked feature of Minitab.)

The output in Figure 6.6 will now be considered in detail.

- The histogram of the 100 (25 subgroups of 4 bottles) bottle weight values has super-imposed on it reference lines indicating the target of 490 and the LSL and USL of 485 and 495, respectively. Also superimposed are two fitted normal distributions labelled Within (solid curve) and Overall (dashed curve). Further reference will be made to these distributions below.

- In the top left-hand corner of the output the text box labelled Process Data includes two standard deviations – within and overall. These are two estimates of the process standard deviation σ. The value 2.039 15 was obtained using the pooled standard deviation method. Since this method of estimating the process standard deviation is based on the 25 subgroup standard deviations, and since standard deviation measures variability *within* subgroups, it is natural to refer to the estimate in this way. If the subgroups are stacked into a single column then **Descriptive Statistics** gives the standard deviation of the overall data set as $s = 2.093\,59$. However, although sample variance s^2 provides an unbiased estimate of σ^2, sample standard deviation s provides a biased estimate of σ. An unbiased estimate is obtained by dividing s by c_4, a constant whose value depends on sample size. For sample size 100 the value of c_4 is 0.997 48 and division of 2.093 59 by this value yields 2.098 88, the value referred to as the *overall* estimate of the process standard deviation.

- The within normal curve corresponds to the $N(489.754, 2.039\,15^2)$ distribution and the overall to the $N(489.754, 2.098\,88^2)$ distribution.

- Three bottles from the 100 bottles measured had weight less than the LSL and none had weight above the USL. This is equivalent to a total of 30 000 ppm failing to meet the specifications as recorded in the text box labelled Observed Performance at the bottom left of the display.

- The text boxes to the right of the one recording Observed Performance give the expected performances calculated using the within and overall normal distributions. Since the overall standard deviation exceeds the within standard deviation the predicted fall-out is greater for the former.

- The C_p and C_{pk} indices have been calculated using the within estimate of the process standard deviation and agree with the calculations given above in Box 6.3.

- The P_p and P_{pk} indices are referred to as *process performance indices*. They are analogous to the C_p and C_{pk} indices but are calculated using the overall standard deviation. The reader is invited to confirm the given values of P_p and P_{pk} as an exercise.

In this example the values of C_p and P_p (0.82 versus 0.79) are similar and the values of C_{pk} and P_{pk} (0.78 versus 0.76) are similar. This is typical of scenarios where the quality characteristic of interest is normally distributed and the process is behaving in a stable and predictable manner as evidenced by monitoring using control charts.

In discussing process performance indices, Montgomery refers to the recommendation that the capability indices C_p and C_{pk} should be used when a process is in a state of statistical control and that the process performance indices P_p and P_{pk} should be used when a process is not in a state of statistical control. He comments 'if the process is not in control the indices P_p and P_{pk} have no meaningful interpretation relative to process capability because they cannot predict process performance' (Montgomery, 2009, p. 363).

Step-by-step assessment of process capabilty in the above example involved: -

- Xbar and R control charts (Figure 6.2);

- a normal probability plot (Figure 6.3);

- a histogram with superimposed normal distribution and reference lines indicating the specification limits (Figure 6.4);

- the capability indices (Figure 6.6).

Use of **Stat > Quality Tools > Capability Sixpack > Normal...** essentially provides all of the output from these steps plus a run chart of the data for the last 25 subgroups and what is referred to as a capability plot. (Sixpack™ is a trademarked feature of Minitab.) Under **Estimate...**, the default **Pooled standard deviation** was chosen as the means of estimating process standard deviation and the option **Use unbiasing constants to calculate overall standard deviation** was checked. Defaults were accepted otherwise. **Perform all eight tests** was selected under **Tests...**, and **Options...** was used to specify the **Target** as 490. The source of the data and the specification limits were indicated as before. The output for the data set considered above is shown in Figure 6.7. Note that in this case there were only 25 subgroups so all of the 100 data values are displayed in the run chart.

The capability plot consists of three line segments. The lower (labelled Specs) indicates the customer tolerance range from the lower specification limit to the upper specification limit and has a tick at its midpoint, representing the target. The middle segment (labelled Overall) represents the natural tolerance range obtained using the overall estimate of standard deviation

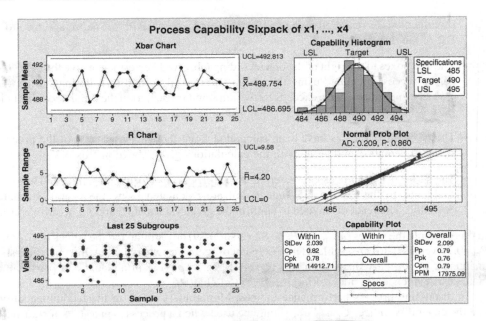

Figure 6.7 Process Capability Sixpack output.

while the upper (labelled Within) represents the natural tolerance range obtained using the within estimate of standard deviation. The index C_p is the ratio of the length of the lower segment (labelled Specs) to that of the upper segment (labelled Within); the plot gives an immediate visual indication that C_p is less than 1.

The index C_{pm} given in the output is defined as:

$$C_{pm} = \frac{USL - LSL}{6\sqrt{\sigma^2 + (\mu - T)^2}}.$$

T is the target value, normally the mid-point of the specification range. With two-sided specification limits, the value of the C_{pk} index does not give any indication of the location of the process mean μ in relation to the specification limits LSL and USL. The C_{pm} index was developed in order to deal with this inadequacy of the C_{pk} index. If the process is centred 'on target' with the process mean, μ, equal to the target, T, then the C_{pm} index is identical to C_p. Minitab gives this index computed using the overall estimate of standard deviation. The main point to note is that the closer the value of C_{pm} is to the value of C_p, the closer is the process to being centred on target.

6.1.2 Process capability indices and sigma quality levels

Consider now an alternative version of the output from **Capability Analysis > Normal...** to that displayed in Figure 6.6. In order to obtain this alternative one proceeds with the dialog displayed in Figure 6.5, except that under **Options...** one selects **Benchmark Z's (sigma level)** and checks **Include confidence intervals**. The output is shown in Figure 6.8.

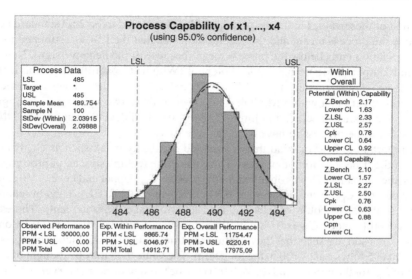

Figure 6.8 Alternative process capability output.

The expected process performance, calculated using the within estimate of the process standard deviation, is a fall-out of 14 913 nonconforming bottles per million as far as specifications for weight are concerned. Reference to the table in Appendix 1 indicates that this corresponds to a sigma quality level of around 3.67. The formula in Box 6.4 may be used to convert the Z.Bench value quoted under Potential (Within) Capability to an estimate of the sigma quality level of the process. Thus the sigma quality level of the process is estimated to be $2.17 + 1.5 = 3.67$. Readers interested in the technical details of how Z.Bench is computed in Minitab should consult the Help facility.

When dealing with random variable X having mean μ and standard deviation σ, the corresponding random variable Z given by $Z = (X - \mu)/\sigma$ is referred to as the standardized random variable. Using the overall mean weight, 489.754, of the 100 bottles measured and the within estimate of standard deviation, the standardized values corresponding to specification limits, Z_1 and Z_2 are calculated in Box 6.5.

$$\textbf{Sigma quality level} = \textbf{Z.Bench} + \textbf{1.5}$$

Box 6.4 Formula for sigma quality level.

$$Z_1 = \frac{\text{USL} - \mu}{\sigma} = \frac{495 - 489.754}{2.03915} = 2.57$$

$$Z_2 = \frac{\text{LSL} - \mu}{\sigma} = \frac{485 - 489.754}{2.03915} = -2.33$$

Box 6.5 Calculation of Z_1 and Z_2.

The first of these values indicates that the USL is estimated to be 2.57 standard deviation units above the process mean, and the second indicates that the LSL is estimated to be 2.33 standard deviation units below the process mean (because of the negative sign). Note that, in the Minitab output in Figure 6.8, under Potential (Within) Capability the numerical values of these Z-values are quoted as Z.LSL = 2.33 and Z.USL = 2.57. They are used in the computation of Z.Bench within Minitab. The lower of these two values, i.e. 2.33, gives the distance to the nearest specification limit (DNS) in standard deviation units. The DNS must be at least 3 for the process to have a C_{pk} of at least 1.

It is important to bear in mind that any quoted capability index such as C_{pk} is in fact an estimate of the 'true' C_{pk} for the process. Thus, using the within estimate of process standard deviation, the estimate 0.78 was obtained for the process C_{pk}. Upper and lower 95% confidence limits for the true C_{pk} of the process are 0.64 (LCL) and 0.92 (UCL). Thus in reporting the capability analysis for bottle weight it is advisable to make the statement: 'The estimated process capability index C_{pk} is 0.78 with 95% confidence interval (0.64, 0.92).' Confidence intervals of this sort are such that they capture the true value of that which is being estimated from the data 95 times out of 100 in the long term. The value 0.78 may be thought of as a *point estimate* of C_{pk} for the process and (0.64, 0.92) may be thought of as an *interval estimate* of C_{pk} for the process. Confidence intervals will be considered in more detail in Chapter 7.

Note that in the case of Z.Bench only a Lower CL value of 1.63 is quoted. A Z.Bench value of 1.64 corresponds to a sigma quality level of 1.63 + 1.5 = 3.13. Thus in reporting the sigma quality level for bottle weight it is advisable to make the statement: 'The estimated sigma quality level is 3.67 and it can be stated with 95% confidence that the sigma quality level is at least 3.13.'

There are situations where an assessment of process capability is required from data obtained from a single sample of product. In such situations the customer would be wise to seek assurance from the supplier that the sample to be used was taken while the process was operating in a stable and predictable manner and that the sample is representative of the population of product. Montgomery (2009, p. 348) gives such data for the burst strength (psi) of a sample of 100 bottles. The data are provided in the worksheet Burst.MTW and are reproduced by permission of John Wiley & Sons, Inc., New York. This data set will be used to illustrate a situation where there is only one specification limit, in this case a lower specification limit of 200 psi. The output from use of **Stat > Quality Tools > Capability Analysis > Normal...** with subgroup size specified as 100 is shown in Figure 6.9. Clicking on **Estimate...**, the default **Pooled standard deviation** method was checked under **Methods of estimating within subgroup standard deviation** and the option **Use unbiasing constants to calculate overall standard deviation** also selected. Under **Options...**, both **Benchmark Z's (sigma level)** and **Include confidence intervals** were selected.

The C_{pk} value is 0.67 with 95% confidence interval (0.55, 0.78). (It should be noted that in this case the within and overall estimates of the process standard deviation are identical as there is only a single sample.) The predicted fall-out based on a normal distribution of burst strength is 22 983 ppm. Appendix 1 indicates that this corresponds to a sigma quality level of around 3.5 for the process. Addition of 1.5 to the Z.Bench value of 2.0 confirms the sigma quality level of 3.5.

6.1.3 Process capability analysis with nonnormal data

Consider now data stored in columns C1 to C5 of the worksheet Density.MTW giving density measurements (g/m^2) for 80 consecutive hourly samples of size $n = 5$ from a process for the

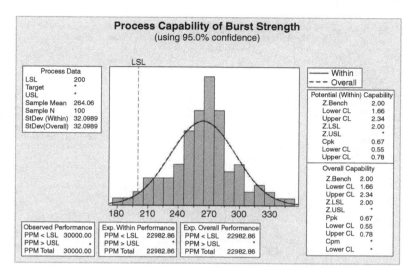

Figure 6.9 Capability analysis of burst strength.

fabrication of plastic sheeting. The specification limits are 45 to 55 g/m². Naïve use of **Stat > Quality Tools > Capability Analysis > Normal...** yields a C_{pk} of 1.29, which borders on the widely recommended minimum of 1.33. The predicted fall-out is of the order of 90 ppm. However, use of **Stat > Quality Tools > Capability Analysis > Capability Sixpack > Normal...** reveals a normal probability plot with strong evidence that the distribution of density is nonnormal (P-value less than 0.005). In situations such as this one can employ **Capability Sixpack (Nonnormal Distribution)** to investigate alternative probability distributions to the normal, such as the Weibull. In this case the Weibull probability distribution provides a satisfactory model, with a P-value of 0.195. Subsequent use of **Capability Analysis (Nonnormal Distribution)**, with selection of Weibull, gives a predicted fall-out of the order of 3000 ppm based on the fitted Weibull distribution (see Figure 6.10).

Reference to Table 6.4 indicates that this level of fall-out for a scenario where the distribution was normal, and the process was stable, predictable and centred, would correspond to C_p and C_{pk} values less than 1. Thus, in general parlance, the data indicate that the process is not capable.

Another method for dealing with data that do not have a normal distribution is to seek a transformation that will yield a new variable that is at least approximately normally distributed. Minitab provides a facility for implementing Box–Cox transformations in which the original random variable Y is transformed to $W = Y^{\lambda}$ when $\lambda \neq 0$ and to $W = \ln(Y)$, the natural logarithm of Y, when λ is zero. The user may either specify a value for λ or implement a procedure within the software to select an optimum value for λ.

Consider the data in the worksheet Roughness.MTW which gives roughness measurements (nm), for a sample of 200 machined automotive components. The upper specification limit is 800 nm. Naïve use of **Capability Analysis (Normal Distribution)**, with the data considered as a single subgroup of 200, yields a C_{pk} of 2.03 and a sigma quality level of 7.59. However, scrutiny of the histogram in the output suggests that the distribution of roughness is nonnormal. In order to carry out a capability analysis of the data following a Box–Cox

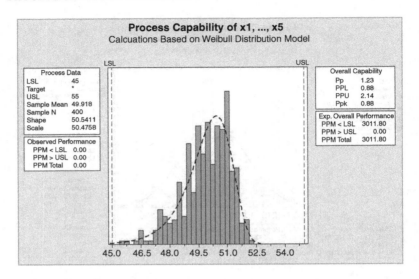

Figure 6.10 Capability analysis of density using a Weibull distribution.

transformation, use **Capability Analysis (Normal Distribution)**, select **Transform...** and check both **Box-Cox power transformation** and **Use optimal lambda** as indicated in Figure 6.11.

We have already seen that, in the case of a single subgroup, the within and overall analyses are identical (Figure 6.9). Thus under **Options...** one can uncheck **Overall analysis**. **Benchmark Z's (sigma level)** and **Include confidence intervals** were checked under **Options...** in order to obtain the output in Figure 6.12.

The heading in the output indicates that the Box–Cox transformation selected employed $\lambda = -1$. This means that the roughness values Y were replaced by $W = Y^{-1}$, i.e. by their reciprocals. Thus, for example, the roughness values 200, 500 and 800, the USL, would be

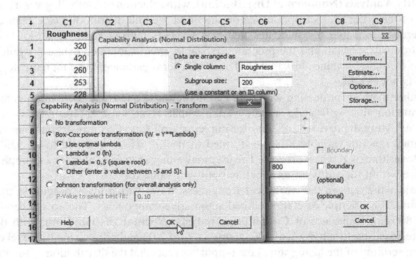

Figure 6.11 Capability analysis using a Box–Cox transformation.

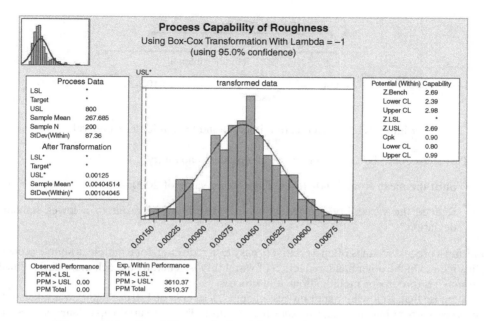

Figure 6.12   Capability analysis of the transformed roughness data.

replaced by 0.005, 0.002 and 0.00125 respectively. The USL of 800 for Y corresponds to an LSL of 0.00125, denoted by USL* in the output, for the transformed random variable W. The histogram of the transformed data appears to indicate a normal distribution and should be compared with the histogram of the raw data shown in the top left-hand corner of the output. (The reader is encouraged to use the **Calc** menu to compute a column of the reciprocals, W, of the roughness values, Y, to perform a normality test in order to confirm that the transformation has indeed been effective and to perform the capability analysis directly on the transformed data. It has to be borne in mind that roughness less than 800 is equivalent to the reciprocal of roughness exceeding 0.00125.) However, the key information from the output is that the C_{pk} is 0.90 with 95% confidence interval (0.80, 0.99). Also the sigma quality level predicted by the analysis is $2.69 + 1.5 = 4.19$ with 95% confidence interval given by $(2.39 + 1.5, 2.98 + 1.5) = (3.89, 4.48)$.

6.1.4   Tolerance intervals

Consider a process for the production of an electronic component with a target capacitance of 2000 nF and specification limits of 1900 and 2100 nF. Suppose that the process currently yields components with capacitances that are normally distributed with mean $\mu = 2025$ nF and standard deviation $\sigma = 50$ nF. For a normal distribution 99% of values lie between $\mu - 2.58\sigma$ and $\mu + 2.58\sigma$, which in this case would be 1896 and 2154 nF. The interval (1896, 2154) is the 99% *tolerance interval* for capacitance, and we may refer to the interval (1900, 2100) as the *specification interval*. These intervals are displayed in Figure 6.13. The fact that the tolerance interval is wider than the specification interval gives an immediate indication of poor process capability. The reader may readily verify that C_{pk} is 0.50 and C_p is 0.67.

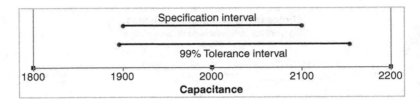

Figure 6.13   Specification interval and 99% tolerance interval.

Two steps could be taken to improve the process capability:

- Shift the mean from 2180 closer to the target value of 2000.

- Reduce the variability – a reduction in variability will result in a lower standard deviation.

Shifting a process mean is often a relatively easy task requiring, for example, a simple process adjustment. On the other hand, reduction of variability is generally a much more difficult task and may require major modifications to a process.

Suppose that, following major process changes, a sample of 150 capacitors was taken with the process operating in a state of statistical control. The measured capacitance values are provided in Capacitance.MTW. The mean and standard deviation are $\bar{x} = 1999.4$ and $s = 13.0$, respectively. The natural thing to do would be to estimate the 99% tolerance interval for the modified process as $\bar{x} - 2.58s$ and $\bar{x} + 2.58s$, i.e. (1966, 2033). However, since we are now using estimates of the population mean and standard deviation, a factor greater than 2.58 should be used. In order to estimate the 99% tolerance interval with 95% confidence the factor 2.86 should be used, with sample size 150, which yields the interval (1962, 2037). The intervals for the modified process are displayed in Figure 6.14. Clearly there has been a dramatic improvement – the reader is invited to verify that the estimated C_{pk} and C_p are 2.54 and 2.55, respectively.

Use of 95% confidence means that in the long term, when 99% tolerance intervals are calculated for samples from a normal distribution, 95 out of 100 calculated tolerance intervals will cover at least 99% of the population. The required factors for these calculations may be obtained from tables such as those in Hogg and Ledolter (1992, p. 453). Alternatively Minitab may be used.

Use of **Stat** > **Quality Tools** > **Tolerance Intervals. . .** is required. With **Samples in columns:** Capacitance, clicking on **Options. . .** and selecting **Confidence level: 95, Minimum percentage of population in interval: 99, Tolerance interval:** Two-sided and defaults otherwise, the output in Figure 6.15 is obtained. The display includes a histogram of the data, a normal probability plot that indicates that a normal distribution is a reasonable model,

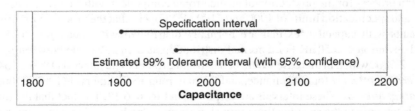

Figure 6.14   Specification interval and estimated 99% tolerance interval for modified process.

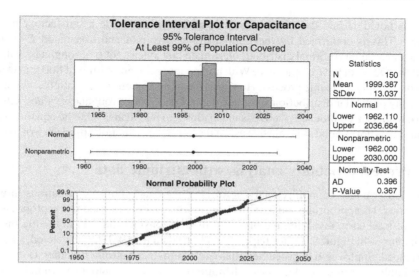

Figure 6.15   Estimated 99% tolerance interval from Minitab.

summary statistics and a 99% normal tolerance interval, after rounding, of (1962, 2037) and a 99% nonparametric tolerance interval, after rounding, of (1962, 2030). In cases where a normal distribution is an appropriate model these two intervals will be similar.

It is instructive to apply **Stat > Quality Tools > Tolerance Intervals...** to the roughness data considered at the end of the previous section. With **Samples in columns:** Roughness, clicking on **Options...** and selecting **Confidence level: 95, Minimum percentage of population in interval: 99, Tolerance interval:** Upper bound and defaults otherwise, the output in Figure 6.16 is obtained.

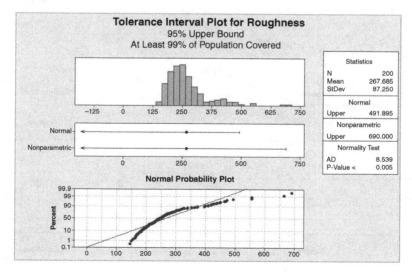

Figure 6.16   Tolerance intervals for roughness.

As was noted earlier, the normal distribution does not provide a satisfactory model for roughness. This is confirmed by the normal probability plot and associated *P*-value. The calculations based on a normal distribution estimate that at least 99% of roughness values will be less than 492, with 95% confidence. With an upper specification limit of 800 this would, on the face of it, imply a capable process. However, the nonparametric calculations, which are not based on any particular distribution, estimate that at least 99% of roughness values will be less than 690, with 95% confidence. As 690 is much closer to 800 than is 492, the conclusion is that the capability is not as good as erroneous use of the normal distribution approach would suggest.

6.1.5   Process capability analysis with attribute data

Minitab also provides capability analysis for attribute data – both for situations in which a binomial model is appropriate and for situations in which a Poisson model is appropriate. As an example of a scenario involving the binomial model, consider the data in the file Invoices1. MTW displayed in Figure 2.15 in Chapter 2, with the data for 10/01/2000 deleted, since a new inexperienced employee had processed many of the invoices during that day. Use of **Stat** > **Quality Tools** > **Capability Analysis** > **Binomial...** yields the output shown in Figure 6.17. In the dialog, **Defective:** was specified as No. Incomplete and **Use sizes in:** was specified as No. Invoices. Under **Tests...** the option to **Perform all four tests** was checked.

There are no signals of any special cause behaviour on the control chart of the data in the top left-hand corner of the output. Additional evidence that the process is behaving in a stable, predictable manner is provided by the display of the cumulative proportion of nonconforming invoices shown in the bottom left-hand corner of the output. It shows the cumulative proportion of nonconforming invoices levelling off at around 16% as more and more data became available. (This display is similar to Figure 4.1.) The histogram displays the proportions

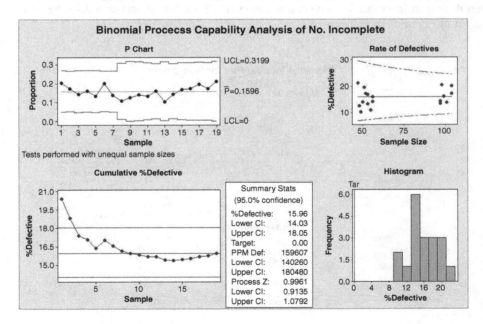

Figure 6.17   Binomial capability analysis of invoice data.

(expressed as percentages) of nonconforming invoices for the 19 samples analysed. The Rate of Defectives plot is essentially a funnel plot as described in Chapter 5. It is provided in this context in order to give insight into whether or not the proportion of nonconforming invoices is influenced by the subgroup size. Here the subgroups comprised all the invoices processed on each day. Had the group of points on the right been located at a higher level than the group of points on the left, then this could possibly have indicated a higher proportion of nonconforming invoices occurring on days when staff were working under greater pressure dealing with higher numbers of invoices. Finally, the Summary Stats table indicate that the overall proportion of nonconforming invoices is estimated at 15.96% or 159 607 ppm. This converts, via Appendix 1, to a sigma quality level of around 2.5. Alternatively, addition of 1.5 to the Process Z of 0.996 1 given in the output yields a sigma quality level of 2.5, to one decimal place. Confidence intervals are also given. (Note that Minitab refers to defectives rather than to nonconforming items.)

As an example of a situation where the Poisson model is potentially appropriate consider the data from work cell B referred to in Exercise 11 in Chapter 5. The worksheet PCB2.MTW gives counts of nonconformities on a series of samples of printed circuit boards. The subgroup size is given in the first column and the number of nonconformities found in the sample in the second. Use of **Stat** > **Quality Tools** > **Capability Analysis** > **Poisson...** yields the output shown in Figure 6.18. In the dialog **Defects:** was specified as Nonconformities and **Use Sizes in:** was specified as Boards. Under **Tests...** the option to **Perform all four tests** was checked.

The U chart of the data provides no evidence of any special cause behaviour affecting the process. The plot of Cumulative DPU (defects per unit) stabilizes at around 0.35, which is indicated by the middle of the three horizontal reference lines on the chart. The upper and lower reference lines correspond to upper and lower 95% confidence limits for defects per unit for the process of 0.31 and 0.40. All three values are given in the Summary Stats table. The

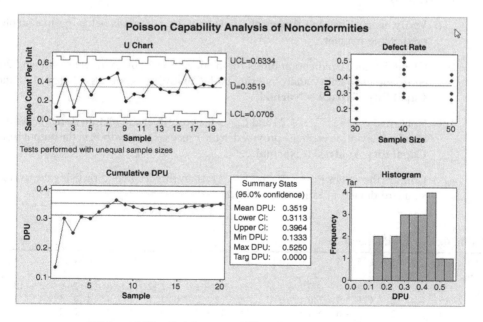

Figure 6.18   Poisson capability analysis of PCB data.

histogram shows the distribution of DPU for the subgroups, the scatterplot shows DPU plotted against subgroup size. There is a suggestion from the arch-shaped scatterplot that DPU may be influenced by subgroup size – further investigation of this would clearly be of interest to those responsible for the identification of nonconformities.

6.2   Exercises and follow-up activities

1. The specification limits for the amount of mineral water delivered by a bottling process to nominal 1.5 litre bottles are 1.480 and 1.520 litres, respectively. Forty samples of five fill volumes taken at 30-minute intervals from a bottling process are given in Volumes. MTW. Assess the capability of the process.

   Stack the data into a single column and obtain a two-sided tolerance interval that covers at least 99.9% of fill volumes with 95% confidence. Move the mouse pointer to the horizontal scale in the central plot so that Y-scale is displayed in a text box. Double-click and change the scale range minimum and maximum to 1.47 and 1.53, respectively. Right-click the display and use **Add > Reference Lines...** to superimpose labelled specification limits on the plot. Note the insight into process capability that the display now provides.

2. The outside diameter (OD) of the pilot on an output shaft is an important quality characteristic. Table 6.5 gives data obtained from 25 subgroups of size 4 and records the diameter measured as deviation from nominal (micrometres). The data are also provided in the worksheet OD.MTW and are reproduced by permission of the Statistics and Actuarial Science Department, University of Waterloo, Canada from Steiner *et al.* (1997, p. 6).

   (i) Verify, using Xbar and R control charts, that the process is behaving in a stable, predictable manner.

   (ii) Given that the specification limits for the deviation from nominal are –20 and 20, carry out a capability analysis using both **Capability Analysis > Normal...** and **Capability Sixpack > Normal....**.

   (iii) Confirm the values $C_p = 1.29$ and $C_{pk} = 1.26$ by direct calculation using the within standard deviation estimate provided in the Process Data textbox in the output from **Capability Analysis > Normal....**.

   (iv) Confirm the values $P_p = 1.19$ and $P_{pk} = 1.16$ by direct calculation using the overall standard deviation estimate provided.

Table 6.5   Pilot diameter data.

Sub group	1	2	3	4	5	6	7	8	9	10	11	12	13	14	15	16	17	18	19	20	21	22	23	24	25
x_1	−10	−14	−2	−3	12	0	2	0	2	−8	4	−8	−10	−8	−4	2	12	2	−6	−2	2	0	−2	−10	6
x_2	−6	−4	12	−5	6	0	−6	6	4	0	2	4	2	2	6	2	6	2	−4	4	4	4	4	4	8
x_3	0	−6	−2	−5	2	−6	8	4	6	−4	2	−14	−10	−4	0	0	0	0	2	0	2	2	−2	−12	−4
x_4	0	4	8	−1	2	−8	−6	8	8	2	6	6	4	4	8	2	2	−8	0	4	6	4	4	4	2

(v) Using the estimate of process 'fall–out' in ppm based on the within standard deviation estimate, obtain an estimate of the sigma quality level of the process using Appendix 1. Use the **Benchmark Z's (sigma level)** option with **Capability Analysis > Normal...** to confirm your estimate.

(vi) Change the data set by *subtracting* 1 from each observation in subgroup 4 and *adding* 10 to each observation in subgroup 15. Repeat the capability analysis and observe that the within estimate of standard deviation is unchanged since the sample ranges are unchanged, but that the overall estimate has increased. As a consequence, C_p is unchanged but P_p is reduced. Some authors argue that this means that P_p and P_{pk} measure how the process *actually* performed, while C_p and C_{pk} measure how the process *could* perform. However, scrutiny of Xbar and R charts of the modified data reveals that they correspond to a process that is not stable and predictable.

3. Door to needle time (DTN) is the time from arrival at hospital when the ambulance stops outside the hospital (door) to the start of the thrombolytic treatment (needle) for patients with an acute myocardial infarction. The health authority responsible for management of the hospital has an upper specification limit of 30 minutes for DTN for such patients. Given the sample of door to needle times in the worksheet DTN1.MTW, carry out a capability analysis. Following process changes, a further sample of door to needle times was recorded and is available in the worksheet DTN2.MTW. Assess the impact of the changes on process capability. Obtain appropriate before and after tolerance intervals that cover 99% of DTN times with 95% confidence and note the insight that these provide into the impact of the process changes.

4. The worksheet Sand.MTW contains the percentage of sand by weight in samples of aggregate material for use in the construction industry. Given that the upper specification limit is 10%, carry out a capability analysis of the data. You should verify that direct use of Capability Analysis (Normal) overestimates capability compared with that obtained using a Box–Cox transformation and with that obtained using a Weibull distribution.

5. A large call centre, which operates from 08.00 until 20.00 Monday to Friday, is staffed by teams A, B and C, with A responsible for 08.00 until 12.00, B for 12.00 until 16.00 and C for the remaining period each day. For one particular week recordings of samples of 40 of the calls received during each half–hour period were analysed by supervisors for conformance to specifications. The worksheet Calls.MTW gives summary data.

   (i) Carry out a binomial capability analysis of the complete data set without taking team into account. Note the signals on the P chart and the oscillatory behaviour of the Cumulative %Defective (nonconforming) plot.

   (ii) Unstack the data by team and carry out a binomial capability analysis for each team. What do you conclude?

6. Set up the funnel plot data in Table 5.7 in Minitab and, viewing the data as a series of samples from a process, carry out a binomial capability analysis. Compare the funnel plot in the output with that in Figure 5.29.

7

Process experimentation with a single factor

Experiment, and it will lead you to the light. (Cole Porter, 'Experiment' from *Nymph Errant*, 1933)

Overview

This chapter deals with statistical tools that are relevant to the improve phase of Six Sigma projects. Having made changes to a process, how do we formally assess data from the modified process for evidence of improvement? Statistical inference techniques may be use to address questions such as:

- Has the change of supplier of the lens coating fluid led to a reduction in the proportion of nonconforming lenses?

- Has the appointment of specialist nurses, empowered to administer thrombolytic treatment to acute myocardial infarction patients on admission to hospital, led to a reduction in the mean door to needle time?

Estimation techniques provide point estimates of the population proportion and the population mean, respectively, for the modified processes in the above scenarios, i.e. of population parameters that are of interest. Estimation techniques provide intervals in which we can have confidence that the values of the parameters are located.

Some of the techniques are based on the normal distribution while others make no such assumption. Minitab is well equipped to deal with both classes of technique.

Six Sigma Quality Improvement with Minitab, Second Edition. G. Robin Henderson.
© 2011 John Wiley & Sons, Ltd. Published 2011 by John Wiley & Sons, Ltd.

7.1   Fundamentals of hypothesis testing

In Chapter 1, the description of the improve phase in a Six Sigma project given by Roger Hoerl included the statement 'determine how to intervene in the process to significantly reduce the defect levels' (Hoerl, 1998, p. 36). Process experimentation may be thought of as a formal approach to the question of determining how to intervene in the process in order to improve it. Wheeler (1993, p. 21) writes:

- Before one can improve any system one must listen to the voice of the system (the voice of the process).

- Then one must understand how the inputs affect the outputs of the system.

- Finally, one must be able to change the inputs (and possibly the system) in order to achieve the desired results.

- This will require sustained effort, constancy of purpose, and an environment where continual improvement is the operating philosophy.

Wheeler (2007, p. 7) distinguishes between *observational* and *experimental* studies. The routine collection of data in order to monitor, using control charts, a process running under normal conditions is an example of an observational study. The collection of data on a process when it is being run under special conditions, with a view to learning how the process might be improved, is an example of an experimental study. He states: 'when we analyze experimental data we are looking for differences that we have paid good money to create and that we believe are contained within the data'.

Consider the process of administering thrombolytic treatment to acute myocardial infarction patients at a hospital. Records show that the process has been behaving in a stable, predictable manner, with door to needle time (DTN) being adequately modelled by the normal distribution with mean 19 minutes and standard deviation 6 minutes. An experiment was conducted over a period of 1 month during which one of a team of specialist nurses, empowered to administer the thrombylotic treatment, was on duty at all times in the accident and emergency department of the hospital. The DTN times for the 25 patients treated during the experimental period are shown in Table 7.1 and are available in the worksheet DTNTime.MTW. The mean DTN for acute myocardial infarction patients during the experimental period was 16.28 minutes. Does this sample mean represent a 'real' improvement to the process in the sense of a reduction of the population mean DTN for acute myocardial infarction patients from 19 minutes to a new, lower population mean? Can we infer from the data that the regular deployment of the specialist nurses would ensure process improvement?

Table 7.1   DTN (minutes) during experimental period.

26	7	24	3	12
17	24	4	5	16
16	22	14	15	14
19	21	18	14	20
29	20	17	9	21

The various steps involved in performing the appropriate statistical inference will now be detailed under the headings Hypotheses, Experimentation, Statistical model and Conclusion.

Hypotheses.   Denoting the *null hypothesis* by H_0 and the *alternative hypothesis* by H_1, our hypotheses are:

$$H_0 : \mu = 19, \quad H_1 : \mu < 19.$$

The null hypothesis represents 'no change' – were the introduction of the specialist nurses to have no impact on DTN, the population mean would remain at 19 minutes. Thus μ represents the population mean DTN with the specialist nurses deployed. The alternative hypothesis represents what might be referred to as the *experimental hypothesis* – the objective of the experiment is to determine whether or not there is evidence that the specialist nurse input improves the process by leading to a mean DTN time which is less than 19 minutes.

Experimentation.   During an experimental period of 1 month, with specialist nurse input, the mean DTN for the 25 patients treated was 16.28 minutes.

Statistical model.   Three assumptions are made:

1. The variability of DTN time is assumed to be unaffected by the process change, i.e. it is assumed that the standard deviation continues to be $\sigma = 6$ minutes.

2. The null hypothesis is assumed true, i.e. it is assumed that the process mean continues to be $\mu = 19$ minutes. (This is analogous to the situation whereby the defendant in a trial in a court of law is considered innocent until there is evidence to the contrary.)

3. The sample of 25 DTNs, obtained during the experimental period, is regarded as a random sample from the population of normally distributed times with mean 19 and standard deviation 6 minutes.

The statistical model is detailed in Box 7.1.

The question asked of the statistical model is 'What is the probability of observing a sample mean for 25 patients which is 16.28 minute or less?' Minitab readily provides the answer using **Calc > Probability Distributions > Normal** to obtain the Session window output in Panel 7.1. **Cumulative probability** must be selected, with **Mean:** 19, **Standard deviation:** 1.2 and **Input constant:** 16.28 specified. Thus, if the null hypothesis was true, i.e. if the deployment of the specialist nurses had no impact on mean DTN, then the probability of observing a mean for a sample of 25 patients as low as 16.28, or lower, would be 0.012 (to three decimal places). This probability of 0.012 is the *P*-value for testing the hypotheses specified above.

Conclusion.   It is conventional in applied statistics to state that a *P*-value less than 0.05 provides evidence for rejection of the null hypothesis, in favour of the alternative hypothesis, at

---

Door to needle time, Y, is normally distributed with mean 19 and standard deviation 6, i.e. $Y \sim N(19, 6^2)$. Mean DTN for samples of $n = 25$ patients, $\bar{Y}$, will be normally distributed with mean $\mu_{\bar{Y}} = \mu = 19$ and standard deviation $\sigma_{\bar{Y}} = \sigma/\sqrt{n} = 6/\sqrt{25} = 1.2$, i.e. $\bar{Y} \sim N(19, 1.2^2)$.

Box 7.1   Statistical model.

```
Cumulative Distribution Function

Normal with mean = 19 and standard deviation = 1.2

     x   P( X <= x )
  16.28    0.0117053
```

Panel 7.1 Probability that sample mean is 16.28 minutes or less.

the 5% level of significance. A P-value less than 0.01 would be said to provide evidence for rejection of the null hypothesis, in favour of the alternative hypothesis, at the 1% level of significance. (The value 0.001, corresponding to the 0.1% level of significance, is also widely used and 0.1, corresponding to the 10% level of significance is sometimes used.) In addition to the highly technical statement that 'the P-value of 0.012 provides evidence for rejection of the null hypothesis, in favour of the alternative hypothesis, at the 5% level of significance', it is important to state that 'the experiment provides evidence that mean DTN is significantly reduced through the deployment of the specialist nurses' and that 'a point estimate of the new mean DTN is a little over 16 minutes'. (It should be noted that although the result of the experiment is *statistically* significant, a reduction of the mean DTN of around 3 minutes might not be of any *practical* significance from a medical point of view. Statistical significance does not equate to practical significance.)

From now on the use of the word 'evidence' will imply that the evidence is convincing, where the word 'convincing' can be further qualified by the significance level. In teaching this topic the author has taught his students to think in terms of a P-value less than 0.05 providing evidence for rejection of the null hypothesis, a P-value less than 0.01 providing strong evidence and a P-value less than 0.001 providing very strong evidence. A P-value less than 0.1 might also be regarded as providing slight evidence for rejection of the null hypothesis.

Of course the mean DTN might have remained at 19 minutes with the introduction of the specialist nurses and the experimenters might have been unlucky enough to obtain a sample of times with a low enough mean to provide evidence, in the sense discussed above, of a reduction in the population mean time. Two types of error can occur in the performance of a test of hypotheses as indicated in Table 7.2.

The probability of a Type I error is denoted by the Greek letter α (alpha) and is the significance level of the test. Thus, if one decides to perform a test of hypotheses at the 5% level of significance, $\alpha = 0.05$ and there is a probability of 0.05 that the null hypothesis will be rejected when it is in fact true. In the case of the DTN scenario this means that, were the

Table 7.2 Possible errors in testing hypotheses.

Errors possible in testing hypotheses		True state	
		H_0 true	H_0 false
Conclusion reached	Accept H_0	Correct decision	Type II error probability $= \beta$
	Reject H_0	Type I error probability $= \alpha$	Correct decision

```
Inverse Cumulative Distribution Function

Normal with mean = 19 and standard deviation = 1.2

P( X <= x )          x
         0.05   17.0262
```

Panel 7.2 Determining the cut-off mean DTN.

introduction of the specialist nurses to have no impact whatsoever on the mean of 19 minutes, there is a probability of 0.05 that the conclusion would be the erroneous one that there was evidence of a decrease. The lower the significance level selected then the lower the risk of committing a Type I error.

Having decided, say, on a significance level of $\alpha = 0.05$ for the DTN experiment, one can use **Calc** > **Probability Distributions** > **Normal…** to complete the final stage of reaching a conclusion in a different way. The statistical model indicates that the means of samples of 25 times follow the $N(19, 1.2^2)$ distribution. Panel 7.2 displays the Session window output obtained using the **Inverse cumulative probability** function for this normal distribution via **Calc** > **Probability Distributions** > **Normal…** and specifying **Input constant:** 0.05. Thus the cut-off between acceptance and rejection of the null hypothesis occurs at the value 17.026 2 for the sample mean. The conclusion would therefore be:

- Do not reject H_0 if the sample mean is greater than 17.026 2.

- Reject H_0 if the sample mean is less than or equal to 17.026 2.

From the data it was established that the sample mean was 16.28. Since this is less than 17.026 2 the conclusion reached was to reject the null hypothesis at the significance level of $\alpha = 0.05$.

A graphical representation of the test is displayed in Figure 7.1. The use of the 5% significance level corresponds to an area of 0.05 in the shaded left-hand tail of the normal curve

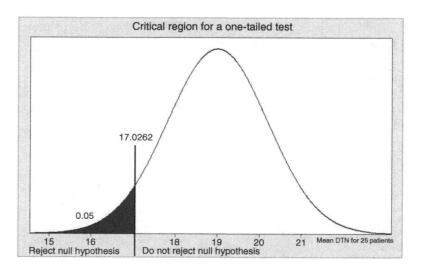

Figure 7.1 Critical region for the statistical test.

```
Cumulative Distribution Function

Normal with mean = 16 and standard deviation = 1.2

       x   P( X <= x )
 17.0262      0.803771
```

Panel 7.3   Probability of rejecting null hypothesis when new population mean is 16.

that specifies the statistical model for the distribution of sample means in this case. As only one tail of the distribution is involved, this test may be referred to as a *one-tailed test*. Values of the sample mean DTN less than 17.026 2 comprise the critical region for the test.

Suppose now that the deployment of the specialist nurses had actually led to a reduction in the mean of the population of DTNs from 19 to 16 minutes. What is the probability that, once data are available for a random sample of 25 patients, the conclusion reached – to reject H_0 in favour of H_1 – will be the correct one? We require the probability of observing a sample mean of 17.0262 or less when the population mean is actually 16. Again **Calc** > **Probability Distributions** > **Normal...** provides the answer – see Panel 7.3.

Thus there is a probability of 0.80 (to two decimal places) of the test of hypotheses providing evidence of a three-minute reduction in the population mean DTN. This probability is the power of the statistical test to detect the change from a population mean of 19 minutes to a population mean of 16 minutes. The other side of the coin is that there is probability $1 - 0.80 = 0.20$ of the experiment failing to provide evidence of the reduction, i.e. there is probability $\beta = 0.20$ of committing a Type II error with a test based on a sample of 25 patients when the mean actually drops from 19 to 16 minutes. (The Greek letter β is beta.)

Figure 7.2 show a plot of the power of the test against the population mean after the process change. The greater the drop in the population mean the more likely it is that the test will provide evidence of the change, i.e. the more powerful is the test. For example, were the mean to drop by 1 minute, so that the population mean after the process change became 18 minutes,

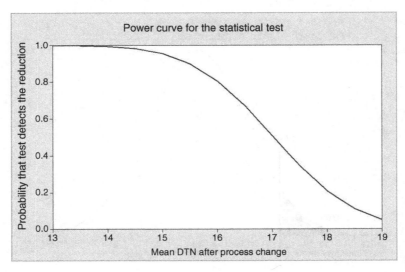

Figure 7.2   Power curve for the statistical test.

then the probability is 0.21 that the conclusion would be to reject the null hypothesis. On the other hand, were the mean to drop by 5 minutes, so that the population mean after the process change became 14 minutes, then the probability is 0.99 that the conclusion would be to reject the null hypothesis. The power of a test may be increased through use of a larger sample.

Now consider the lens coating process referred to in the previous chapter. Records show that prior to the change of coating fluid supplier the process was yielding 4.5% nonconforming lenses. In an experimental run with the new coating fluid there were 80 nonconforming lenses in a batch of 2400, i.e. 3.3% nonconforming. Can we infer from these data that there has been process improvement?

Hypotheses.   In this case, using p to represent the population proportion of nonconforming lenses,

$$H_0 : p = 0.045, \quad H_1 : p < 0.045.$$

The null hypothesis represents 'no change' – were the switch to a new supplier of the coating fluid to have no impact on nonconformance, the proportion of nonconforming lenses would remain at 0.045. The alternative hypothesis represents the experimental hypothesis – the objective of the experiment is to determine whether or not there is evidence that the change of supplier improves the process by leading to a reduction in the proportion of nonconforming lenses.

Experimentation.   From an experimental run, during which 2400 lenses were processed, 80 were found to be nonconforming.

Statistical model.   Three assumptions are made:

1. The null hypothesis is assumed to be true, i.e. that the proportion of nonconforming lenses is assumed to be unaffected by the process change and remains at 0.045.

2. The conditions for the binomial distribution apply, i.e. that there is constant probability of 0.045 that a lens is nonconforming and that the status of each lens is independent of that of all other lenses.

3. The set of 2400 lenses, manufactured using coating fluid from the new supplier, may be considered as a random sample from a population of lenses in which the proportion 0.045 is nonconforming.

The statistical model is detailed in Box 7.2.

The question asked of the statistical model is: 'What is the probability of observing 80 or fewer nonconforming lenses in a batch of 2400?' Minitab readily provides the answer using **Calc > Probability Distributions > Binomial...** to obtain the output in Panel 7.4. **Cumulative probability** must be selected, with **Number of trials:** 2400, **Event probability:** 0.045 and **Input constant:** 80 specified. Thus, if the null hypothesis were true, i.e. if the change of supplier had no impact on the proportion of nonconforming lenses, then the probability of

---

The number of nonconforming lenses, Y, in a batch of 2400 will have the binomial distribution with parameters $n = 2400$ and $p = 0.045$, i.e. $Y \sim B(2400, 0.045)$.

Box 7.2   Statistical model.

```
Cumulative Distribution Function

Binomial with n = 2400 and p = 0.045

  x   P( X <= x )
 80     0.0024365
```

Panel 7.4   Probability that sample includes 80 or fewer nonconforming lenses.

observing 80 or fewer nonconforming lenses in a batch of 2400 would be 0.0024. Thus 0.0024 is the P-value for testing the hypotheses specified above.

Conclusion.   Since the P-value is less than $0.01 = 1\%$ the null hypothesis would be rejected in favour of the alternative hypothesis at the 1% significance level. Thus the data from the experiment provide strong evidence that the change of supplier has led to a significant reduction in the proportion of nonconforming lenses from the previous level of 4.5%. A point estimate of the new proportion of nonconforming lenses is $80/2400 = 0.033 = 3.3\%$.

Having decided, say, on a significance level of $\alpha = 0.01 = 1\%$ for the lens coating experiment, one can use **Calc > Probability Distributions > Binomial...** to look at the final stage of reaching a conclusion in a different way. The statistical model indicates that the number of nonconforming lenses in a batch will have the $B(2400, 0.045)$ distribution. Panel 7.5 displays the Session window output obtained using the **Inverse cumulative probability** function for this binomial distribution via **Calc > Probability Distributions > Binomial...** and specifying **Input constant: 0.01**. Thus the cut-off between acceptance and rejection of the null hypothesis occurs at the value 84 for the number of nonconforming lenses in the batch when the significance level is $\alpha = 0.01 = 1\%$. With the discrete binomial distribution it has not been possible to determine a value x such that $P(X \leq x)$ is precisely 0.01, so the value 84 is used since $P(X \leq 84)$ is closest to, but less than, 0.01. The conclusion would therefore be:

- Do not reject H_0 if the number nonconforming is greater than 84.

- Reject H_0 if the number nonconforming is less than or equal to 84.

It was established that there were 80 nonconforming lenses in the batch. Since this is less than 84, the conclusion reached was to reject the null hypothesis at the significance level of $\alpha = 0.01 = 1\%$.

Suppose now that change of supplier had actually led to a reduction in the proportion of nonconforming lenses from 0.045 to 0.030. What is the probability that, once the nonconforming lenses had been counted in a batch of 2400, the conclusion reached – to reject H_0 in favour of H_1 – will be the correct one? Again **Calc > Probability Distributions > Binomial...** provides the answer – see Panel 7.6.

```
Inverse Cumulative Distribution Function

Binomial with n = 2400 and p = 0.045

  x   P( X <= x )         x   P( X <= x )
 84     0.0085010        85     0.0112890
```

Panel 7.5   Determining the cut-off number of nonconforming lenses.

```
Cumulative Distribution Function

Binomial with n = 2400 and p = 0.03

   x   P( X <= x )
  84      0.929871
```

Panel 7.6   Probability of rejecting null hypothesis when new population proportion is 0.030.

Thus there is a probability of 0.93 (to two decimal places) of the test of hypotheses providing the evidence of a reduction from 0.045 to 0.030 in the population proportion of nonconforming lenses. This probability is the power of the statistical test to detect the change from a population proportion of 0.045 to a population proportion of 0.030. The other side of the coin is that there is probability $1 - 0.93 = 0.07$ of the experiment failing to provide evidence of the reduction, i.e. there is probability $\beta = 0.07$ of committing a Type II error with a test based on a sample of 2400 lenses.

Figure 7.3 shows a plot of the power of the test against the population proportion nonconforming after the process change. Note that the larger the drop in the proportion the more likely it is that the test will provide evidence of the change. For example, were the population proportion of nonconforming lenses to drop by 0.02, so that the population proportion after the process change was 0.025, then the probability is 0.999 that the conclusion would be to reject the null hypothesis, i.e. it is virtually certain that the test would lead to making the correct decision.

In discussing the above two tests of hypotheses the statistical models used employed specific probability distributions, the normal distribution in the case of the DTNs and the binomial distribution in the case of the lens coating process. Other tests of hypotheses are available which do not require use of specific probability distributions. These are referred to as *distribution-free* or *nonparametric* tests. Specific cases of such tests will be introduced later in the chapter.

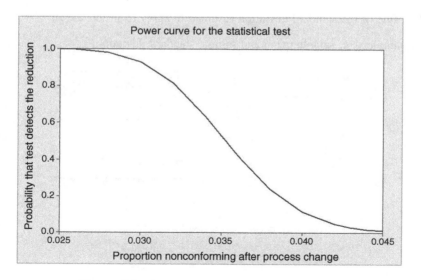

Figure 7.3   Power curve for the statistical test.

Both the above tests of hypotheses were performed from first principles. In the next section we will see how Minitab can be used to streamline performance of the tests.

7.2   Tests and confidence intervals for the comparison of means and proportions with a standard

7.2.1   Tests based on the standard normal distribution – z-tests

Consider again the thrombolytic treatment example and the DTN data in Table 7.1 and worksheet DTNTime.MTW. The standard DTN could be thought of as the mean $\mu = 19$ minutes of the normal distribution of DTNs. When the data for the 25 patients treated with the specialist nurses available are to hand we wish to compare these data with the standard via the formal test of the hypotheses:

$$H_0 : \mu = 19, \quad H_1 : \mu < 19.$$

Recall, too, that the variability was assumed to remain unchanged, with the standard deviation being 6 minutes. The dialog involved in performing the test in Minitab using **Stat > Basic Statistics > 1-Sample Z...** is shown in Figure 7.4.

Here the sample of DTNs is available in column C1. **Standard deviation:** 6 is specified and, with **Perform hypothesis test** checked, **Hypothesized mean:** 19 indicates the null hypothesis. Under **Options...** the alternative hypothesis is specified by use of the scroll arrow to select **less than** in the **Alternative:** window. Finally, under **Graphs...** one can select to display the data in the form either of a histogram, an individual values plot or a boxplot; in this case the **Histogram of data** option was selected. The Session Window output is shown in Panel 7.7 and the graphical output is shown in Figure 7.5.

The Session window output includes the following:

- a statement of the hypotheses under test in the first line;

- the value of the standard deviation assumed to apply (6 in this case);

- the sample size, sample mean and sample standard deviation;

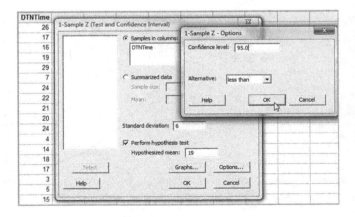

Figure 7.4   Dialog for performing a one-sample z-test.

```
One-Sample Z: DTNTime

Test of mu = 19 vs < 19
The assumed standard deviation = 6


                                           95% Upper
Variable    N     Mean   StDev   SE Mean     Bound      Z        P
DTNTime    25    16.28    6.83    1.20      18.25    -2.27    0.012
```

Panel 7.7   Session window output for z-test.

- the standard error of the mean which is the standard deviation of the sample mean given by $\sigma_{\bar{x}} = \sigma/\sqrt{n} = 6/\sqrt{25} = 1.2$;

- a 95% upper bound of 18.25 which will be explained later in this section;

- a Z-value of -2.27, which is explained in Box 7.3;

- the P-value of 0.012 for the test.

The P-value of 0.012 was calculated in the previous section in the 'grass-roots' version of the test. Since it is less than 0.05 one can immediately conclude that the null hypothesis $H_0: \mu = 19$ would be rejected at the $\alpha = 0.05$ significance level in favour of the alternative hypothesis $H_1: \mu < 19$. In other words, the experiment provides evidence of a significant reduction in the population mean DTN at the 5% level of significance.

In the previous section it was noted that the cut-off between acceptance and rejection of the null hypothesis, when using the 5% significance level, occurs at the value of 17.026 2 for the sample mean. The value of Z corresponding to 17.026 2 is -1.64. In terms of the standardized variable Z the conclusion would therefore be:

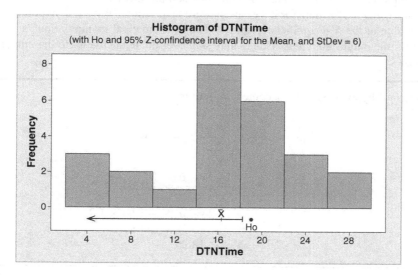

Figure 7.5   Histogram of door to needle times with z-test annotation.

If V is a random variable with mean μ_V and standard deviation σ_V, then the random variable $Z = (V - \mu_V)/\sigma_V$ is the standardized variable which has mean 0 and standard deviation 1. If V is normally distributed then so is Z. In the case of door to needle time, Y, in this example the sample mean, $\bar{Y}$, is, under the null hypothesis, normally distributed with mean 19 and standard deviation 1.2, so the corresponding standardised variable is given by $Z = (\bar{Y} - 19)/1.2$. The mean of the sample of 25 times was $\bar{y} = 16.28$ with corresponding $z = (16.28 - 19)/1.2 = -2.27$ This is the value of Z given in the Session window output.

Box 7.3   Calculation of the z-statistic given in Session window output.

- Do not reject H_0 if Z is greater than -1.64.

- Reject the null hypothesis H_0 if Z is less than or equal to -1.64.

From the data it has been established that the value of Z corresponding to the sample mean of 16.28 was -2.27. Since this is less than -1.64 the conclusion reached would be to reject the null hypothesis at the significance level of $\alpha = 0.05$.

Prior to the availability of statistical software packages, such as Minitab, tests of this type were typically conducted by calculating Z using the formula

$$Z = \frac{\bar{Y} - \mu}{\sigma/\sqrt{n}}.$$

The conclusion regarding acceptance or rejection of the null hypothesis would then be made by reference to tables of critical values of Z, taking into account the significance level of interest and the nature of the alternative hypothesis. (The creation of such a table was set as Exercise 7 in Chapter 4.) The key involvement of Z in such tests of hypotheses involving a single sample gives rise to the nomenclature one-sample Z-test.

Finally, using the formula for Z we can answer the question: 'What null hypotheses would be acceptable at the 5% significance level?' The mathematical manipulations are given in Box 7.4 for the interested reader. Others may skip over the mathematics to the interpretation that follows.

For the null hypothesis to be accepted we require

$$Z > -1.64 \Rightarrow \frac{\bar{Y} - \mu}{\frac{\sigma}{\sqrt{n}}} > -1.64$$

$$\Rightarrow \frac{16.28 - \mu}{1.2} > -1.64$$

$$\Rightarrow 16.28 - \mu > -1.64 \times 1.2$$

$$\Rightarrow \mu < 16.28 + 1.968$$

$$\Rightarrow \mu < 18.25.$$

Box 7.4   Calculation of range of acceptable population means.

The value of 18.25 is given in the Session window output as the 95% upper bound. In applied statistics 95% confidence level goes hand in hand with the 5% significance level. Had the null hypothesis been $H_0 : \mu = 17$, with alternative $H_1 : \mu < 17$, then one can deduce immediately that the null hypothesis would not be rejected at the 5% level of significance since 17 is less than 18.25, the 95% confidence upper bound. Similarly, $H_0 : \mu = 19$, with alternative $H_1 : \mu < 19$ would be rejected at the 5% level of significance (as we have already seen) since 19 is greater than 18.25. Thus one could report the results of the experiment by stating that 'on the basis of the data collected, the population mean DTN was estimated to be 16.28 minutes and that, with 95% confidence, it could be claimed that the true population mean was at most 18.25 minutes'.

The histogram created as the graphical output in Figure 7.5 is annotated with a point labelled H_0 indicating the value 19 specified in the null hypothesis. The arrowed line segment has a tick mark labelled $\bar{X}$ on it, indicating the sample mean. The segment extends from the value 18.25 downwards and indicates the 95% confidence interval for the population mean following the process change. The fact that the point corresponding to H_0 does not lie on the line segment indicates rejection of the null hypothesis $H_0 : \mu = 19$ in favour of the alternative $H_1 : \mu < 19$ at the 5% level of significance.

Had the data been presented in summary form, i.e. that a sample of 25 times following the process change had mean 16.28, then the test could still be performed using **Stat > Basic Statistics > 1-Sample Z...** by checking **Summarized data** and entering **Sample size: 25**, **Mean: 16.28** and **Standard deviation: 6**. With **Perform hypothesis test** checked, **Hypothesized mean: 19** indicates the null hypothesis. Under **Options...** the alternative hypothesis is specified by use of the scroll arrow to select **less than** in the **Alternative:** window. Without the raw data no graphical output is possible. The reader is invited to check that the output in Panel 7.8 results and that it is identical to that obtained by performing the test using the column of raw data, except that it is not possible to deduce the standard deviation of the sample from the summary information provided to the software.

As a second example, consider a type of glass bottle for which burst strength (psi) could be adequately modelled by the normal distribution with mean 480 and standard deviation 64. During a Six Sigma project with the aim of increasing the burst strength of the bottles, a new glass formulation was used in a large production run of the bottle. The burst strength data for a random sample from the batch are given in Table 7.3 and also as a single column in the worksheet Burst_Strength.MTW. Does the data provide evidence of increased mean burst strength?

```
One-Sample Z

* NOTE * Graphs cannot be made with summarized data.

Test of mu = 19 vs < 19
The assumed standard deviation = 6

                          95% Upper
 N    Mean   SE Mean       Bound       Z      P
25   16.28     1.20        18.25    -2.27   0.012
```

Panel 7.8 Session window output for z-test using summarized data.

Table 7.3 Burst strength (psi) for a sample of 50 bottles.

535	476	439	541	526	523	465	476	468	524
449	444	431	582	580	447	503	498	488	467
545	528	538	570	453	700	535	454	403	498
573	558	442	490	503	476	609	483	484	443
535	476	439	541	526	523	465	476	468	524

Here the hypotheses are:

$$H_0 : \mu = 480, \quad H_1 : \mu > 480.$$

Proceeding as in the previous case, use of **Stat > Basic Statistics > 1-Sample Z...** is required with the **Standard deviation** of 64 specified, **Perform hypothesis test** checked and **Hypothesized mean:** 480 entered. Under **Options...** the alternative hypothesis is specified by use of the scroll arrow to select **greater than** in the **Alternative:** window. Under **Graphs:** the **Boxplot of data** option was selected. The Session window output is shown in Panel 7.9 and the graphical output is shown in Figure 7.6.

One-Sample Z: Burst Strength

```
Test of mu = 480 vs > 480
The assumed standard deviation = 64


                                         95% Lower
Variable          N    Mean   StDev  SE Mean    Bound      Z      P
Burst Strength   50  504.08   55.55     9.05   489.19   2.66  0.004
```

Panel 7.9 Session window output for z-test on burst strength data.

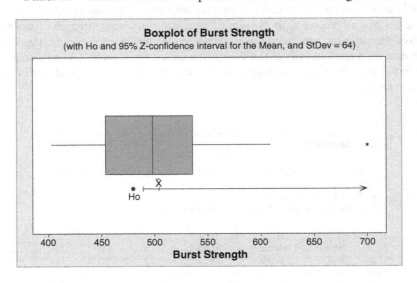

Figure 7.6 Boxplot of burst strength data with z-test annotation.

The conclusion would be that the data provide evidence that the new glass formulation has led to an increase in population mean burst strength – the null hypothesis $H_0 : \mu = 480$ being rejected in favour of the alternative $H_1 : \mu > 480$ at the significance level $\alpha = 0.01$. The P-value for the test is 0.004. The data also enable one to state with 95% confidence that, with the new glass formulation, the population mean burst strength will be at least 489 psi. (Of course a decision as to whether or not to change to the new glass formulation would be likely to involve cost and other considerations.)

As a third example, consider the following scenario. Before revising staffing arrangements at a busy city branch, a major bank determined that the service time (seconds) for business customers could be adequately modelled by a normal distribution with mean 453 and standard deviation 38 seconds. Following implementation of the revision, the mean service time for a random sample of 62 business customers was 447 seconds. The data are stored in Service. MTW. Do these data provide any evidence of a change (either an increase or a decrease) in the mean service time for business customers?

Proceeding as in the previous two cases, but with **Alternative:** set to **not equal**, we can test the null hypothesis $H_0 : \mu = 453$ against the alternative hypothesis $H_1 : \mu \neq 453$. The third available graphical option of an **Individual value plot** (dotplot) was selected in this case. The Session window output is shown in Panel 7.10 and the graphical output in Figure 7.7.

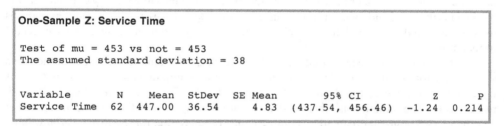

Panel 7.10   Session window output for z-test on service time data.

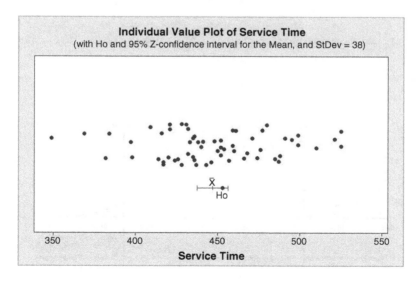

Figure 7.7   Individual value display of service time data with z-test annotation.

(On obtaining the graph the author double-clicked on a data point to access the **Edit Individual Symbols** menu. Under the **Identical Points** tab the **Jitter** option was selected. On clicking **OK** the display shown was obtained. Use of jitter reveals overlapping points.)

The conclusion would be that the data provide no evidence that the revised staffing arrangements have led to a change in population mean service time since the P-value for the test is 0.214. Since the P-value exceeds 0.05 the null hypothesis cannot be rejected at the significance level $\alpha = 0.05$. The data also enable one to state with 95% confidence that, following the revision of staffing arrangements, the population mean service time lies in the interval 438 to 456 seconds, rounded to the nearest integer. This 95% confidence interval (CI) for the population mean service time has been taken from the Session window output and rounded. The fact that the 95% confidence interval includes the mean of 453 specified in the null hypothesis indicates that $H_0 : \mu = 453$ cannot be rejected at the 5% level of significance. This is evident in the graphical display also, as the point representing the value of the population mean specified in the null hypothesis lies on the line segment that represents the 95% confidence interval.

The first and second examples involved one-tailed tests. In terms of z, the criterion for rejection of the null hypothesis at the 5% level of significance was z less than or equal to -1.64 in the first case, z greater than or equal to 1.64 in the second case, and in the third case either z less than -1.96 or z greater than 1.96. (The fact that 1.96 rounds to 2.00 may explain why the 5% level is the most widely used significance level in applied statistics – for a null hypothesis to be rejected at the 5% level in a two-tailed z-test, it is easy to remember that z must exceed 2 in magnitude.)

The test could have been carried out in the third case by obtaining the values of the mean service time for a random sample of 62 customers, on the assumption that the null hypothesis is true, that correspond to the z-values of -1.96 and 1.96. The values are 443.5 and 462.5, respectively. The situation is illustrated in Figure 7.8. Values of the sample mean either less than 443.5 or greater than 462.5 comprise the critical region for the *two-tailed test* here.

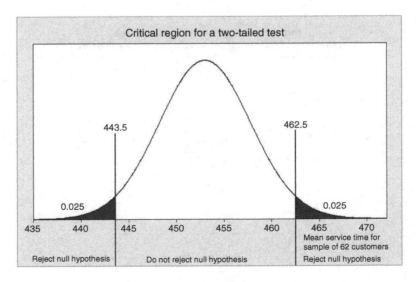

Figure 7.8   Critical region for a two-tailed test.

The sample mean obtained was 447.0, which is not in the critical region – hence the decision not to reject the null hypothesis at the 5% level of significance. Thus there is no evidence from the data to suggest that the revised staffing arrangements have had any impact on the mean time taken to serve business customers.

7.2.1.1   Some comments on tests of hypotheses and P-values

Vickers (2010) states that 'the p-value is the probability that the data would be at least as extreme as those observed, if the null hypothesis were true'. His book gives much sound advice on hypothesis testing via both light-hearted contexts and real applications. Ronald Fisher, who played a key role in the development of statistical inference, wrote on P-values in *Statistical Methods for Research Workers*, first published in 1925: 'We shall not often be astray if we draw a conventional line at 0.05' (Fisher, 1954, p. 80) This comment will undoubtedly have contributed to the level of significance 0.05 becoming the most widely used in applied statistics.

Statistical tests of hypothesis are controversial. Sterne and Davey Smith (2001, p. 226) make the following summary points:

> P values ... measure the strength of the evidence against the null hypothesis; the smaller the P value, the stronger the evidence against the null hypothesis.

> An arbitrary division of results, into 'significant' or 'nonsignificant' according to the P value, was not the intention of the founders of statistical inference.

> A P value of 0.05 need not provide strong evidence against the null hypothesis, but it is reasonable to say that $P < 0.001$ does. In the results ... the precise P value should be presented, without reference to arbitrary thresholds.

> Confidence intervals should always be included in reporting the results of statistical analyses, with emphasis on the implications of the ranges of values in the intervals.

The value of z quoted in the output is known as the test statistic. Montgomery (2009, p. 117) comments:

> It is customary to call the test statistic (and the data) significant when the null hypothesis H_0 is rejected; therefore, we may think of the P-value as the smallest level α at which the data are significant. Once the P-value is known, the decision maker can determine for himself or herself how significant the data are without the data analyst formally imposing a pre-selected level of significance.

This author strongly recommends that a display of the data should be incorporated into the reporting whenever it is possible to do so.

7.2.1.2   Some comments on confidence intervals

In order to aid the reader's understanding of confidence intervals a series of 80 random samples of size 62 was generated from the $N(453, 38^2)$ distribution using Minitab. (The sample size and parameters chosen relate to the earlier example on customer service time in a bank, but one

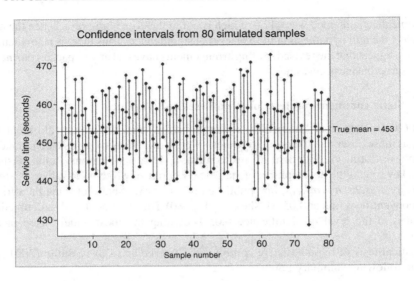

Figure 7.9   Display of 95% confidence intervals from simulated samples.

could choose arbitrary values.) On the assumption that the standard deviation of the population was known to be 38, two-sided 95% confidence intervals for the population mean were computed and displayed using the **1-Sample Z...** facility in Minitab – see Figure 7.9.

The reference line indicates the true population mean of 453. The first vertical line segment represents the 95% confidence interval based on the first simulated sample which was (440.1, 459.0) and captures within it the true population mean of 453 – capture is indicated by the line segment crossing the reference line. The 63rd line segment represents the 95% confidence interval based on the 63rd sample which was (453.7, 472.6) and fails to capture within it the true population mean of 453 – failure to capture is indicated by the segment not crossing the reference line. Of the 80 confidence intervals, the 63rd, 74th and 79th fail to 'capture' the true population mean. This corresponds to 77 captures from 80 attempts, which is equivalent to 96.2%. This capture rate is close to the long-term capture rate for such intervals of 95%.

The formula for calculating a 95% two-sided confidence interval for the mean of a normally distributed population, with known standard deviation σ, based on a sample of size n, is

$$\bar{x} \pm 1.96 \frac{\sigma}{\sqrt{n}}.$$

Suppose that a machine for filling jars with instant coffee granules delivers amounts that are normally distributed with standard deviation 0.3 g. The process manager wishes to know how big a sample of jars is required in order to estimate, with 95% confidence, the mean amount delivered to within 0.1 g of its true value. This means that we require

$$1.96 \frac{0.3}{\sqrt{n}} < 0.1 \Rightarrow n > \left(\frac{1.96 \times 0.3}{0.1} \right)^2 = 34.5.$$

Thus a random sample of at least 35 jars would be required.

The value 0.1 g may be referred to as the margin of error, E. In general, in order to estimate, with 95% confidence, the mean of a population to within E of its true value requires a sample size n of at least $3.84\sigma^2/E^2$. If the standard deviation, σ, of the population is unknown then an estimate of the standard deviation from a pilot sample may be used in the calculation. If the pilot sample is small then caution should be exercised in applying the formula. For estimation with 99% confidence, the factor of 3.84 is replaced by 6.63 in the above formula, and for 99.9% confidence it is replaced by 10.83. A point to note is that halving the margin of error quadruples the sample size required.

7.2.2 Tests based on the Student t-distribution – t-tests

At the core of the tests we have considered so far in this section is the test statistic given by

$$z = \frac{\bar{y} - \mu}{\sigma/\sqrt{n}}.$$

In each of the examples considered it was assumed that process variability was unaffected by the process changes so that the standard deviation, σ, was known.

Hence the tests are called z-tests.

What do we do if this assumption is suspect? If we have modified a process. might not the changes made affect variability as well as location? A natural thing to do is to calculate the test-statistic value:

$$t = \frac{\bar{y} - \mu}{s/\sqrt{n}}$$

This formula may be obtained from the previous one by using the sample standard deviation s in place of the population standard deviation, σ. If the underlying random variable Y has a normal distribution then the random variable T has a Student's t-distribution. The distribution is named in honour of William S. Gosset who was appointed to a post in the Guinness brewery in Dublin in 1899 and made a major contribution to the development of applied statistics. He developed the t-test to deal with small samples used for quality control in brewing. He wrote under the pseudonym 'Student' because his company had a policy against work done for the company being made public. A sample of n values has $\nu = n - 1$ degrees of freedom. (The Greek letter ν is nu.) There is a separate t-distribution for each number of degrees of freedom 1, 2, 3, ...

As an example suppose that initially assembly of P87 modules took on average 48.0 minutes with standard deviation 3.7 minutes. At a later date the following sample of eight assembly times was obtained:

$$46 \quad 48 \quad 45 \quad 48 \quad 46 \quad 47 \quad 43 \quad 48.$$

We wish to evaluate the evidence from this sample of assembly times for a reduction in the population mean assembly time.

In order to perform the t-test of the null hypothesis $H_0: \mu = 48$ against the alternative hypothesis $H_1: \mu < 48$, first set up the data in a column named Assembly time. Use of **Stat > Basic Statistics > 1-Sample t...** is required with 'Assembly time' selected under

One-Sample T: Assembly time

Test of mu = 48 vs < 48

Variable	N	Mean	StDev	SE Mean	95% Upper Bound	T	P
Assembly time	8	46.375	1.768	0.625	47.559	-2.60	0.018

Panel 7.11   Session window output for *t*-test on assembly time data.

Samples in columns:, **Perform hypothesis test** checked, **Hypothesized mean:** 48 entered and **less than** specified via **Options...** under **Alternative:**. With such a small sample an **Individual value plot**, rather than a histogram or boxplot, is recommended under **Graphs....**. The Session window output is shown in Panel 7.11 and the graphical output in Figure 7.10.

The output follows the same pattern as for a *z*-test. A statement of the hypotheses of interest is followed by summary statistics for the sample. Finally, the confidence interval, test statistic and *P*-value are given. The data provide evidence via the *t*-test of a reduction in the population mean assembly time, the null hypothesis that the mean is 48 minutes being rejected at the 5% significance level (*P*-value = 0.018). A point estimate of the new population mean is 46.4 minutes and, with 95% confidence, it can be stated that the new population mean is at most 47.6 minutes. Note how, in Figure 7.10, the point representing the value of the mean specified in the null hypothesis does not lie on the line segment that represents the 95% confidence interval. This provides visual confirmation of the rejection of the null hypothesis at the 5% significance level.

The *t*-test requires the random variable of interest to be normally distributed. The normal probability plot in Figure 7.11 provides no evidence of nonnormality of the data (*P*-value = 0.237).

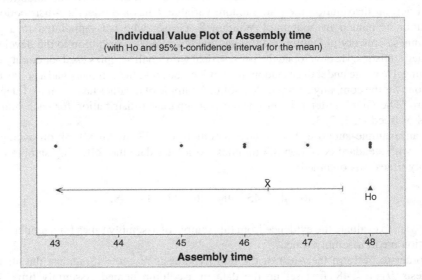

Figure 7.10   Individual value display of assembly time with *t*-test annotation.

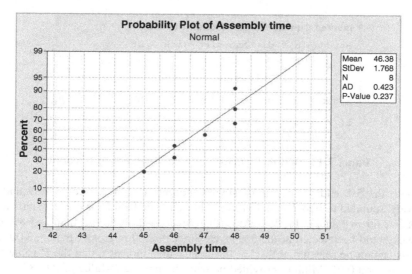

Figure 7.11 Probability plot of assembly time data.

If the normality test provided strong evidence of nonnormality of the distribution of the random variable of interest, then one possible approach would be to seek an appropriate transformation of the data. Box–Cox transformation may be explored via **Stat > Control Charts > Box-Cox Transformation....**. An alternative approach would be to perform a nonparametric test that will be introduced later in the chapter.

The probability density functions for the standard normal distribution, i.e. the N(0,1) distribution, and the t distribution with parameter $\nu = 7$ degrees of freedom are displayed in Figure 7.12. One of Gosset's major contributions was the creation of tables of critical values of to enable tests such as the above to be performed. In Panel 7.12 the critical values for the above

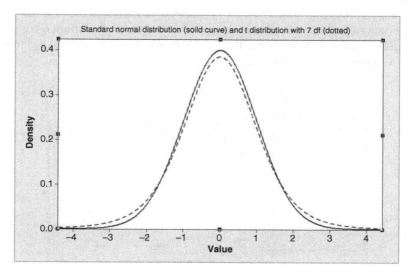

Figure 7.12 Probability density functions for the $N(0,1)$ and t_7 distributions.

```
Inverse Cumulative Distribution Function
Student's t distribution with 7 DF

P( X <= x )          x
         0.05   -1.89458

Student's t distribution with 7 DF

P( X <= x )          x
         0.01   -2.99795
```

Panel 7.12   Critical values of t with 7 degrees of freedom.

test at both the 5% and 1% levels of significance are displayed, i.e. -1.89 and -3.00 respectively, rounded to two decimal places. The calculated value of the test statistic, t, was given as -2.60 in Panel 7.11. It follows that, since -2.60 is less than -1.89, the null hypothesis would be rejected at the 5% level. Since -2.60 is greater than -3.00, the null hypothesis cannot be rejected at the 1% level. This also indicates that the P-value must lie between 0.05 and 0.01. Minitab provided us with the precise P-value of 0.018.

7.2.2.1   Power and sample size

Recall that, in the DTN example, the population mean time before the process change was 19 minutes with population standard deviation 6 minutes. Suppose that the project team decided that, were population mean time to decrease by 3 minutes due to the introduction of the specialist nurses, then they would like to be 95% certain to detect the decrease, using a significance level of $\alpha = 0.05$. Thus they would be specifying power 0.95 for the z-test to detect the change from population mean 19 to population mean 16. The question to be answered is therefore that of how big a sample should be taken. Minitab provides the answer via **Stat > Power and Sample Size > 1-Sample Z…**. The required dialog is shown in Figure 7.13.

The change of interest, from 19 to 16, is specified as **Differences:** -3; the population standard deviation is assumed to remain unchanged so **Standard deviation:** 6 is entered. The Session window output is shown in the Panel 7.13. In addition, a power curve similar to that displayed in Figure 7.2 is created by default.

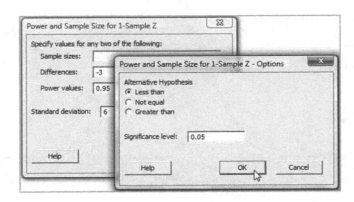

Figure 7.13   Dialog for sample size calculation.

```
Power and Sample Size

1-Sample Z Test

Testing mean = null (versus < null)
Calculating power for mean = null + difference
Alpha = 0.05   Assumed standard deviation = 6

            Sample  Target
Difference    Size   Power  Actual Power
        -3      44    0.95      0.952715
```

Panel 7.13 Session Window output for z-test on Burst Strength data.

Panel 7.13 shows that the required sample size is 44 patients. (Note the actual power is 0.953 – Minitab computes the lowest sample size that gives power greater than the target power specified by the user.) Minitab also provides power and sample size calculations for t-tests. An exercise will be provided.

7.2.3 Tests for proportions

In the lens coating example the nonconformance rate was 4.5% prior to the introduction of the new supplier. In a trial run with the coating fluid from the new supplier there were 80 nonconforming lenses in a batch of 2400. Here the test is of null hypothesis $H_0 : p = 0.045$ versus the alternative hypothesis $H_1 : p < 0.045$. To perform the test directly via Minitab, use **Stat > Basic Statistics > 1 Proportion...** with the dialog shown in Figure 7.14.

The Session window output is given in Panel 7.14. The P-value of 0.002 was obtained in Section 7.1 using grass-roots computation and the binomial distribution – see Panel 7.4. Thus the data provide evidence of a reduction in the proportion of nonconforming lenses at the 1% significance level since the P-value is less than 0.01. The estimated proportion nonconforming following the process change is 3.3%. The process owner can be 95% confident that the proportion nonconforming is at worst 4.0% (the upper bound expressed as a percentage and rounded to one decimal place) following the process change. Of course a decision on whether or not to continue with the new supplier would typically involve consideration of the costs involved.

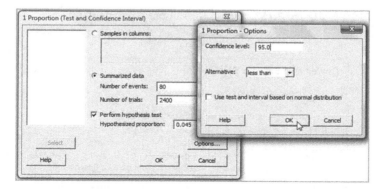

Figure 7.14 Dialog for testing a proportion.

```
Test and CI for One Proportion

Test of p = 0.045 vs p < 0.045


                                  95% Upper    Exact
Sample    X      N   Sample p        Bound    P-Value
1        80   2400   0.033333      0.040008     0.002
```

Panel 7.14 Session window output for test of proportion nonconforming.

Suppose that the process team had wished to detect a reduction in the proportion of nonconforming lenses from 4.5% to 3.0% with power 0.99, using a significance level of 0.01. Use of **Stat > Power and Sample Size > 1 Proportion...** with the dialog shown in Figure 7.15 provides the sample size required in the Session window output shown in Panel 7.15. In addition, a power curve similar to that displayed in Figure 7.3 is created by default. The size of sample required is 3435. Thus in order to ensure a probability of 0.99 of detecting a reduction in the proportion of nonconforming lenses from 4.5% to 3.0%, using a significance level of 0.01, the process team would require a sample size of the order of 3500.

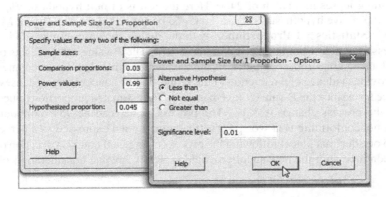

Figure 7.15 Dialog for sample size calculation.

```
Power and Sample Size

Test for One Proportion

Testing p = 0.045 (versus < 0.045)
Alpha = 0.01


                Sample   Target
Comparison p      Size    Power   Actual Power
        0.03      3435     0.99       0.990004
```

Panel 7.15 Session window output for sample size computation.

Table 7.4   Waiting time (seconds) data.

471	36	55	185	88
188	58	72	18	47
42	34	41	67	28

7.2.4   Nonparametric sign and Wilcoxon tests

Following alterations to a checkout facility at a supermarket, a random sample of customers was observed and the time they waited for service recorded. The data are displayed in Table 7.4 and available in Wait.MTW.

Prior to the alterations extensive monitoring had shown that the median waiting time was 120 seconds. For the symmetrical normal distribution the mean and median are identical, so were we to evaluate the evidence for a reduction in the median following the changes by performing a t-test of null hypothesis $H_0 : \mu = 120$ versus the alternative hypothesis $H_1 : \mu < 120$ then the output in Panel 7.16 would be obtained from Minitab.

With a P-value of 0.212 there would appear to be no evidence of a reduction in the mean, and by implication, in the median. However, a normal probability plot of the data provides strong evidence of the distribution of waiting time being nonnormal. As an alternative to seeking a suitable transformation of waiting time one may proceed as follows. Order the data and record a negative sign if an observed value is less than 120; record a positive sign if an observed value is greater than 120. (Any values of exactly 120 would be excluded from the analysis.) The results are shown in Table 7.5.

By definition, the probability that an observed value is less than the median of a population is 0.5; likewise, the probability that it is greater than the median of the population is 0.5. For the above data we have observed a value less than the hypothesized median of 120 for 12 observations from a total of 15 observations (corresponding to the 12 negative signs from a total of 15 signs). We therefore ask what is the probability of observing this pattern, or a more extreme one. This is equivalent to asking what is the probability of obtaining 12 or more tails from 15 tosses of a coin. Consider therefore the random variable X having the $B(15, 0.5)$ distribution. We require

```
One-Sample T: Wait

Test of mu = 120 vs < 120

                                          95%
                                         Upper
Variable    N      Mean       StDev   SE Mean    Bound       T       P
Wait       15   95.3333   115.9136   29.9288   148.0471   -0.82   0.212
```

Panel 7.16   Session window output for t-test on waiting time data.

Table 7.5   Coded data for waiting time.

Wait	18	28	34	36	41	42	47	55	58	67	72	88	185	188	471
Sign	−	−	−	−	−	−	−	−	−	−	−	−	+	+	+

```
Cumulative Distribution Function

Binomial with n = 15 and p = 0.5

  x   P( X <= x )
  11      0.982422
```

Panel 7.17   Calculation of probability of 11, or fewer, $+$ signs.

$P(X \geq 12) = 1 - P(X \leq 11)$. Use of **Calc** > **Probability Distributions** > **Binomial...** yields the result in Panel 7.17. Hence, $P(X \geq 12) = 1 - P(X \leq 11) = 1 - 0.982\,422 = 0.017\,578$. Since this probability is less than 0.05 we have evidence of a reduction in the median waiting time. This test of hypotheses has made no appeal to any particular probability distribution of the waiting time. It is known as the *sign test* and is an example of a distribution-free or nonparametric test.

Use of **Stat** > **Nonparametrics** > **1-Sample Sign...** enables the test to be performed directly. The completed dialog is shown in Figure 7.16. The Session window output is shown in Panel 7.18. The output includes a statement of the hypotheses under test together with counts of the numbers of observations in the sample that are less than, equal to and greater than the

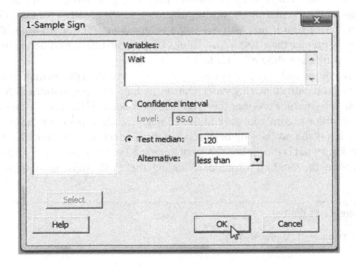

Figure 7.16   Dialog for sign test.

```
Sign Test for Median: Wait

Sign test of median =  120.0 versus  < 120.0

          N  Below  Equal  Above       P  Median
Wait     15     12      0      3  0.0176   55.00
```

Panel 7.18   Session window output for sign test.

Sign CI: Wait

Sign confidence interval for median

			Achieved	Confidence Interval		
Wait	N	Median	Confidence	Lower	Upper	Position
Wait	15	55.00	0.8815	41.00	72.00	5
			0.9500	**37.87**	**82.02**	**NLI**
			0.9648	36.00	88.00	4

Panel 7.19   Session window output giving confidence interval for median.

median specified in the null hypothesis. The *P*-value, computed as above using the binomial distribution, and the median of the sample are also given. In situations where observations are equal to the median specified in the null hypothesis these are discounted from the calculations of the *P*-value for the test.

The sign test procedure in Minitab offers as default the option to obtain a two-sided confidence interval for the median instead of performing a test of hypotheses. The Session window output for the waiting time data is shown in Panel 7.19. The key element is the 95% confidence interval (37.87, 82.02) for the population median waiting time following the alterations to the checkout facility. The fact that this interval does not include 120 formally provides evidence at the 5% level of significance of a change in the population median result of the alterations. (The reader will recall that 95% confidence intervals go hand in hand with 5% significance.)

Another nonparametric test that may be used in the type of scenario discussed in this chapter, where comparison is being made with a standard, is the one-sample Wilcoxon signed-rank test. With **Stat > Nonparametrics > 1-Sample Wilcoxon...** one can test hypotheses concerning the median or obtain the corresponding point estimate and confidence interval. As with the sign test no appeal is made to any particular underlying distribution for the random variable of interest, but the assumption that the data constitute a random sample from a continuous, symmetric distribution is required. The reader is invited to verify that this test yields a *P*-value of 0.035 so the conclusion, at the 5% level of significance, is the same as that from the sign test. The median of the sample of 15 values of waiting time is 55 seconds. The output from the Wilcoxon test includes an estimate of 58.75 for the population median. As with the sign test, the default output from the Wilcoxon procedure is computation of a two-sided 95% confidence interval for the population median.

In all the scenarios discussed in this section we have been concerned with processes as depicted in Figure 1.3. Typically there has been a well-established current parameter for a process performance measure, *Y*, of interest. The performance measure has been usually either a mean or a proportion. The author has chosen to refer to the current level of performance as a standard in the heading for section 7.2. The methods introduced are useful for assessing the impact, if any, of a change to an input, *X*, or factor on *Y*. In the next section we will look at techniques which can be applied in situations where two choices are available for the levels of the factor, e.g. where we wish to compare two processes for dealing with the administration of thrombolytic drugs in a hospital accident and emergency department or to compare two potential suppliers of lens coating fluid where there has are no well-established current parameters for the process performance measure, *Y*, of interest.

7.3   Tests and confidence intervals for the comparison of two means or two proportions

7.3.1   Two-sample t-tests

An assembly operation requires a 6-week training period for a new employee to reach maximum efficiency. A new method of training was proposed and an experiment was carried out to compare the new method with the standard method. A group of 18 new employees was split into two groups at random. Each group was trained for 6 weeks, one group using the standard method and the other the new method. The time (in minutes) required for each employee to assemble a device was recorded at the end of the training period. Here the X is the training method and the Y is the assembly time. The two levels of the factor training method are 'standard' and 'new'.

If the data can be considered as independent random samples from normal distributions with means μ_1 and μ_2 with common variance σ^2, then a two-sample t-test is available via **Stat > Basic Statistics > 2-Sample t...**. The completed dialog is shown in Figure 7.17. The data are available in Assembly.MTW.

Training method is indicated in the text column named Method, with entries New and Standard. Minitab treats these identifying labels, New and Standard, in alphabetical order so that it takes μ_1 to refer to the new method and μ_2 to refer to the standard method of training. The null hypothesis is $H_0 : \mu_1 = \mu_2$, i.e. that there is no difference in mean assembly time for employees trained by the new and the standard method. The alternative hypothesis is $H_1 : \mu_1 < \mu_2$, since it was of interest to determine whether or not there was evidence that new training method led to a reduction in the mean assembly time. Thus under **Options...** less than has to be selected as **Alternative:. Assume equal variances** was checked. One should always choose one of the display options under **Graphs.... Boxplots of data** was selected in

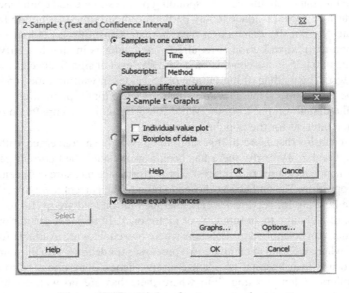

Figure 7.17   Dialog for two-sample t-test.

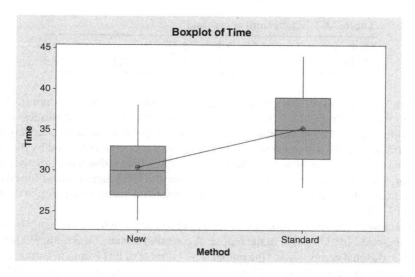

Figure 7.18 Boxplots of assembly time data.

this case. The graphical output is shown in Figure 7.18. The Session window output is shown in Panel 7.20.

The box sections of the boxplots are of similar length, indicating that the assumption of equal variances for the two populations is reasonable. Note that the sample means are also displayed and connected by a line segment. Normal probability plots of the two samples provide no evidence of nonnormality. Thus a two-sample t-test, with the assumption of equal variances, would appear to be a sound method of analysis.

The Session window output gives the summary statistics sample size, mean, standard deviation and standard error of the mean for each sample. The null and alternative hypotheses,

$$H_0 : \mu_1 = \mu_2 \quad \text{and} \quad H_1 : \mu_1 < \mu_2,$$

may be written in terms of the *difference* between the population means as

$$H_0 : \mu_1 - \mu_2 = 0 \quad \text{and} \quad H_1 : \mu_1 - \mu_2 < 0.$$

```
Two-Sample T-Test and CI: Time, Method

Two-sample T for Time

Method     N    Mean   StDev   SE Mean
New        9   30.33    4.15      1.4
Standard   9   35.22    4.94      1.6


Difference = mu (New) - mu (Standard)
Estimate for difference:   -4.89
95% upper bound for difference:   -1.13
T-Test of difference = 0 (vs <): T-Value = -2.27   P-Value = 0.019   DF = 16
Both use Pooled StDev = 4.5659
```

Panel 7.20 Session window output for two-sample t-test.

Table 7.6 Magnesium assay data.

Method A	3.6	3.5	3.4	3.5				
Method B	2.6	2.9	3.5	2.7	3.8	3.2	2.8	3.3

Minitab states the hypotheses in this format in the Session window output in the rather cryptic shorthand 'difference $= 0$ (vs $<$)'. The P-value is 0.019 so, since this is less than 0.05, there is evidence, at the 5% level of significance, that the population mean assembly time for operators trained by the new method is lower than that for operators trained by the standard method. The statement 'Estimate for difference: -4.89' indicates that the estimated reduction in the mean assembly time is 4.89 minutes. The statement '95% upper bound for difference: -1.13' indicates that, with 95% confidence, it may be stated that the reduction in the mean is at least 1.13 minutes.

Each sample of nine observations has 8 degrees of freedom, yielding a total of 16 degrees of freedom, indicated by 'DF $= 16$' in the output. A common variance was assumed for the two populations; the final component of the output is an estimate of this common variance. Further detail may be found in Montgomery (2009, pp. 132–134) or Hogg and Ledolter (1992, p. 236).

As a second example consider the data in Table 7.6 on determinations of the percentage of magnesium in a batch of ore by two chemical assay procedures. Performance of a two-sample t-test of the hypotheses $H_0 : \mu_A = \mu_B$ and $H_1 : \mu_A \neq \mu_B$, with equal variances for the two populations of determinations assumed, yields the Session window output shown in Panel 7.21. The data are available in Magnesium.MTW in two separate columns named A and B. In this case, as the data appear in separate columns, the **Samples in different columns** option is required with **First:** A and **Second:** B specified. **Assume equal variances** was checked. **Individual value plot** was selected under **Graphs. . ..**

The null hypothesis $H_0 : \mu_A = \mu_B$ may be stated in the form $H_0 : \mu_A - \mu_B = 0$, i.e. that the difference in the population means is zero. The alternative hypothesis $H_1 : \mu_A \neq \mu_B$ may be stated in the form $H_1 : \mu_A - \mu_B \neq 0$, i.e. that the difference in the population means is nonzero. In the Session window output the hypotheses are indicated by 'difference $= 0$ (vs not $=$)'. The 95% confidence interval $(-0.084, 0.884)$ for $\mu_A - \mu_B$ includes the value 0, which indicates that the null hypothesis cannot be rejected at the 5% significance level. This conclusion is confirmed by the P-value of 0.096 being in excess of 0.05. This analysis suggests that both methods of determining the magnesium content of the ore would yield the same mean value from many repeated assays.

```
Two-Sample T-Test and CI: A, B

Two-sample T for A vs B

    N     Mean    StDev   SE Mean
A   4    3.5000   0.0816    0.041
B   8    3.100    0.421     0.15


Difference = mu (A) - mu (B)
Estimate for difference:   0.400
95% CI for difference:   (-0.084, 0.884)
T-Test of difference = 0 (vs not =): T-Value = 1.84   P-Value = 0.096   DF = 10
```

Panel 7.21 Session window output for two-sample t-test – equal variances assumed.

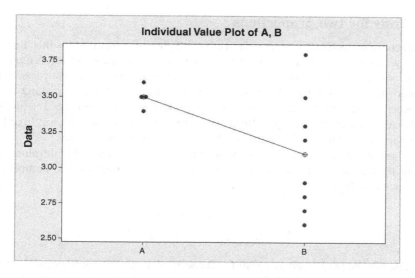

Figure 7.19   Dotplots of magnesium assay data.

Although normal probability plots provide no evidence of nonnormality, the individual value plots of the data displayed in Figure 7.19 cast doubt on the assumption of equal variances, the spread of the method A data being much greater than that of the method B data. Performing the test again, with **Assume equal variances** unchecked, yields the Session window output in Panel 7.22. This version of the test provides evidence, at the 5% level of significance, that the population means differ for the two methods of assay. Later in the chapter we will look at tests of hypotheses concerning variances.

In the examples considered the null hypothesis has been that the two population means are equal, i.e. that the difference between the population means is 0. It is possible to test the null hypothesis that the difference between the population means is some value other than 0. For example, if it is claimed that the use of high octane fuel would improve the fuel consumption of a type of vehicle by 5 mpg on average over that obtained with regular octane fuel then the null hypothesis would be

$$H_0 : \mu_{\text{High}} = \mu_{\text{Low}} + 5, \quad \text{i.e. } \mu_{\text{High}} - \mu_{\text{Low}} = 5.$$

```
Two-Sample T-Test and CI: A, B

Two-sample T for A vs B

      N     Mean    StDev   SE Mean
A     4   3.5000   0.0816     0.041
B     8    3.100    0.421      0.15


Difference = mu (A) - mu (B)
Estimate for difference:   0.400000
95% CI for difference:   (0.035131, 0.764869)
T-Test of difference = 0 (vs not =): T-Value = 2.59  P-Value = 0.036  DF = 7
```

Panel 7.22   Session window output for two-sample *t*-test – equal variances not assumed.

7.3.2 Tests for two proportions

A manufacturer of laptop computers claims that a higher proportion of his machines will be operating without any hardware faults after 1 year than those of a competitor. A multinational company which had purchased a large number of machines from both manufacturers established that, of a sample of 200 machines from the manufacturer making the claim, 13 had experienced hardware faults during the first year while, of a sample of 150 produced by the rival manufacturer, 19 had experienced hardware faults during the first year. In order to put the manufacturer's claim to the test formally one can proceed as follows.

Let p_1 represent the proportion of the manufacturer's machines that develop hardware faults during the first year and let p_2 represent the proportion for the competitor. Our hypotheses are as follows:

$$H_0 : p_1 = p_2 \quad \text{or} \quad p_1 - p_2 = 0,$$
$$H_1 : p_1 < p_2 \quad \text{or} \quad p_1 - p_2 < 0.$$

The test may be performed using **Stat > Basic Statistics > 2 Proportions. . . .** The dialog required is shown in Figure 7.20. It is recommended that the pooled estimate of a common proportion be used for the test (Montgomery, 2009, p. 139; Hogg and Ledolter, 1992, p. 242). Thus **Use pooled estimate of p for test** should be checked under **Options. . .**, together with **Alternative: less than**.

The Session Window output is shown in Panel 7.23. A summary of the data provided is given indicating that, to two decimal places, 6.5% of the manufacturer's machines developed hardware faults within a year while 12.7 % of the competitor's machines developed hardware faults within a year. The P-value of 0.024 provides evidence, at the 5% level of significance, that the manufacturer's claim is true. The point estimate of the difference is 6.1% fewer machines developing hardware faults within a year for the manufacturer compared with the competitor. The confidence interval for the difference indicates that at least 0.9% fewer of the manufacturer's machines developed hardware faults within a year (after rounding –0.008 587 15 to three significant figures, converting to a percentage and

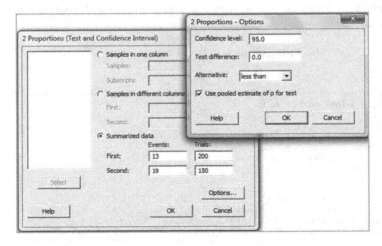

Figure 7.20   Dialog for hypothesis test of two proportions.

```
Test and CI for Two Proportions

Sample   X     N   Sample p
1       13   200   0.065000
2       19   150   0.126667


Difference = p (1) - p (2)
Estimate for difference:  -0.0616667
95% upper bound for difference:  -0.00858715
Test for difference = 0 (vs < 0):  Z = -1.98  P-Value = 0.024

Fisher's exact test: P-Value = 0.037
```

Panel 7.23   Session window output for test of proportions.

interpreting the negative to imply fewer). The z-value quoted is the test statistic used. The P-value of 0.037 results from application of the alternative test of the hypotheses provided by Fisher's exact test and leads to the same conclusion.

7.3.2.1 Power and sample size

Minitab enables power and sample size calculations to be performed for the two-sample t-test and for tests concerning two proportions. Suppose that yields from a batch chemical process are known to be normally distributed and to vary with a standard deviation of the order of 5 kg under a wide variety of operating conditions. Suppose also that discussions with the process team reveal that a switch to a new catalyst would be viable from an economic point of view if the mean yield per batch were to increase by 4 kg. Using **Stat > Power and Sample Size > 2-Sample t-test...**, once can determine the sample size required to perform a two-sample t-test of the null hypothesis $H_0: \mu_{Standard} = \mu_{New}$, i.e. $\mu_{Standard} - \mu_{New} = 0$, at significance level $\alpha = 0.05$ and with power $= 0.9$. The dialog is shown in Figure 7.21. Note that the difference is specified as -4 and that the alternative hypothesis is $H_1: \mu_{Standard} < \mu_{New}$. i.e. $\mu_{Standard} - \mu_{New} < 0$, that **Less than** is checked.

The Session window output is shown in Panel 7.24. The calculated sample size, for each group, is 28. Thus a sample of 28 yields from the process run with the standard catalyst and a sample of 28 yields from the process run with the new catalyst would be required.

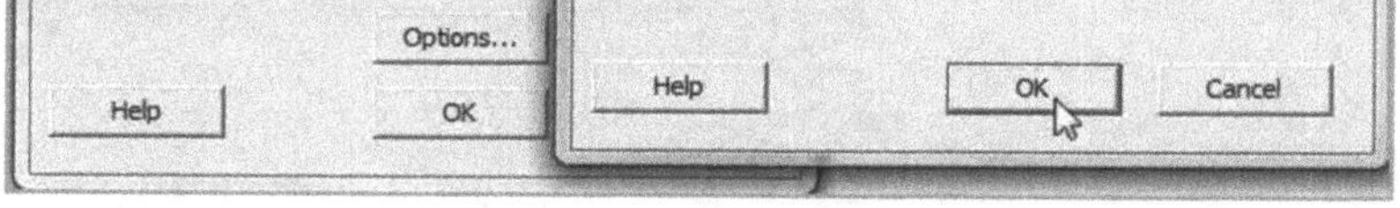

Figure 7.21   Dialog for sample size calculation.

```
Power and Sample Size

2-Sample t Test

Testing mean 1 = mean 2 (versus <)
Calculating power for mean 1 = mean 2 + difference
Alpha = 0.05   Assumed standard deviation = 5

                Sample  Target
Difference       Size   Power   Actual Power
      -4          28     0.9       0.905010
```

Panel 7.24   Sample size calculation for two-sample *t*-test.

Suppose that a procurement manager wishes to ascertain whether there is evidence that the proportion of nonconforming items from supplier A is 2% lower than that obtained from supplier B, for whom records indicate that about 8% of items are nonconforming. Suppose that random samples of 500 items from each supplier were to be checked. The manager would like to know the power of a test performed at the 5% level of significance, based on these sample sizes, to detect superior performance by supplier A, by 2%, in the proportion of nonconforming items. Use of **Stat > Power and Sample Size > 2 Proportions...** provides the answer. The dialog is displayed in Figure 7.22.

The Session window output is shown in Panel 7.25. It gives the power of the test to detect superiority of supplier A, by 2%, as 0.34. This means that the probability of committing a Type II error is $1 - 0.34 = 0.66$, which indicates that if supplier A is actually operating with a nonconformance rate of 6%, as compared with 8% for supplier B, then there is probability of approximately 0.66 that the test would fail to provide evidence of the difference. If you decide that you would like the power to be 0.9 then, by specifying this value in the dialog shown in Figure 7.22 and clearing the box **Sample sizes:**, the procedure returns a sample size of 2786. The reader is invited to verify this as an exercise. People involved in quality improvement often fail to realize the size of sample required to formally detect changes or differences of a magnitude that is of practical significance to their organization.

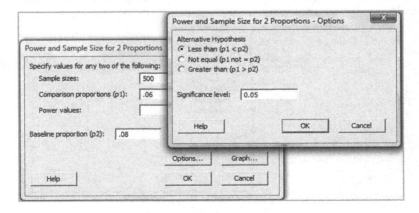

Figure 7.22   Dialog for power calculation.

```
Power and Sample Size

Test for Two Proportions

Testing proportion 1 = proportion 2 (versus <)
Calculating power for proportion 2 = 0.08
Alpha = 0.05

                Sample
Proportion 1     Size      Power
        0.06      500    0.342455

The sample size is for each group.
```

Panel 7.25   Sample size calculation for test of two proportions.

7.3.3 Nonparametric Mann–Whitney test

This nonparametric test provides an alternative to the two-sample t-test. It is based on allocating ranks to the combined data from both samples. It is also referred to as the two-sample rank test or the two-sample Wilcoxon rank-sum test. The null hypothesis is that the two population medians are equal. The assumptions required are that the data are independent random samples from two distributions that have the same shape.

Consider the data in Table 7.7 on the length of drive (in metres) achieved on striking golf balls of two different types with a mechanical club device used by a golf manufacturer for product testing purposes. The manufacturer's quality manager wishes to know if the data provide evidence that the median length of drive achieved with type A is less than that achieved with type B.

The test involves ranking the combined data is shown in Table 7.8. In this case the actual observed rank sum for type A is $W = 1 + 2 + 3 + 4 + 8 = 18$. Had all the type A distances been less than all the type B distances then the rank sum for type A would have been

Table 7.7   Distance driven (m) for two types of golf ball.

Type A	181	183	176	221	180
Type B	215	197	229	222	195

Table 7.8   Ranked data for distance.

Distance	Type	Rank
176	Type A	1
180	Type A	2
181	Type A	3
183	Type A	4
195	Type B	5
197	Type B	6
215	Type B	7
221	Type A	8
222	Type B	9
229	Type B	10

$W = 1 + 2 + 3 + 4 + 5 = 15$. In this case one might feel that there was no need for a formal test of hypotheses! The reader might wish to take a few moments to convince him/herself that the possible ranks for type A that would give rise to a rank sum of 18 or less are as follows:

$$W = 1 + 2 + 3 + 4 + 5 = 15,$$
$$W = 1 + 2 + 3 + 4 + 6 = 16,$$
$$W = 1 + 2 + 3 + 4 + 7 = 17,$$
$$W = 1 + 2 + 3 + 5 + 6 = 17,$$
$$W = 1 + 2 + 4 + 5 + 6 = 18,$$
$$W = 1 + 2 + 3 + 4 + 8 = 18,$$
$$W = 1 + 2 + 3 + 5 + 7 = 18.$$

Thus there are seven possible rankings for type A yielding a rank sum of 18 or less. Combinatorial mathematics indicates that there are 252 possible rankings for type A that can arise when two samples of size 5 are tested. Were the null hypothesis true then each ranking for type A would have equal probability of occurring in the experiment and $P(W \leq 18) = 7/252 = 0.0278$. Since this probability is less than 0.05, the null hypothesis that the medians are equal would be rejected in favour of the alternative hypothesis that the median for type A is less than that for type B. (The Greek letter η (eta) may be used to denote a population median.)

With the data for each type in separate columns in a Minitab worksheet the test may be performed using **Stat > Nonparametrics > Mann-Whitney...** with **Alternative: less than** selected. The Session window output is shown in Panel 7.26. It begins by giving the sample sizes and sample medians by way of data summary. Next follows the point estimate for the difference in the population medians – note that this is not the difference between the sample medians. (Interested readers will find details of the estimation procedure employed in Minitab via the Help facility.) Even though a one-sided alternative hypothesis was specified here, Minitab gives an approximate two-sided 96.3% confidence interval. (Since it includes zero we have an indication that the null hypothesis of equal population medians would not be rejected in favour of the alternative hypothesis of unequal population medians at the $(100 - 96.3)\% = 3.7\%$ significance level.) Then the rank sum $W = 18$ found earlier is stated. (If there are two or more equal values in the combined data set then the mean of the associated ranks is allocated to these equal values. Thus noninteger values of W may occur.) Finally, the null and alternative hypotheses are stated and the P-value is given as 0.0301 rather than 0.0278,

```
Mann-Whitney Test and CI: A, B
    N   Median
A   5   181.00
B   5   215.00


Point estimate for ETA1-ETA2 is -21.00
96.3 Percent CI for ETA1-ETA2 is (-48.00,5.98)
W = 18.0
Test of ETA1 = ETA2 vs ETA1 < ETA2 is significant at 0.0301
```

Panel 7.26   Session Window output for Mann-Whitney test.

the value calculated above. This is because Minitab uses an approximate method, based on the normal distribution, to compute the probability, and not because the author has made an error!

In discussing the Mann–Whitney test Daly *et al.* (1995, pp. 372–373) write:

> The idea of using ranks instead of the data values is an appealing one. Furthermore, it has an obvious extension to testing two groups of data when a two-sample *t*-test may not be applicable because of lack of normality. The test itself was first proposed by H.B. Mann and D.R. Whitney in 1947, and modified by Wilcoxon; it turns out to be very nearly as powerful as the two-sample *t*-test, which tests for equal means. However it is nevertheless a test of the equality of locations of the two groups and using it as an alternative to the two-sample *t*-test is an approximation often made in practice.

7.4 The analysis of paired data – *t*-tests and sign tests

A finance company gave a group of employees a test before and after a refresher course on tax legislation. The scores obtained are displayed in Table 7.9 and are available in the worksheet TaxTest.MTW.

In order to evaluate the evidence for an improvement in the knowledge of the employees there are two approaches. The first approach is to form the differences obtained by subtracting, for each employee, the score obtained before the course from the score obtained after the course. If the differences may reasonably be regarded as a random sample from the normal distribution $N(\mu, \sigma^2)$ then the evidence may be evaluated by using a one-sample *t*-test of the null hypothesis $H_0 : \mu = 0$ versus the alternative hypothesis $H_1 : \mu > 0$. Given the two columns of before and after scores, **Calc > Calculator** may be used to form the differences and subsequently **Stat > Basic Statistics > 1-Sample t ...** may be used to perform the *t*-test. However, Minitab provides **Stat > Basic Statistics > Paired t ...** to enable the test to be performed via a single dialog as displayed in Figure 7.23.

Table 7.9 Test scores before and after refresher course.

Employee number	Score before	Score after	Difference x	Sign
1	48	58	10	+
2	87	91	4	+
3	82	81	−1	−
4	44	55	11	+
5	56	60	4	+
6	71	68	−3	−
7	60	66	6	+
8	66	82	16	+
9	84	89	5	+
10	48	55	7	+
11	63	73	10	+
12	48	49	1	+

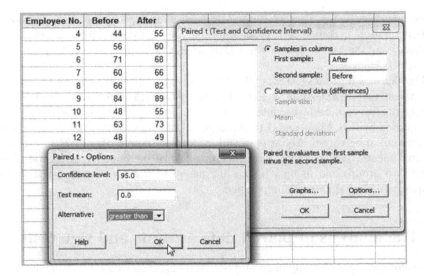

Figure 7.23 Dialog for paired *t*-test.

It is important to note the statement in the main dialog box: **Paired t evaluates the first sample minus the second sample**. Thus care has to be taken in specifying which column is deemed to contain the first sample data and which deemed to contain the second sample data and in specifying the hypothesis of interest in relation to that choice. It could be argued that the natural thing to do would be to specify the pre-course scores as the first sample; the reason the author chose the reverse was that positive differences then correspond to improvement. The alternative hypothesis is specified in the usual way using the **Options...** subdialog box and, as ever, creation of some form of display of the data using **Graphs...** is recommended. Here an individual value plot was selected. (The author edited the symbols so that an employee whose score after was higher that his/her score before is represented by an upward pointing triangle and so that an employee whose score after was lower that his/her score before is represented by an downward pointing triangle. To edit all graph symbols click on a symbol, pause and double-click to access the **Edit Individual Symbols** menu. To edit a single graph symbol click on it, pause, click again, pause and double-click to access the **Edit Individual Symbols** menu.)

The individual value plot of the differences in Figure 7.24 indicates that the scores increased for 10 of the 12 employees but decreased for the remaining two. The fact that the value of 0, specified under the null hypothesis H_0, lies outwith the line segment representing the one-sided 95% confidence interval for the mean of the population of differences indicates that the null hypothesis would be rejected in favour of the alternative at the 5% level of significance.

The Session Window output is shown in Panel 7.27. Summary statistics are given for the two sets of scores and for the differences. The *P*-value of 0.002 indicates that the null hypothesis would be rejected in favour of the alternative at the 1% level of significance. Rounded to the nearest whole number, the point estimate of the mean increase in score is 6 and, with 95% confidence, it can be stated that the mean increase in score is at least 3 points. A normal probability plot of the difference data provides no evidence of nonnormality, so the *t*-test is an appropriate method of analysis.

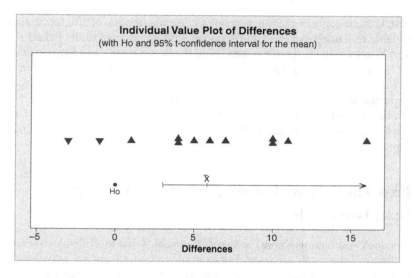

Figure 7.24   Display of differences with *t*-test annotation.

```
Paired T-Test and CI: After, Before

Paired T for After - Before

              N    Mean   StDev   SE Mean
After        12   68.92   14.20      4.10
Before       12   63.08   15.21      4.39
Difference   12    5.83    5.41      1.56


95% lower bound for mean difference: 3.03
T-Test of mean difference = 0 (vs > 0): T-Value = 3.74  P-Value = 0.002
```

Panel 7.27   Session window output for paired *t*-test.

Alternatively, the nonparametric sign test may be used to test the null hypothesis that the median of the population of differences is 0 against the alternative hypothesis that the median is greater than 0. The Session window output from this test is shown in Panel 7.28. With *P*-value 0.0193 the null hypothesis that the median population difference is 0 would be rejected in favour of the alternative hypothesis that the median is greater than 0 at the 5% level of significance. Thus the sign test also provides evidence that the refresher course has improved the employees' knowledge of the tax legislation. The sign test is a less powerful test than the

```
Sign Test for Median: Difference

Sign test of median =  0.00000 versus > 0.00000

                 N  Below  Equal  Above       P  Median
Difference      12      2      0     10  0.0193   5.500
```

Panel 7.28   Session window output for sign test.

paired *t*-test. If one is concerned about normality of the distribution of differences then the sign test is available as a nonparametric, but less powerful, alternative to the paired *t*-test.

Had one erroneously analysed the data using the two-sample *t*-test then no evidence of a significant impact of the refresher course would have been found. This emphasizes the fact that the two sets of scores do not constitute independent random samples from two normal populations and also the need for care in the selection of methods for the analysis of data. Paired experiments of the type discussed here are a special case of the use of blocking in the design of experiments. This powerful technique will be discussed in detail later in the chapter.

7.5   Experiments with a single factor having more than two levels

We will now look at situations where the factor of interest, X, has more than two levels. In some cases the effects of the factor are *fixed*. For example, suppose there are only three adhesives available on the market that may be used to bond components to a substrate in the fabrication of a particular type of electronic circuit. For an experiment in which bond strength, Y, was measured for 10 components bonded to substrate with each available adhesive, the factor adhesive would be said to be fixed. In some cases the effects of the factor are said to be *random*. Were there many adhesives available then, for an experiment in which bond strength, Y, was measured for 10 components bonded to substrate with each adhesive from a sample of three adhesives, selected at random from the available adhesives, the factor adhesive would be said to be random.

In analysing data from fixed effect scenarios, interest centres on testing hypotheses concerning means and making comparisons between means. Thus in the case of there being only three adhesives, all involved in the experiment, the questions being addressed would be:

- Is there evidence that the population mean bond strengths differ for the three available adhesives?

- If the answer to the first question is an affirmative, then what is the extent of the differences?

In analysing data from random effects scenarios, interest centres on partitioning the variation observed into components. Thus in the case of there being a random sample of three adhesives involved in the experiment, the questions being addressed would be:

- How much of the variation observed is attributable to real differences between means in the population of adhesives from which the three used in the experiment were selected?

- How much of the variation observed is attributable to random variation about these population means?

7.5.1   Design and analysis of a single-factor experiment

In order to introduce key concepts and techniques data for a fictitious golfer, Lynx Irons, will be used. Lynx is interested in improving the process of driving a golf ball from the tee at holes where she needs to use a driving club in order to achieve maximum distance. She wishes to determine the effect of ball type on the length of her drives. She will hit a number of drives with each of a *fixed* set of ball types she is prepared to use – Exoset, Flyer and Gutty (E, F and G for short).

Coleman *et al.* (1996, pp. 137–141) refer to requirements of good experimentation including:

- reliable measurement,

- randomization,

- replication.

Let us assume that we can reliably measure the response, Y, of interest – the length of drive in metres. Suppose that it has been decided to incorporate replication by having Lynx hit five balls of each type. (Were she to hit only a single ball of each type there is a risk that the distance achieved with one particular type might be atypically low and lead to failure to identify an opportunity for process improvement.) Finally, let us assume that all balls of a particular type are absolutely identical – unrealistic, of course, but necessary to make the example tractable. To achieve randomization the 15 balls could be put into a bag, given a thorough mix and a ball selected in turn for each drive. Were Lynx asked to hit all five balls of type E first, then all five of type F second and finally all five of type G then, for example, fatigue might lead to lower drive length for type G than might otherwise be obtained.

Minitab may also be used to carry out the randomization. Having decided to hit five balls of each type, this can be achieved by setting up columns as shown in the background in Figure 7.25. **Calc > Make Patterned Data > Simple Set of Numbers...** and **Calc > Make**

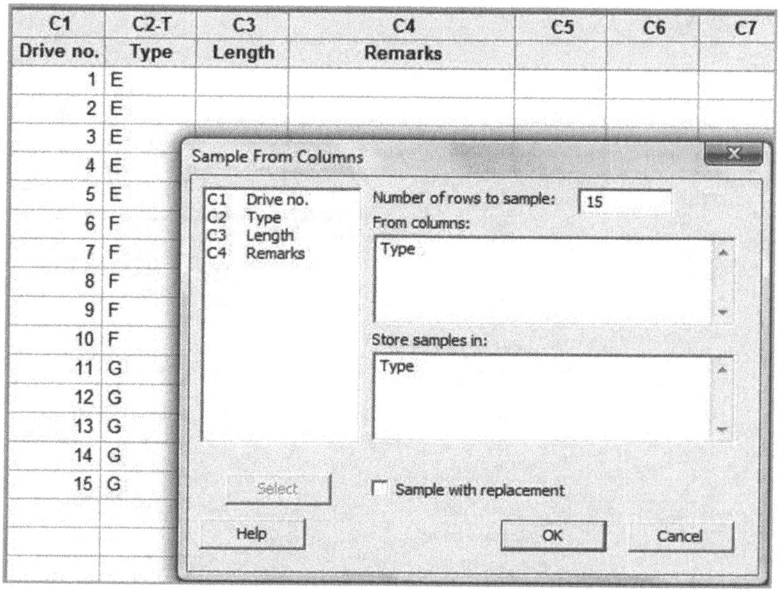

Figure 7.25 Initial worksheet for golf ball experiment with dialog for randomization.

Patterned Data > Text Values... may be used to set up the first two. (It is a simple matter to make the required entries via the keyboard, but experience of using the facilities for creating patterned data is worth having. It should be noted that in creating a column named Drive no. the entry 'Drive no.' is required in the **Store patterned data in:** window.) The third will be use to record the length of each drive. The final column may be used to note any unusual occurrences during the conduct of the experiment or any information that might be relevant.

To achieve randomization, use may be made of **Calc > Random Data > Sample From Columns...** via the dialog shown in Figure 7.25. By sampling, without replacement, 15 values at random from the entries in column 2 (**Sample with replacement** must not be checked) and using the same column to store the results, the original entries in column 2 are rearranged into random order. (If the reader tries this for her/himself it is unlikely, but possible, that the same sequence will be obtained as that obtained by the author in Figure 7.26!)

The resulting worksheet may then be stored in a project file and also printed off as a pro forma for the recording of the length of the 15 drives at the golf range where the experiment is to be performed. The data are displayed in Figure 7.26 and are available in Types.MTW.

We could analyse these data formally using three two-sample *t*-tests - one to compare E with F, one to compare F with G, and a final one to compare G with E.

Had Lynx wished to investigate seven ball types this approach would have required 21 two-sample *t*-tests in total. Apart from the tedium, there is a problem with this approach. When employing a 5% significance level there is a 5%, or 1 in 20, probability of a Type I error, i.e. of rejecting a null hypothesis when in fact that hypothesis is true. Thus with seven ball types, which in reality have identical mean distances for Lynx, we would expect the *t*-test approach to throw up spurious evidence of a significant difference between two of the ball types. We are now going to look at a *single* analysis that will seek evidence from the data of a real difference between ball types as far as mean length for Lynx is concerned. The technique is *analysis of variance* (ANOVA).

Figure 7.26   Data from golf ball experiment and dialog for ANOVA.

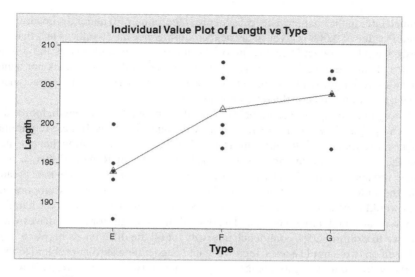

Figure 7.27   Individual value plot of length by type.

The null hypothesis is that the population means are equal and the alternative is that not all the means are identical:

$$H_0 : \mu_E = \mu_F = \mu_G,$$
$$H_1 : \text{Not all } \mu\text{s are identical.}$$

The ANOVA can be performed using **Stat > ANOVA > One-Way…**, the 'one-way' indicating that there is only a single factor of interest in this experiment. The dialog required is also displayed in Figure 7.26. The **Response:** (Y) is length and the **Factor:** (X) is type of ball. **Store residuals** and **Store fits** have both been checked. The default **Confidence level:** of 95% has been accepted. In order to display the data **Individual value plot** was selected under **Graphs…** together with **Four in one** for the **Residual plots**. Finally under **Comparisons… Tukey's, family error rate:** 5 (the default 5%), was selected. The output will now be discussed step by step.

The individual value plot of the data is shown in Figure 7.27. The plot suggests that ball types F and G are on a par (no pun intended!) as far as length performance for Lynx Irons is concerned, while E is inferior to both. The triangular symbols denote the mean lengths of the five drives with each ball type. The author edited the crossed circle symbols obtained by default. The mean value for both E and G equals an observed value.

The Session window output enables a formal test of the null hypothesis of equal population means to be tested against the alternative specified above. The relevant section of the output is presented in Panel 7.29. This section of the output is the ANOVA table. Its construction will

```
One-way ANOVA: Length versus Type

Source   DF     SS     MS      F      P
Type      2   280.0  140.0   7.30  0.008
Error    12   230.0   19.2
Total    14   510.0
```

Panel 7.29   ANOVA table for golf ball experiment.

be explained later in the chapter. The test statistic is the value 7.30 under the heading F and the associated P-value of 0.008 is given in the next column. Since the P-value is less than 0.01 null hypothesis would be rejected in favour of the alternative hypothesis at the 1% level of significance. Thus the experiment provides strong evidence that Lynx does not achieve the same mean drive length with the three types of golf ball. In view of the appearance of the individual value plot, this conclusion is not surprising.

The theory underlying the test requires that the three populations of length, for the three ball types driven by Lynx, have normal distributions with equal variances. If these requirements are met and the null hypothesis is true then the test statistic has an F-distribution that may be used to compute the P-value. The distribution is named in honour of Ronald Fisher, a pioneer in developing the application of statistical methods to experimentation. Since the spreads of the points in the individual value plot are similar for each ball type, the assumption of equal variances would appear to be a reasonable one. Some descriptive statistics for length and for length by type are shown in Panel 7.30. The overall mean length for all 15 drives was 200 m.

Were we to examine 95% confidence intervals for all the differences between population means, based on two-sample t-tests, then we would encounter a similar problem to that which would arise were we to compare means using a series of two-sample t-tests. If we require a statement with 95% confidence for *all* possible differences between population means then Tukey's multiple comparison provides this. The corresponding section of the Session window output is shown in Panel 7.31. Thus with overall 95% confidence we can state the following:

- The population mean length achieved with type F is greater than that achieved with type E by 8 m, with confidence interval (0.6, 15.4)

- The population mean length achieved with type G is greater than that achieved with type E by 10 m, with confidence interval (2.6, 17.4)

- The population mean length achieved with type G is greater than that achieved with type F by 2 m, with confidence interval (−5.4, 9.4).

The first two confidence intervals do not include 0, while the third does. Thus, with overall 95% confidence, we can state that there is evidence from the experiment that F is superior to E and that G is superior to E, as far as the mean length of drive for Lynx is concerned. (We can make this statement since the corresponding confidence intervals do not include the value 0 and the intervals cover ranges of positive values.) There is no evidence of any difference between F and G in this respect. (The corresponding confidence interval includes 0.) Thus we use ANOVA to

```
Descriptive Statistics: Length

Variable     Mean   StDev
Length      200.00   6.04


Descriptive Statistics: Length

Variable  Type     Mean   StDev
Length     E      194.00   4.30
           F      202.00   4.74
           G      204.00   4.06
```

Panel 7.30   Descriptive statistics for length and for length by ball type.

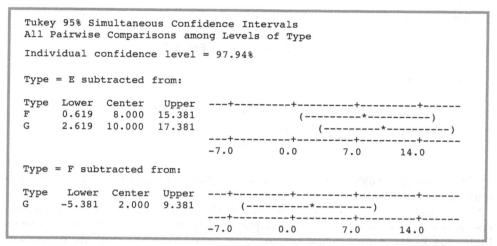

```
Tukey 95% Simultaneous Confidence Intervals
All Pairwise Comparisons among Levels of Type

Individual confidence level = 97.94%

Type = E subtracted from:

Type   Lower   Center   Upper    ---+---------+---------+---------+------
F      0.619   8.000    15.381                (---------*----------)
G      2.619   10.000   17.381                   (---------*----------)
                                  ---+---------+---------+---------+------
                                  -7.0       0.0       7.0      14.0

Type = F subtracted from:

Type   Lower   Center   Upper    ---+---------+---------+---------+------
G      -5.381  2.000    9.381         (----------*----------)
                                  ---+---------+---------+---------+------
                                  -7.0       0.0       7.0      14.0
```

Panel 7.31   Session window output for Multiple Comparison procedure.

look for evidence that the factor of interest (type of ball) influences the response (length). If such evidence is found then the follow-up using multiple comparisons establishes evidence of where the differences lie.

These conclusions are summarized by the software in the component of the Session window output displayed in Panel 7.32. The rows for types F and G have the letter A in common, indicating the earlier conclusion of no difference in mean length for these two types. The rows for E and F and the rows for E and G have no letters in common, indicating significant differences between E and F on the one hand and between E and G on the other. In essence group A consists of types F and G and group B consists of type E on its own.

The overall mean for the experiment was 200. The mean for type E was 194, so we can think of the effect for type E as being − 6, i.e. using type E reduced mean length by 6 m from the overall mean. The mean for type F was 202, so we can think of the effect for type F as being 2, i.e. using type F increased mean length by 2 m from the overall mean. Similarly, for type G the effect was 4. Note that the three effects sum to 0. In fitting a statistical model to data it is usual to write

$$\text{Observed data value} = \text{Value fitted by model} + \text{Residual},$$

or, more succinctly,

$$\text{Data} = \text{Fit} + \text{Residual}.$$

```
Grouping Information Using Tukey Method

Type   N     Mean    Grouping
G      5     204.000  A
F      5     202.000  A
E      5     194.000     B

Means that do not share a letter are significantly different.
```

Panel 7.32   Session window summary of Tukey Multiple Comparisons procedure.

In this case we take

$$\text{Fit} = \text{Overall mean} + \text{Effect}.$$

Thus we have:

$$\text{type E,}\quad \text{Fit} = 200 + (-6) = 194;$$
$$\text{type F,}\quad \text{Fit} = 200 + 2 = 202;$$
$$\text{type G,}\quad \text{Fit} = 200 + 4 = 204.$$

Thus knowing the data and fit values we can compute the residual values as the difference Data − Fit.

The fit and residual values were computed by Minitab by checking **Store fits** and **Store residuals** in the ANOVA dialog. They are displayed in Figure 7.28 – Minitab assigns the names FITS1 and RESI1 to the columns containing the fitted values and residuals, respectively. The first drive was of length 200 with a ball of type F for which the fit is 202. Hence, Residual = Data − Fit = 200 − 202 = −2. The reader is invited to check the remaining residual values displayed in Figure 7.28.

The **Four in one** plot facility under **Graphs...** yields four plots that will be discussed in turn. They are displayed in Figure 7.29. Theoretically the ANOVA methods used in this chapter require the assumptions of independence, random samples and normal distributions with equal variances. The four plots can often indicate when these assumptions are suspect.

1. *Normal probability plot of residuals.* In this case the plot is reasonably linear so the normality assumption underlying the valid use of the F-distribution for testing the null hypothesis appears reasonable.

C1	C2-T	C3	C4-T	C5	C6
Drive no.	Type	Length	Remarks	RESI1	FITS1
1	F	200	Tape measure works well	-2	202
2	G	206	OK	2	204
3	E	195	OK	1	194
4	E	188	OK	-6	194
5	G	207	OK	3	204
6	E	194	OK	0	194
7	F	208	Wind dropped slightly	6	202
8	E	193	OK	-1	194
9	F	206	OK	4	202
10	F	197	OK	-5	202
11	G	206	OK	2	204
12	E	200	OK	6	194
13	F	199	OK	-3	202
14	G	197	OK	-7	204
15	G	204	Impressive driving!	0	204

Figure 7.28   Worksheet with columns of fits and residuals Columns.

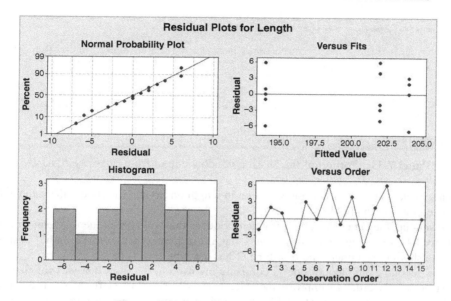

Figure 7.29   Four in one residual plots.

2. *Histogram of residuals*. A histogram of residuals that does not exhibit reasonable symmetry would suggest that the assumption of normality was suspect.

3. *Residuals versus the fitted values*. In addition to normality, the valid use of the F-distribution requires that the populations of lengths obtained with each type of ball have equal variances. Support for the assumption of equal variances is provided by similar vertical spread in the three sets of points.

4. *Residuals versus the order of the data*. In this case the data are in time order, so any unusual features in this run chart of the residuals could alert the experimenters to some factor other than type that might be having an influence on length.

These checks of the assumptions underlying valid use of the F-distribution are often referred to as *diagnostic checks*. Having the fits and residuals stored in the worksheet means that one can carry out one's own diagnostic checks, e.g. one could use **Stat > Basic Statistics > Normality Test...** in order to obtain a normal probability plot of the residuals with associated P-value. However visual scrutiny of the four plots discussed above will often be sufficient. The use of the F-test is fairly robust to relatively minor departures from the assumptions of normality and equal variances. However, if there are major concerns from scrutiny of the diagnostic plots one can either seek to transform the data or carry out a nonparametric test.

The portion of the Session window output in Panel 7.33 will now be considered. The number s = 4.378 is an estimate of the common standard deviation, assumed to apply to all three ball types. It is used to obtain, using the appropriate t-distribution, the 95% confidence intervals for the population means for the three ball types that are displayed to the right of the summary statistics. The R-sq (R^2) value of 54.9% is the coefficient of determination for length and fit expressed as a percentage. The reader may readily verify that the correlation between length and fitted value is 0.741 yielding $r^2 = 0.549 = 54.9\%$. It indicates that the model fitted to

```
S = 4.378   R-Sq = 54.90%   R-Sq(adj) = 47.39%

                             Individual 95% CIs For Mean Based on Pooled StDev
Level   N    Mean   StDev    -+---------+---------+---------+--------
E       5  194.00    4.30    (--------*--------)
F       5  202.00    4.74                       (--------*--------)
G       5  204.00    4.06                           (--------*--------)
                             -+---------+---------+---------+--------
                            190.0     195.0     200.0     205.0

Pooled StDev = 4.38
```

Panel 7.33 Portion of the Session window output from one-way ANOVA.

the data explains just over half the variation in length observed. The R-sq (adj) value will be discussed later in the book.

To sum up, the experiment has provided evidence that ball type influences length of drive for Lynx. The follow-up analysis indicates that, if the greatest achievable length is desirable, then she should use either type F or type G but not type E.

As a second example, consider a company involved in telesales that was running a Six Sigma project in order to improve the sales performance of its staff. A group of 40 new employees with similar educational backgrounds was split at random into four equal sized groups. The first group received the standard in-house training, while the other three groups were each trained by a different external training provider. The three external training providers comprise the list of accredited trainers for the company. Numbers of sales made by each employee during their first month of telephone contacts with prospective customers was recorded. The data are displayed in Table 7.10 and are available in the worksheet Sales.MTW.

The ANOVA for unstacked data in this form can be carried out using **Stat > ANOVA > One-way (Unstacked)....** The dialog required is shown in Figure 7.30. Here the factor (X) of interest is the training with in-house, trainer P, trainer Q and trainer R as levels. The in-house trained sales staff may be regarded as a control group so, clicking on **Comparisons...**, **Dunnett's** multiple comparison method was selected. As there is no run order in this case Minitab offers, by clicking on **Graphs...**, a **Three in one** plotting option for the residuals from the fitted model, which was selected together with **Boxplots of data**. (These plots are not given

Table 7.10 Telesales data.

In-house	Trainer P	Trainer Q	Trainer R
71	62	55	71
71	62	67	72
59	82	67	90
67	82	71	94
66	64	62	80
45	70	71	80
58	71	72	77
58	71	69	76
55	82	58	75
67	93	53	80

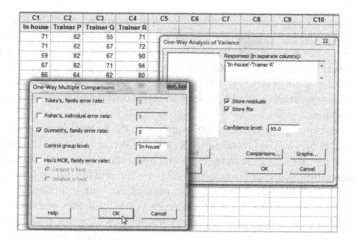

Figure 7.30   Dialog for ANOVA with unstacked data.

in the text.) Note that the default family error rate of 5% has been accepted and **Control group level:** 'In-house' indicates that employees trained in-house comprise the control group.

The reader is invited to verify that a P-value of 0.000 is obtained, which indicates that the actual P-value is less than 0.0005 (in fact it is 0.000 075). Thus the experiment provides strong evidence for rejection of the null hypothesis of equal mean sales performance for populations of employees trained by the four methods. The residual plots were deemed satisfactory. It should be noted that the data here are actually discrete but the residual plots indicate that the assumption of approximate normal distributions with equal variances is reasonable.

The Session window output for Dunnet's multiple comparison procedure is shown in Panel 7.34. The family error rate of $\alpha = 0.05$ means that there is an overall probability of 5% of a Type I error, which means in turn that we can have 95% confidence in the group of three confidence intervals provided. This can be interpreted to mean that, were we to repeat the experiment over and over again, on 5 occasions out of 100 in the long term the three confidence intervals would fail to capture all three true differences between the trainer means and the in-house mean. Conversely, on 95 occasions out of 100 in the long term the three

```
Dunnett's comparisons with a control

Family error rate = 0.05
Individual error rate = 0.0192

Critical value = 2.45

Control = In-house

Intervals for treatment mean minus control mean

Level       Lower   Center   Upper    ------+---------+---------+---------+---
Trainer P   3.036   12.200   21.364              (--------*--------)
Trainer Q  -6.364    2.800   11.964   (--------*--------)
Trainer R   8.636   17.800   26.964                   (--------*--------)
                                      ------+---------+---------+---------+---
                                            0        10        20        30
```

Panel 7.34   Dunnett's multiple comparisons for training experiment.

confidence intervals would capture all three true differences between the trainer means and the in-house mean. The individual error rate of $\alpha = 0.0192$ means that, were we to repeat the experiment over and over again, on 192 occasions out of 10 000 in the long term an individual confidence interval would fail to capture the true difference between the trainer mean and the in-house mean. Conversely, on 9808 occasions out of 10 000 in the long term an individual confidence interval would capture the true difference between the trainer mean and the in-house mean.

Thus from the experiment we estimate (rounded to the nearest integer) that the trainer P population would achieve 12 more sales in a month on average than the in-house population, with confidence interval $(3, 21)$, the trainer Q population would achieve 3 more on average than the in-house population, with confidence interval $(-6, 12)$, and the trainer Q population would achieve 18 more on average than the in-house population, with confidence interval $(9, 27)$. Since the confidence interval for trainer Q includes 0, this indicates that there is insufficient evidence to conclude that employees trained by trainer Q will perform any better than those trained in-house. Since the confidence intervals for trainers P and R do not include 0 and cover positive ranges of values, this indicates that there is evidence that employees trained by trainer P and R will perform better than those trained in-house. This information is of potential value to the Six Sigma project team. The summary provided by the software is displayed in Panel 7.35.

7.5.2   The fixed effects model

Consider again the golf ball experiment where there was a *fixed* number, $a = 3$, of ball types of interest. Thus the factor, X, of interest had $a = 3$ levels. The response, Y, of interest was the length of drive achieved. There were $n = 5$ replications, i.e. the response was measured for five drives with a ball of each type. The underlying statistical model assumed was that, for each type, the response Y was normally distributed with the same variance for all three types. Thus the model states that:

$$\text{for type E,} \quad Y \sim N(\mu_E, \sigma^2);$$
$$\text{for type F,} \quad Y \sim N(\mu_F, \sigma^2);$$
$$\text{for type G,} \quad Y \sim N(\mu_G, \sigma^2).$$

With this formulation of the model the null and alternative hypotheses are stated as follows:

$$H_0 : \mu_E = \mu_F = \mu_G, \quad H_1 : \text{Not all } \mu s \text{ are identical.}$$

```
Grouping Information Using Dunnett Method

Level                  N     Mean  Grouping
In-house (control)    10   61.700  A
Trainer R             10   79.500
Trainer P             10   73.900
Trainer Q             10   64.500  A

Means not labeled with letter A are significantly different from control level mean.
```

Panel 7.35   Summary of conclusions from experiment.

Table 7.11 ANOVA table for golf ball experiment with expected mean squares (fixed effects).

Source of variation	Degrees of freedom (DF)	Sum of squares (SS)	Mean square (MS)	Expected mean square (EMS)
Type	2	280	140	$\sigma^2 + \dfrac{n\sum_{i=1}^{a}\alpha_i^2}{a-1}$
Error	12	230	19.2	σ^2
Total	14	510		

It is customary to write μ_E as $\mu + \alpha_1$, μ_F as $\mu + \alpha_2$ and μ_G as $\mu + \alpha_3$, where μ is referred to as the *overall mean* and the *effects* α_1, α_2, and α_3 are such that $\alpha_1 + \alpha_2 + \alpha_3 = 0$. Thus the model states that:

$$\text{for type E,} \quad Y \sim N(\mu + \alpha_1, \sigma^2)$$
$$\text{for type F,} \quad Y \sim N(\mu + \alpha_2, \sigma^2)$$
$$\text{for type G,} \quad Y \sim N(\mu + \alpha_3, \sigma^2).$$

With this formulation of the model the null and alternative hypotheses are stated as follows:

$$H_0 : \alpha_1 = \alpha_2 = \alpha_3 = 0, \quad H_1 : \text{Not all } \alpha\text{s are zero.}$$

Table 7.11 shows the ANOVA table for the golf ball experiment together with the expected values of the mean squares. The analysis of variance partitions the total variation as represented by the *total* sum of squares (510 with 14 degrees of freedom) into a component attributable to the source *type* of ball (280 with 2 degrees of freedom) and a component attributable to the source random variation or random *error* (230 with 12 degrees of freedom). The component degrees of freedom and the component sums of squares add up to the corresponding totals ($2 + 12 = 14$ and $280 + 230 = 510$). In total the experiment yielded a sample of 15 values of length and a sample of 15 values has 14 degrees of freedom. We can think also of a sample of three means corresponding to the data for the three ball types involved in the experiment and a sample of three has two degrees of freedom. The mean square corresponding to the type and error components is obtained by dividing the sum of squares by degrees of freedom. Montgomery (2009, pp. 142–146) gives general formulae for the calculation of degrees of freedom and sums of squares.

The test statistic is the ratio of the mean squares, i.e. $140/19.2 = 7.30$. If the null hypothesis is true then all the αs would be zero and the expected value of both mean squares would be σ^2. Thus, if the null hypothesis is true the test statistic would be expected to have a value around 1. If the null hypothesis is false then not all the αs are zero and the expected value of the numerator of the ratio yielding the test statistic would be greater than the expected value of the denominator. Thus if the null hypothesis is false the test statistic would be expected to have a value greater than 1. When the null hypothesis is true, with the model specified above, the test statistic has the F distribution with parameters 2 and 12, i.e. the degrees of freedom for type and error respectively. **Calc > Probability > F...** may be used to confirm the P-value for the test.

```
Cumulative Distribution Function

F distribution with 2 DF in numerator and 12 DF in denominator

  x P (X <= x)
7.3    0.991571
```

Panel 7.36 Calculation of the P-value for the golf ball experiment.

The Session window output in Panel 7.36 indicates that the probability of obtaining a value for F of 7.3 or greater is $1 - 0.991\,571 = 0.008\,429$ so the P-value for the test is 0.008, to three decimal places, as displayed in the Session window output in Panel 7.29.

The fixed effects model may also be specified as detailed in Box 7.5. The number of levels of the factor is a and the number of replicates is n. With $i = 2$ and $j = 3$ we have, for example,

$$Y_{23} = \mu + \alpha_2 + \varepsilon_{23}.$$

In terms of the golf ball experiment, where there were just three levels of the factor type of interest, this equation states that the length, Y, for level 2 of the factor (ball type F) with drive 3 is made up of the overall mean, μ, plus the effect, α_2, for level 2 (ball type F) plus a random error, ε_{23} (a value from the normal distribution with mean 0 and variance σ^2).

7.5.3 The random effects model

Let us examine the data from the golf ball experiment again, but now with one major difference. Instead of a fixed set of three ball types of interest, let us consider the three types used in the experiment to have been a *random sample* of types from the myriad available on the market. In this scenario a random effects model is appropriate. The questions to be addressed by the analysis would thus be: How much of the variation observed is attributable to real differences between mean length, in the population of types from which the three used in the experiment were selected? How much of the variation observed is attributable to random variation about these population means?

The *random* effects model may be specified as detailed in Box 7.6. As in the case of the fixed effects model, the number of levels of the factor is a and the number of replicates is n. With $i = 2$ and $j = 3$ we have, for example,

$$Y_{23} = \mu + \alpha_2 + \varepsilon_{23}.$$

Observed data value $=$ Overall mean $+$ Effect $+$ Random Error,

$$Y_{ij} = \mu + \alpha_i + \varepsilon_{ij} \qquad i = 1, 2, \ldots, a, \qquad j = 1, 2, \ldots, n$$

$$\sum_{i=1}^{a} \alpha_i = 0, \qquad \varepsilon_{ij} \sim N(0, \sigma^2)$$

Box 7.5 Fixed effects model.

In terms of the golf ball experiment, where there were just three levels of the factor type of interest, this equation states that the length, Y, for level 2 of the factor (ball type F) with drive 3 is made up of the overall mean, μ, plus the effect, α_2 (a value from the normal distribution with mean 0 and variance σ_a^2) for level 2 (ball type F) plus a random error, ε_{23} (a value from the normal distribution with mean 0 and variance σ^2).

The ANOVA table is exactly as for the fixed effects model in the case where there is a single factor of interest. However, the null and alternative hypotheses are:

$$H_0 : \sigma_a^2 = 0, \quad H_1 : \sigma_a^2 \neq 0.$$

In order to be able to make a full analysis in the random effects case we will obtain the ANOVA table using **Stat** > **ANOVA** > **Balanced ANOVA...** as indicated in the dialog in Figure 7.31.

The design or plan used for the experiment was such that there were equal numbers of drives made with each level of the factor type. This makes the design a balanced one. The following points should be noted concerning the dialog:

- Under **Graphs...** here there is no facility to create boxplots or an individual plot as in Figure 7.27 by way of initial display of the data, but these may – and, in the author's view, one or other should – always be created separately using the **Graphs** menu.

- Here with a single factor, type, involved the model is:

  Observed data value = Overall mean + Effect of type + Random error.

  This information is communicated to the software by inserting or selecting Type under **Model:**. There is always an overall mean and a random error term on the right-hand side of the equation that specifies models of the sort employed here, so in this case of a single factor this entry conveys the key information.

- The information that the factor Type is random is communicated by inserting Type under **Random factors:**.

- Under **Graphs...** the **Four in one** option for **Residual Plots** is strongly recommended.

- Finally, under **Results...** the option **Display expected mean squares and variance components** should be checked and **Display means corresponding to the terms:** should specify the factor Type.

Observed data value = Overall mean + Effect + Random error,

$$Y_{ij} = \mu + \alpha_i + \varepsilon_{ij}, \qquad i = 1, 2, \ldots, a, \qquad j = 1, 2, \ldots, n$$
$$\alpha_i \sim N(0, \sigma_a^2), \quad \varepsilon_{ij} \sim N(0, \sigma^2)$$

(the random variables α_i and ε_{ij} are independent)

Box 7.6 Random effects model.

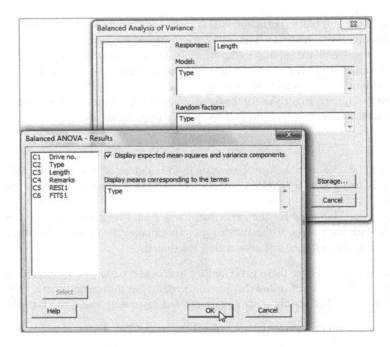

Figure 7.31   Dialog for balanced ANOVA.

The individual value plot in Figure 7.27 suggests that there is a component of variation in the response length of drive (*Y*) that may be attributed to factor ball type (*X*). Formal confirmation is obtained from the portion of the Session window output displayed in Panel 7.37. The null hypothesis $H_0 : \sigma_a^2 = 0$ would be rejected in favour of the alternative hypothesis $H_1 : \sigma_a^2 \neq 0$ at the 1% level of significance, since the *P*-value is 0.008. Thus the experiment provides strong evidence that σ_a^2, the component of variance in drive length attributable to ball type, is nonzero.

Table 7.12 gives the ANOVA table, together with the expected values of the mean squares for a random effects scenario. Here the number of replicates, *n*, was 5. We can take the observed mean squares as estimates of the corresponding expected mean squares. Thus σ^2 is estimated by 19.17 so σ is estimated by the square root of 19.17, which is 4.378. Also $\sigma^2 + 5\sigma_a^2$

```
ANOVA: Length versus Type

Factor   Type      Levels   Values
Type     random         3   E, F, G

Analysis of Variance for Length

Source   DF       SS       MS      F       P
Type      2   280.00   140.00   7.30   0.008
Error    12   230.00    19.17
Total    14   510.00
```

Panel 7.37   ANOVA table for golf ball experiment.

Table 7.12   ANOVA table for golf ball experiment with expected mean squares (random effects).

Source of variation	Degrees of freedom (DF)	Sum of squares (SS)	Mean square (MS)	Expected mean square (EMS)
Type	2	280	140	$\sigma^2 + n\sigma_a^2$
Error	12	230	19.17	σ^2
Total	14	510		

is estimated by 140, so $5\sigma_a^2$ is estimated by the difference $140 - 19.17 = 120.83$. Hence, σ_a^2 is estimated by $120.83/5 = 24.17$. The two components of variance are both given in the annotated section of the Session window output shown in Panel 7.38. The shorthand (2) represents the component of variance due to ball type i.e. to σ_a^2. The shorthand (1) represents the random error variance σ^2. Thus Minitab does all the calculations of the components of variance. The value s in the top left corner is the estimate of σ. The residuals and fitted values are the same as in the case of the fixed effects model so the R-sq value is as before.

The final portion of the Session window output gives the means for the three ball types that were selected at random from the population of available types. The fits and residuals are exactly as before and therefore the diagnostic plots are as before.

Imagine that Lynx goes to a driving range and selects a bucket of golf balls that constitute a random sample from the large population of golf ball types used in the random effects experiment. We can use the model to predict the distribution of length that will be achieved:

$$\text{Observed data value} = \text{Overall mean} + \text{Effect of type} + \text{Random error.}$$

Since the observed data value is a constant plus the sum of two independent random variables the result in Box 4.2 in Section 4.3.1 may be applied to calculate the mean and variance of the random variable length as detailed in Box 7.7. (A constant may be considered as a random variable with variance zero!) Thus the estimated total variance is 43.34 and the estimated proportion of total variance accounted for by ball type is $24.17/43.34 = 55.8\%$. Since a sum of independent normally distributed random variables is also normally distributed we can finally predict that the distribution of length on the driving range would be $N(200, 6.58^2)$. Were Lynx to opt, for example, to use a balls of type G only then the distribution of length would be estimated to be $N(204, 4.38^2)$.

```
S = 4.37798    R-Sq = 54.90%    R-Sq(adj) = 47.39%


                                    Expected Mean
                                    Square for Each
                                    Term (using
                     Variance  Error unrestricted
          Source    component  term  model)
     1    Type        24.17      2    (2) + 5 (1)   [corresponding to σ²+nσ²ₐ]

     2    Error       19.17           (2)           [corresponding to σ²]
```

Panel 7.38   Components of variance for golf ball experiment.

$Y_{ij} = \mu + \alpha_i + \varepsilon_{ij}, \quad \alpha_i \sim N(0, \sigma_a^2); \, \varepsilon_i \sim N(0, \sigma^2).$

Mean of $Y_{ij} = \mu + 0 + 0 = \mu$ which is estimated as 200, the overall mean for the experiment.

Variance of $Y_{ij} = 0 + \sigma_a^2 + \sigma^2$ which is estimated by $0 + 24.17 + 19.17 = 43.34 = 6.58^2$.

Box 7.7 Calculation of the mean and variance of length.

Montgomery (2005a, p. 487) gives an industrial example involving components of variance. A textile company weaves a fabric on a large number of looms. Interest centred on loom-to-loom variability in the tensile strength of the fabric. Four looms were selected at random and four random samples of fabric from each loom were tested, yielding the data in Table 7.13. The data are available in stacked form in the worksheet Looms.MTW and are reproduced by permission of John Wiley & Sons, Inc., New York.

Initial analysis of the data was carried out using **Stat > ANOVA > One-Way...** with the **Individual value plot** option selected under **Graphs...** together with **Normal plot of residuals** and **Residuals versus fits**. Scrutiny of the individual value plot suggests that there is variation attributable to the factor loom; this is confirmed by a *P*-value of 0.000 (to three decimal places) which indicates very strong evidence of such significant variation. The normal probability plot was reasonably straight and the vertical spreads of residuals similar in the plot of residuals versus fits. Thus one can be satisfied that a random effects model of the form used in the previous example is appropriate.

Having established a significant loom effect, **Stat > ANOVA > Balanced ANOVA...** was used to obtain the components of variance shown in Panel 7.39. Hence the mean and variance of Tensile Strength may be estimated as detailed in Box 7.8. Thus the estimated total

Table 7.13 Tensile strength data.

Loom	Tensile strength (psi)			
1	98	97	99	96
2	91	90	93	92
3	96	95	97	95
4	95	96	99	98

```
S = 1.37689    R-Sq = 79.68%    R-Sq(adj) = 74.60%


                                     Expected Mean
                                     Square for Each
                                     Term (using
                 Variance    Error   unrestricted
        Source   component   term    model)
 1  Loom           6.958     2  (2) + 4 (1)
 2  Error          1.896        (2)
```

Panel 7.39 Components of variance for looms experiment.

$$Y_{ij} = \mu + \alpha_i + \varepsilon_{ij}, \quad \alpha_i \sim N(0, \sigma_a^2); \; \varepsilon_i \sim N(0, \sigma^2).$$

Mean of $Y_{ij} = \mu$ which is estimated as 95.438, the overall mean for the experiment.

Variance of $Y_{ij} = \sigma_a^2 + \sigma^2$ which is estimated by $6.958 + 1.896 = 8.854 = 2.976^2$.

Box 7.8 Calculation of the mean and variance of tensile strength.

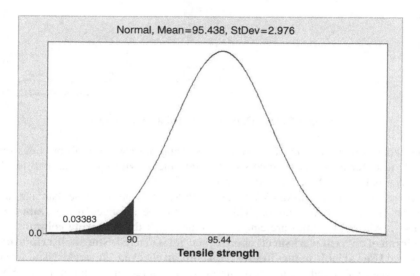

Figure 7.32 Estimated distribution of tensile strength.

variance is 8.854 and the estimated proportion of total variance accounted for by the factor loom is $6.958/8.854 = 78.6\%$. The overall mean for the experiment was 95.438, so we can estimate that the distribution of tensile strength for fabric produced on the population of looms would be $N(95.438, 2.976^2)$. This model, together with a reference line indicating the lower specification limit of 90 psi for tensile strength, is shown in Figure 7.32. The process is therefore operating with a C_{pk} of the order of 0.6 (sigma quality level of around 3.3).

Montgomery comments:

> A substantial proportion of the production is fallout. This fallout is directly related to the excess variability resulting from differences between looms. Variability in loom performance can be caused by faulty set-up, poor maintenance, inadequate supervision, poorly trained operators and so forth. The engineer or manager responsible for quality improvement must remove these sources of variability from the process.

7.5.4 The nonparametric Kruskal–Wallis test

Consider again the telesales data displayed in Table 7.10 and available in worksheet Sales. MTW. The sales figures are actually counts, so it could be argued that an analysis of variance,

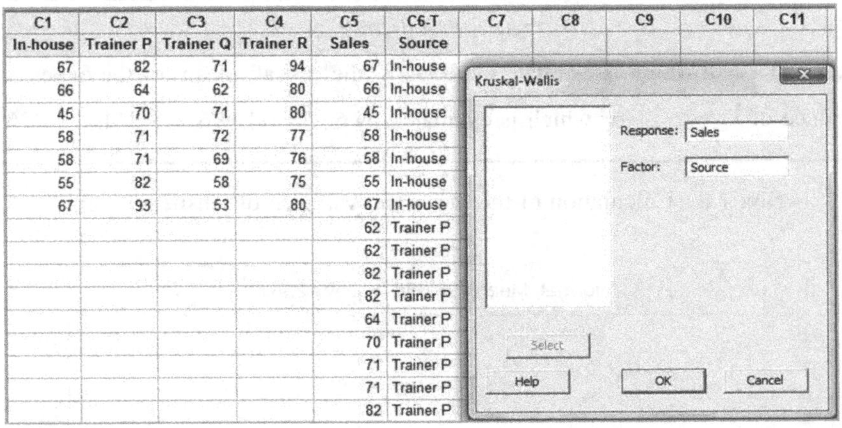

C1	C2	C3	C4	C5	C6-T	C7	C8	C9	C10	C11
In-house	Trainer P	Trainer Q	Trainer R	Sales	Source					
67	82	71	94	67	In-house					
66	64	62	80	66	In-house					
45	70	71	80	45	In-house					
58	71	72	77	58	In-house					
58	71	69	76	58	In-house					
55	82	58	75	55	In-house					
67	93	53	80	67	In-house					
				62	Trainer P					
				62	Trainer P					
				82	Trainer P					
				82	Trainer P					
				64	Trainer P					
				70	Trainer P					
				71	Trainer P					
				71	Trainer P					
				82	Trainer P					

Figure 7.33   Dialog for Kruskal–Wallis test.

with the underlying assumption of normality, is inappropriate. Minitab provides two non-parametric tests for experiments involving a single factor with more than two levels – the Kruskal–Wallis test and Mood's median test.

In order to perform a Kruskal–Wallis test with Minitab the response data and the factor levels must appear in two columns. This can readily be arranged using **Data** > **Stack** > **Columns…**. All four columns are entered into the **Stack the following columns:** window. With **Column of current worksheet:** checked and Sales entered, **Store subscripts in**: Source specified and **Use variable names in subscript column** checked, the stacked data are stored in a column named Sales and the levels of the factor are stored in a column named Source. A portion of the stacked data can be seen in Figure 7.33. Then **Stat** > **Nonparametrics** > **Kruskal-Wallis…** leads to the dialog also shown in Figure 7.33.

The Session window output is shown in Panel 7.40. The sample size and median are given for each level of Source. The test is a generalization of the Mann–Whitney test and is based on ranks. The average rank for each level of the factor is given together with a corresponding z-value. The null hypothesis is that the samples are from identical populations. The test-statistic is denoted by the letter H. The corresponding P value is given and in this case indicates very strong evidence that the populations sampled are not identical, i.e. that the training methods are not equally effective. The procedure does not

```
Kruskal-Wallis Test: Sales versus Source

Kruskal-Wallis Test on Sales

Source       N   Median   Ave Rank        Z
In-house    10    62.50       11.4    -2.83
Trainer P   10    71.00       24.4     1.22
Trainer Q   10    67.00       14.4    -1.91
Trainer R   10    78.50       31.8     3.51
Overall     40                20.5

H = 19.09   DF = 3   P = 0.000
H = 19.24   DF = 3   P = 0.000   (adjusted for ties)
```

Panel 7.40   Session window output for Kruskal–Wallis test.

provide an option to display the data, nor does it provide any facility for comparisons. When used in a situation in which ANOVA could legitimately be used it provides a less powerful test than ANOVA.

7.6   Blocking in single-factor experiments

To introduce the idea of blocking, we will consider an experiment where the single factor of interest is variety of potato and the response of interest is yield in tonnes per hectare (t/ha). Denote the three levels of the variety factor by A, B and C. Suppose that 12 plots, numbered 1 to 12, are available for the experiment, as shown in Figure 7.34, and that random allocation of varieties to the plots led to the design indicated.

Imagine that there is a wood to the west of the plots and a river to the east. This could conceivably lead to a fertility gradient in the direction of the arrow due to greater amounts of both moisture and nutrients in the soil, the further plots are from the wood. A concern with this completely randomized design is that variety A might appear to perform well in terms of yield not because it was superior to the other varieties but because the plots planted with A were favourably placed in terms of a possible fertility gradient.

A superior experimental design in this situation would be achieved through the use of blocking. Each strip of three plots running in a north–south direction would be designated as a block, yielding four blocks as indicated in Figure 7.35. Subsequently the three varieties would be allocated at random within each block. Suppose that the arrangement shown in Figure 7.36 arose. This is a randomized complete block design. The numbers in brackets are the yield values. (As with the golf ball experiment, fictitious integer data have been used in order to make the arithmetic simple when introducing key concepts.)

The term 'block' is a legacy from the early application of designed experiments to agricultural research. 'Block' meant a block of land as in this introductory example. In experimental design it now refers to groups of experimental units that are homogeneous. If the factor of interest has, say, four levels then a block might consist of four plots of land adjacent to each other, four test cubes of concrete from the same batch, four pigs from the same litter etc.

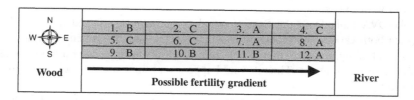

Figure 7.34   Completely randomized design.

	Block 1	Block 2	Block 3	Block 4	
	C (27)	B (28)	B (29)	C (37)	
	A (20)	C (32)	A (25)	B (40)	
	B (31)	A (24)	C (36)	A (31)	

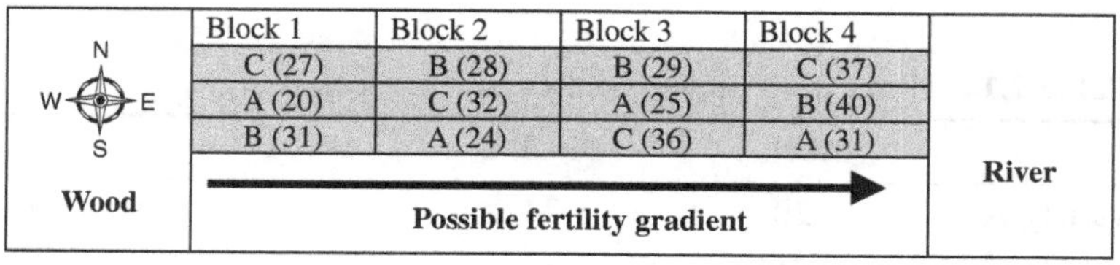

Figure 7.35   Randomized complete block design.

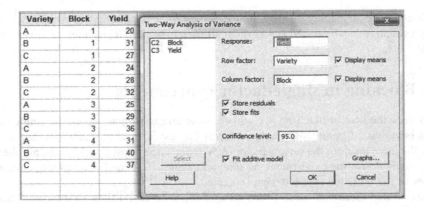

Figure 7.36 Dialog for a two-way analysis of variance.

The data from the experiment are tabulated in Table 7.14 and available in Potato.MTW. In order to analyse the data via Minitab they have to be arranged into three columns specifying variety, block and yield, respectively. Yield is the response and we have two factors, variety and block. Thus we may use **Stat > ANOVA > Two-Way...** to perform an analysis of variance. The dialog is shown in Figure 7.36.

Yield is entered as the **Response:**, Variety as the **Row factor:** and Block as the **Column factor:**. (The levels of variety correspond to the rows, and the levels of block correspond to the columns of Table 7.14.) The **Display means** option was checked for both factors. **Store residuals** and **Store fits** were checked. It is essential that **Fit additive model** be checked when using this procedure to analyse data from a randomized complete block design where there is a single factor of interest (variety in this case).

Under **Graphs...**, the option to display the data using an **Individual value plot** was selected together with **Normal plot of residuals**. The individual value plot given in Figure 7.37 suggests that both variety B and variety C give heavier yield than does variety A. It also suggests that the use of blocking may have been wise since yield generally increases across the blocks from west to east.

The ANOVA table from the Session window output is displayed in Panel 7.41. The P-values for both variety and block are less than 0.05 so the experiment provides evidence, at the 5% level of significance, that variety has a significant influence on the response yield and evidence of block effect. Thus it would appear that the experimenters were justified in using blocking.

The overall mean yield for the experiment was 30. Having checked the **Display means** option for both the row and column factors, the means for variety and block are displayed below the ANOVA table in the Session window output and are given in Table 7.15.

Table 7.14 Data from randomized complete block design.

	Block 1	Block 2	Block 3	Block 4
Variety A	20	24	25	31
Variety B	31	28	29	40
Variety C	27	32	36	37

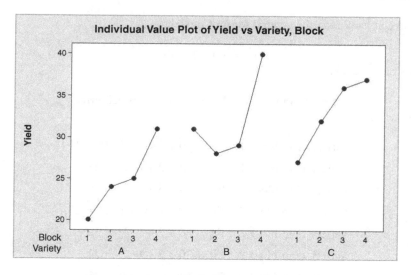

Figure 7.37   Individual value plot of Yield.

```
Two-way ANOVA: Yield versus Variety, Block

Source    DF    SS      MS      F       P
Variety    2   152   76.0000   9.91   0.013
Block      3   168   56.0000   7.30   0.020
Error      6    46    7.6667
Total     11   366

S = 2.769    R-Sq = 87.43%    R-Sq(adj) = 76.96%
```

Panel 7.41   ANOVA for potato experiment.

The mean for variety A was 25. One can therefore think of the effect for variety A as being −5, i.e. variety A reduced mean yield by 5 t/ha from the overall mean of 30 t/ha. The mean for variety B was 32 so we can think of the effect for variety B as being 2, i.e. variety B increased mean yield by 2 t/ha from the overall mean of 30. Similarly, for variety C the effect was 3. Note that the three variety effects sum to 0. The mean for block 1 was 26. One can think of the effect for block 1 as being −4, i.e. block 1 reduced mean yield by 4 t/ha from the overall mean. Similarly, the effects for blocks 2, 3 and 4 were −2, 0 and 6, respectively. Note that the four block effects sum to 0.

Table 7.15   Data with means from randomized complete block experiment.

	Block 1	Block 2	Block 3	Block 4	Mean
Variety A	20	24	25	31	25
Variety B	31	28	29	40	32
Variety C	27	32	36	37	33
Mean	26	28	30	36	30

We have already seen the general form of model:

$$Data = Fit + Residual.$$

In this case we take

$$Fit = Overall\ mean + Variety\ effect + Block\ effect.$$

Thus we have:

variety A in block 1 : Fit $= 30 + (-5) + (-4) = 21$,

variety B in block 1 : Fit $= 30 + 2 + (-4) = 28$,

variety C in block 1 : Fit $= 30 + 3 + (-4) = 29$,

variety A in block 2 : Fit $= 30 + (-5) + (-2) = 23$,

variety B in block 2 : Fit $= 30 + 2 + (-2) = 30$,

variety C in block 2 : Fit $= 30 + 3 + (-2) = 31$,

variety A in block 3 : Fit $= 30 + (-5) + 0 = 25$,

variety B in block 3 : Fit $= 30 + 2 + 0 = 32$,

variety C in block 3 : Fit $= 30 + 3 + 0 = 33$,

variety A in block 4 : Fit $= 30 + (-5) + 6 = 31$,

variety B in block 4 : Fit $= 30 + 2 + 6 = 38$,

variety C in block 4 : Fit $= 30 + 3 + 6 = 39$.

We can now compute the residual values as the differences Data – Fit. The reader is invited to check the calculations in the first few rows of Table 7.16 and to observe that, having checked both **Store residuals** and **Store fits**, both residuals and fits are displayed in the worksheet.

The fixed effects model may also be specified as detailed in Box 7.9. For the potato experiment the number of levels of the factor of interest, variety, is $a = 3$ and the number of

Table 7.16 Fitted values and residuals for potato experiment.

Variety	Block	Data (yield)	Overall mean	Variety effect	Block effect	Fit	Residual
A	1	20	30	−5	−4	21	−1
B	1	31	30	2	−4	28	3
C	1	27	30	3	−4	29	−2
A	2	24	30	−5	−2	23	1
B	2	28	30	2	−2	30	−2
C	2	32	30	3	−2	31	1
A	3	25	30	−5	0	25	0
B	3	29	30	2	0	32	−3
C	3	36	30	3	0	33	3
A	4	31	30	−5	6	31	0
B	4	40	30	2	6	38	2
C	4	37	30	3	6	39	−2

Observed data value = Overall mean + Factor effect + Block effect + Random error

$$Y_{ij} = \mu + \alpha_i + \beta_j + \varepsilon_{ij}, \quad i = 1, 2, \ldots, a, \quad j = 1, 2, \ldots, b,$$

$$\sum_{i=1}^{a} \alpha_i = 0, \quad \sum_{i=1}^{b} \beta_i = 0, \quad \varepsilon_i \sim N(0, \sigma^2)$$

Box 7.9   The fixed effects model

levels of what some refer to as a 'nuisance' factor block is $b = 4$. With $i = 2$ and $j = 3$, for example, we have the specific relationship

$$Y_{23} = \mu + \alpha_2 + \beta_3 + \varepsilon_{23}.$$

This equation therefore states that the yield for the plot planted with variety 2 (B) in block 3 is made up of the overall mean, μ, plus the effect for variety 2, α_2, plus the effect for block 3, β_3, plus a random error (value) from the normal distribution with mean 0 and variance σ^2. At the core of this model we have the addition of the two effects α_2 and β_3, one for variety and one for block – hence the need to check **Fit additive model** in the dialog.

There are two null hypotheses to be tested against alternatives. The first states that all the variety effects are zero, the second that all the block effects are zero. Formally they are stated as follows:

$$H_0 : \alpha_1 = \alpha_2 = \alpha_3 = 0, \quad H_1 : \text{Not all } \alpha\text{s are zero};$$
$$H_0 : \beta_1 = \beta_2 = \beta_3 = \beta_4 = 0, \quad H_1 : \text{Not all } \beta\text{s are zero}.$$

The P-values corresponding to these were 0.013 and 0.020 respectively, so both null hypotheses would be rejected at the 5% level of significance.

Having obtained evidence of variety having leverage in determining yield, it is useful to be able to carry out follow-up analysis using multiple comparisons. This is not available via **Stat > ANOVA > Two-Way...** but is available via **Stat > ANOVA > General Linear Model...**. The dialog is shown in Figure 7.38.

In **Model:** we are communicating to Minitab the nature of the model we are using, i.e. $Y_{ij} = \mu + \alpha_i + \beta_j + \varepsilon_{ij}$. Generally such models always include an overall mean and the random error term so by entering Variety and Block in **Model:** we are indicating the expression $\alpha_i + \beta_j$ at the core of the equation that defines the model in this scenario.

The subdialog for **Comparisons:** is also shown in Figure 7.38. Here **Pairwise comparisons** were selected, by the **Tukey** method with **Terms:** Variety. **Grouping information** and **Confidence interval**, with default **Confidence level:** 95.0, were checked. The latter part of corresponding section of the Session window output is shown in Panel 7.42.

The interpretation of this is as follows:

- On average the yield of variety B is 7 t/ha more than that for variety A, with confidence interval 1 to 13 t/ha (to the nearest integer).

- On average the yield of variety C is 8 t/ha more than that for variety A, with confidence interval 2 to 14 t/ha (to the nearest integer).

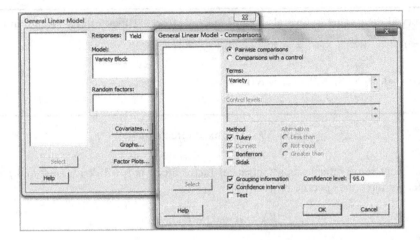

Figure 7.38   General Linear Model dialog.

- On average the yield of variety C is 1 t/ha more than that for variety B, with confidence interval −5 to 7 t/ha (to the nearest integer).

Since the first two confidence intervals do not include 0 we have evidence that the yield of both varieties B and C is superior to that of variety A. The fact that the third confidence interval includes 0 means that we have no evidence of a difference in mean yield for varieties B and C. Thus the experimentation has provided evidence that both varieties B and C give significantly greater mean yield than does variety A. However, it does not provide any evidence of a difference in yield for B and C.

The earlier part of the Session window output from Comparisons is displayed in Panel 7.43. This summarizes the conclusions that stem from scrutiny of the confidence intervals, i.e. that

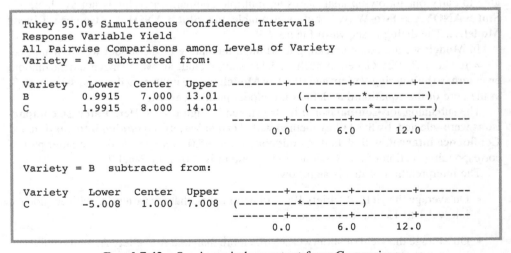

Panel 7.42   Session window output from Comparisons.

```
Grouping Information Using Tukey Method and 95.0% Confidence

Variety  N  Mean  Grouping
C        4  33.0  A
B        4  32.0  A
A        4  25.0     B

Means that do not share a letter are significantly different.
```

Panel 7.43   Session window output from Comparisons.

the means for varieties B and C do not differ significantly whereas the means for both A and B and A and C do.

Iman and Conover (1989, p. 631) give an example of a purchasing agent seeking to obtain word-processing software with which operators have the best production rate. Three candidate packages for purchase were assessed in a randomized complete block experiment in which six operators were treated as blocks. The response was the time taken (minutes) to input a standard document. The data are reproduced by permission of the authors in Table 7.17 and are available, in stacked form, in the worksheet Packages.MTW.

It is left as an exercise for the reader (remember to check **Fit additive model!**) to verify that there is evidence of differences between packages (*P*-value 0.045) and very strong evidence of differences between operators (*P*-value 0.000 to three decimal places). The normal probability plot of the residuals is satisfactory. The Session window output for **Multiple Comparisons** obtained via **Stat > ANOVA > General Linear Model...** using the **Tukey method** is shown in Panel 7.44. Note that in the **Comparisons...** subdialog box **Terms: Package** is required.

For example, the point estimate of the population mean time using package 1 was 47.3 minutes, while that for package 3 was 50.3 minutes. The point estimate of the difference in the means is 3 minutes with confidence interval (0.2, 5.8) minutes (rounded to one decimal place). The fact that this confidence interval does not include 0 indicates that the document can be produced significantly faster with package 1 than with package 3. Since the other confidence intervals both include 0 we cannot claim a significant difference between package 1 and 2 and we cannot claim a significant difference between package 2 and 3. These three confidence intervals together have overall confidence level of 95%.

Table 7.17   Word-processing package assessment data.

	Word-processing package		
Operator	1	2	3
1	42	45	45
2	37	36	40
3	53	56	55
4	68	73	75
5	48	45	47
6	36	39	40

```
Grouping Information Using Tukey Method and 95.0% Confidence

Package  N  Mean  Grouping
3        6  50.3  A
2        6  49.0  A B
1        6  47.3    B

Means that do not share a letter are significantly different.


Tukey 95.0% Simultaneous Confidence Intervals
Response Variable Time
All Pairwise Comparisons among Levels of Package
Package = 1  subtracted from:

Package    Lower   Center  Upper   -------+---------+---------+---------
2         -1.147   1.667   4.480   (-------------*-------------)
3          0.186   3.000   5.814        (-------------*-------------)
                                    -------+---------+---------+---------
                                        0.0       2.0       4.0


Package = 2  subtracted from:

Package    Lower   Center  Upper   -------+---------+---------+---------
3         -1.480   1.333   4.147   (-------------*-------------)
                                    -------+---------+---------+---------
                                        0.0       2.0       4.0
```

Panel 7.44   Session window output from Comparisons.

Some report the grouping information provided at the top of Panel 7.44 by listing the levels of the factor of interest and underlining those for which the mean responses do not differ significantly:

Package 1   Package 2   Package 3

Iman and Conover (1989, p. 638) comment that 'the data are not sufficiently strong to indicate a difference between software package 2 and the other word processing software packages, but they are significantly strong to declare a difference between software packages 1 and 3'.

Some authors refer to experimental designs that involve blocking as 'noise-reducing' experimental designs. 'Designing for noise reduction is based on the single principle of making all comparisons of treatments within relatively homogeneous groups of experimental units. The more homogeneous the experimental units the easier it is to detect a difference in treatments' (Mendenhall *et al.*, 1986, p. 525). The reader is invited to verify that, were we to ignore the blocking and do a one-way ANOVA then a *P*-value of 0.923 would be obtained. Thus naïve analysis of the data, which fails to take operator into account as a blocking factor, provides no evidence of any package effect.

The Friedman test is a nonparametric test that may be used to analyse data from a randomized complete block experiment. It may be implemented using **Stat > Nonparametrics > Friedman. . ..** For the data from the potato experiment the Session window output is shown in Panel 7.45. The reader is invited to check it as an exercise. In some cases where

```
Friedman Test: Yield versus Variety blocked by Block
S = 6.00   DF = 2   P = 0.050

                            Sum of
   Variety  N  Est Median   Ranks
   A        4      24.500     4.0
   B        4      31.167    10.0
   C        4      31.833    10.0

Grand median = 29.167
```

Panel 7.45   Session window output for Friedman test.

identical values or ties occur amongst the values of the response two P-values are given, the second taking ties into account.

In the dialog for this analysis Minitab refers to 'Treatment' rather than 'Factor'. The P-value quoted is for a test of the null hypothesis H_0 : all treatment effects are zero, versus the alternative hypothesis H_1 : not all treatment effects are zero. It is on the brink of being significant at the 5% level. The earlier analysis based on the assumption of underlying normal distributions gave a corresponding P-value of 0.013.

The Friedman test is another non-parametric procedure that is based on ranks. Residuals and fits may be computed using Minitab. Multiple comparisons based on ranks may be made, but this facility is not provided by Minitab – technical details are given in Iman and Conover (1989, p. 658).

7.7   Experiments with a single factor, with more than two levels, where the response is a proportion

A glass bottle manufacturer had been receiving complaints from customers concerning tears in and imperfect sealing of the shrinkwrap used on pallets of bottles. A Six Sigma project team carried out a single-factor experiment in which shrinkwrap from each of three suppliers A, B and C was used to seal 1200 pallets of bottles. Following shipment to a customer all 3600 pallets were checked and the numbers of nonconforming pallets recorded. The data are summarized in Table 7.18.

The null hypothesis of interest here is $H_0 : p_1 = p_2 = p_3$ and the alternative is H_1 : Not all p_i are identical $(i = 1, 2, 3)$, where p_1, p_2 and p_3 represent the population proportion of nonconforming pallets sealed with shrinkwrap from suppliers A, B and C, respectively. We could analyse the above data formally using three tests for equality of two proportions – one to compare A with B, one to compare B with C, and a final one to compare C with A. However, the

Table 7.18   Nonconfoming pallet data.

Status	Supplier			
	A	B	C	Total
Nonconforming	34	57	29	120
Conforming	1166	1143	1171	3480
Total	1200	1200	1200	3600
% Nonconforming	2.8	4.8	2.4	3.3

problem of the increased risk of a Type I error with this approach has already been discussed in Section 7.5.

If the null hypothesis is true then the proportions are homogeneous across suppliers – hence the test to be used is referred to as a test of homogeneity. It is available using **Stat > Tables > Chi-Square Test (Two-Way Table in Worksheet). . . .** The dialog is shown in Figure 7.39. The table required can be seen in the worksheet in the figure and consists of the shaded portion of Table 7.18. Note that the first row of each column gives the number of nonconforming pallets for the supplier and the second row gives the number of conforming pallets for the supplier.

The Session window output is displayed in Panel 7.46. Out of a total of 3600 pallets, 120 were nonconforming. If the null hypothesis is true then $120/3600 = 1/30$ provides an estimate of the common proportion of nonconforming pallets for shrinkwrap from all three suppliers. Thus for shrinkwrap from each supplier we would expect to find one in 30 of the 1200, i.e. 40 pallets, to be nonconforming and the remaining 1160 to be conforming. These expected counts, E_i, have been computed and displayed below the observed counts, O_i, in the table in the output.

The test statistic involves the differences between the observed and expected counts and is calculated as shown in Box 7.10. The central involvement of $O_i - E_i$ in the formula for the chi-square test statistic means that its value is relatively low when there is good agreement between observed and expected counts. Poor agreement arises when the null hypothesis is false, leading to a relatively large value of the test statistic. The P-value may be confirmed using **Calc > Probability Distributions > Chi-square. . . .** The Session window output is given in

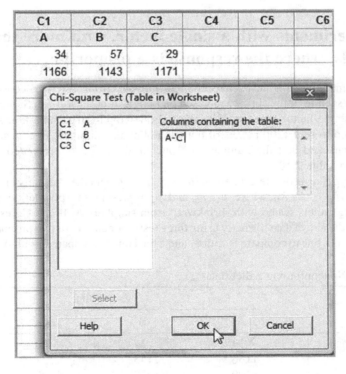

Figure 7.39   Dialog for test of homogeneity of proportions.

```
Chi-Square Test: A, B, C
Expected counts are printed below observed counts
Chi-Square contributions are printed below expected counts

                A         B         C     Total
   1           34        57        29       120
            40.00     40.00     40.00
             0.900     7.225     3.025

   2         1166      1143      1171      3480
           1160.00   1160.00   1160.00
             0.031     0.249     0.104

Total        1200      1200      1200      3600

Chi-Sq = 11.534, DF = 2, P-Value = 0.003
```

Panel 7.46   Session window output for chi-square test.

Panel 7.47. It indicates that the probability of obtaining a chi-square value of 11.534 or greater, were the null hypothesis true, would be $1 - 0.996\,871 = 0.003\,129$, or 0.003 to three decimal places, as stated in Panel 7.46. Thus the data from the experiment provide evidence, at the 1% level of significance, that the suppliers perform differently in terms of proportion of non-conforming. Formally, the null hypothesis $H_0 : p_1 = p_2 = p_3$ would be rejected in favour of the alternative $H_1 :$ Not all p_i are identical ($i = 1, 2, 3$) at the 1% level of significance. It therefore appears that supplier B performs significantly worse than the other two (4.8% nonconforming compared with 2.8% and 2.4%, respectively).

The chi-square test statistic is given by

$$\chi^2 = \sum_{i=1}^{k} \frac{(O_i - E_i)^2}{E_i}$$

where k is the number of cells in the table. Thus

$$\chi^2 = \frac{(34-40)^2}{40} + \frac{(57-40)^2}{40} + \frac{(29-40)^2}{40}$$
$$+ \frac{(1166-1160)^2}{1160} + \frac{(1143-1160)^2}{1160} + \frac{(1171-1160)^2}{1160}$$
$$= 0.900 + 7.225 + 3.025 + 0.031 + 0.249 + 0.104$$
$$= 11.534$$

(note how these six contributions to the test statistic appear in the output). The only parameter of a chi-square distribution is its number of degrees of freedom, which in this case is $a - 1$ where a is the number of levels of the supplier factor, i.e. $3 - 1 = 2$.

Box 7.10   Calculation of the chi-square test statistic.

```
Cumulative Distribution Function

Chi-Square with 2 DF

       x   P( X <= x )
  11.534      0.996871
```

Panel 7.47   Calculation of the *P*-value.

7.8   Tests for equality of variances

Two-sample *t*-tests and one-way ANOVA tests that required the assumption of equal population variances were discussed earlier in the chapter. Support for the assumption may be obtained from scrutiny of displays of the data in the form of individual value plots or boxplots. Minitab provides formal tests of the null hypothesis of equal variances. Consider again the magnesium assay data in Table 7.6, stored in Magnesium.MTW. Use of **Stat > Basic Statistics > 2 Variances...** provides a test of the null hypothesis that the two population variances are the same, i.e. the null hypothesis $H_0 : \sigma_1^2 = \sigma_2^2$, versus the alternative hypothesis that they are not, $H_1 : \sigma_1^2 \neq \sigma_2^2$. The null hypothesis is equivalent to $H_0 : \sigma_1 = \sigma_2$, which in turn is equivalent to $H_0 : \sigma_1/\sigma_2 = 1$. The dialog is shown in Figure 7.40. The alternative hypothesis is equivalent to $H_1 : \sigma_1 \neq \sigma_2$, which in turn is equivalent to $H_1 : \sigma_1/\sigma_2 \neq 1$. In completing the dialog the reader is urged to use the **Graphs...** button to select either an individual value plot or boxplot of the data.

Key components of the Session window output are shown in Panel 7.48. The *F*-test for equality of variances yields a *P*-value 0.021, so the null hypothesis of equal variances would be rejected at the 5% level of significance. Valid application of this test requires the distributions to be normal. Levene's test for equality of variances yields a *P*-value 0.016, so the null

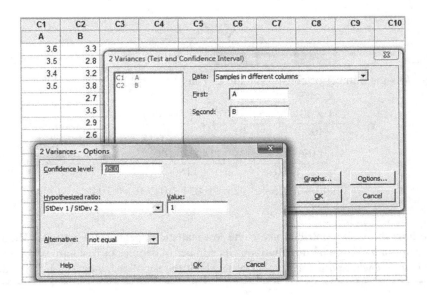

Figure 7.40   Dialog for test of equality of variances.

```
Test and CI for Two Variances: A, B
Method

Null hypothesis         Sigma(A) / Sigma(B) = 1
Alternative hypothesis  Sigma(A) / Sigma(B) not = 1
Significance level      Alpha = 0.05

.............................................................................

Tests

                                                 Test
Method                         DF1   DF2    Statistic   P-Value
F Test (normal)                  3     7         0.04     0.021
Levene's Test (any continuous)   1    10         8.28     0.016
```

Panel 7.48   Session window output for test for equal variances.

hypothesis of equal variances would be rejected at the 5% level of significance. Valid application of this test simply requires the distributions to be continuous.

As a second example, consider the drive length data in Figure 7.26, also available in Types. MTW. Here, in the fixed effects scenario, there were three populations of interest corresponding to the three ball types. Use of **Stat > ANOVA > Test for Equal Variances...** provides a test of the equality of two or more variances – in this case of the null hypothesis $H_0 : \sigma_1^2 = \sigma_2^2 = \sigma_3^2$ versus the alternative hypothesis H_1: not all variances are equal. The dialog is shown in Figure 7.41.

The Session Widow output is shown in Panel 7.49. Graphical output is provided but is not reproduced here. With P-values of 0.956 and 0.851 there is no reason to doubt the null hypothesis of equal variances for the three types. Bonferroni confidence intervals are given for the population standard deviations. As with the Tukey procedure for multiple comparisons, the Bonferroni procedure is designed to yield a set of confidence intervals with an overall joint confidence level – the default of 95% was specified in this example. The user may change the

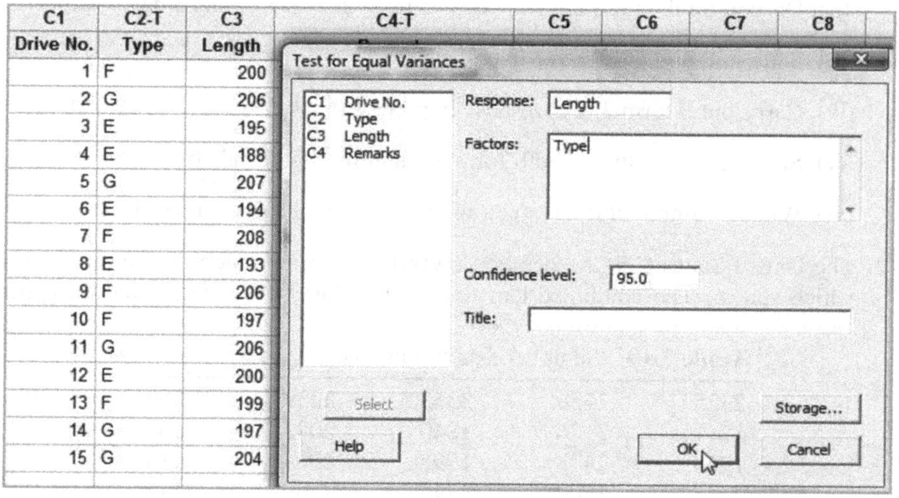

Figure 7.41   Dialog for test of equality of variances.

```
Test for Equal Variances: Length versus Type

95% Bonferroni confidence intervals for standard deviations

Type  N    Lower     StDev    Upper
   E  5  2.32449   4.30116  16.5547
   F  5  2.56350   4.74342  18.2569
   G  5  2.19525   4.06202  15.6343


Bartlett's Test (Normal Distribution)
Test statistic = 0.09, p-value = 0.956

Levene's Test (Any Continuous Distribution)
Test statistic = 0.16, p-value = 0.851
```

Panel 7.49 Session window output for test for equal variances.

confidence level via **Options:**. The assumption required for the valid use of each test is stated in brackets in the Session window output after the name of each test.

In addition to the use of these tests concerning the variability of populations for checking the validity of assumptions underlying the use of tests concerning location, they provide a means of analysing data from an experiment with a single factor where variability is the response of interest.

7.9   Exercises and follow-up activities

1. The output from a manufacturing operation over many weeks was, on average, 2000 units per day with standard deviation 400 units. Following changes to plant configuration the outputs for a sample of 25 days were as displayed in Table 7.19.

   The data are available in the supplied file Output.xls.

   (i) Display the data.

   (ii) Do you think that there has been a 'real' increase in mean daily output?

   (iii) State null and alternative hypotheses.

   (iv) Carry out a formal test of these hypotheses.

   (v) State your conclusion both formally and in plain English.

   (vi) What assumption(s) have you made in order to carry out the test?

2. The United States Golf Association (2008) has a requirement concerning golf balls which states: 'The combined carry and roll of the ball, when tested on apparatus

Table 7.19 Sample of daily outputs.

2325	2756	2358	2239	2660
1711	2176	1840	2082	2008
1760	2473	1998	2304	2215
2145	1900	1779	1709	2145
1596	2278	2482	2426	2160

Table 7.20   Sample of distances (yards) achieved.

316	315	315	317	333	328	330	317
313	329	339	302	324	323	311	296

approved by the United States Golf Association, must not exceed the distance specified under the conditions set forth in the Overall Distance Standard for golf balls on file with the United States Golf Association.' The QC Manager at Acme tested a random sample of 16 balls of a particular brand using similar equipment on the company's range, with the results displayed in Table 7.20. (Experience has shown that a standard deviation of 12 yards is typical for balls manufactured by Acme.)

(i) Display the data.

(ii) Do you think that the data for the brand of ball suggest a mean value that differs from the current Overall Distance Standard of 317 yards?

(iii) Explain why a two-tailed test is appropriate here.

(iv) State null and alternative hypotheses and carry out a formal test of these hypotheses.

(v) State your conclusion both formally and in plain English.

(vi) What assumption(s) have you made in order to carry out the test?

3. Suppose that initially assembly of P86 modules took on average 50.0 minutes with standard deviation 4.2 minutes. At a later date a sample of nine assembly times (minutes) was 44, 48, 45, 45, 46, 49, 48, 51 and 47.

   Evaluate the evidence for a reduction in the mean assembly time using a z-test, a t-test, a sign test and a Wilcoxon test. State any assumptions required for each test and whether or not they are reasonable.

   Estimate the size of sample required to detect evidence, at significance level 0.01 and with power 0.99, of a reduction in the mean of 2 minutes assuming that the standard deviation has remained at 4.2 and also without making this assumption.

4. An accountant believes that a company's cash flow problems are due to outstanding accounts receivable. She claims that 70% of the current accounts receivable are over 3 months old. A sample of 120 accounts receivable revealed 78 over 3 months old. Verify that the accountant's claim cannot be rejected at the at the 5% level of significance. Estimate the size of sample required to provide evidence, at significance level 0.05 and with power 0.9, of a reduction in the proportion from 70% to 60%.

5. A supplier claims that at least 95% of the parts it supplies meet the product specifications. In a sample of 500 parts received over the last 6 months, 36 were defective. Test the supplier's claim at the 5% level of significance.

6. In discussion of the Minitab Pulse.MTW data set earlier in the book it was noted that 35 students from a class of 92 ran on the spot, the decision whether or not to run

supposedly based on the outcome of the flip of a coin by the student. Do these data provide any evidence of 'cheating'?

7. Under the Weights and Measures (Packaged Goods) Regulations 1986, display of the text '330 ml e' on a bottle of beer with nominal content volume of 330 ml means that not more than 2.5% of bottles may be deficient in content volume by more than the tolerable negative error (TNE) specified for the nominal quantity (Trading Standards Net). The TNE for nominal volume of 330 ml is 3% of the nominal volume.

   A brewery has bottling machines which deliver amounts that are normally distributed with known standard deviation 6 ml and which are capable of filling bottles with nominal capacities of both 330 and 500 ml.

   (i) Explain why, when filling 330 ml bottles, the brewery should aim to set up the filling process to operate with a mean of 332 ml (to the nearest ml).

   (ii) Following set-up for a run of 330 ml bottles, after a run of 500 ml bottles, a sample of 24 bottles was checked and found to have the content volumes stored in Beer. xls. Use hypothesis testing to investigate whether or not the machine has been set up correctly.

8. This exercise has been created as an aid to understanding the concept of a confidence interval. The worksheet Strength.MTW contains data for the tensile strengths (N/mm^2) of a sample of 16 components. Assume that the production process for the components is known to operate with a standard deviation of 1.9 N/mm^2.

   Verify, using Minitab, that in a z-test of $H_0 : \mu = 50.0$ versus $H_1 : \mu \neq 50.0$ the decision would be to accept H_0. Repeat the test for all the other null hypotheses listed in Table 7.21 and record your decisions. Note the range of values for the population mean, μ, that would be accepted. By further changing of the mean value specified in the null hypothesis, determine, to 2 decimal places, the range of mean values that would be accepted. Check that the formula $\bar{x} \pm 1.96\sigma/\sqrt{n}$ gives the same results, apart form a small rounding error.

   The range of mean values that would be accepted is a *95% confidence interval for the mean* tensile strength. Check that the 95% confidence interval provided by Minitab confirms your calculations.

   Obtain a 99% confidence interval for the mean tensile strength. Note that with greater confidence we now have a wider range of mean values that would be accepted.

Table 7.21   Decisions from tests of hypotheses.

Null hypothesis H_0	Decision at 5% significance level (two-tailed z-test)
$\mu = 48.5$	
$\mu = 49.0$	
$\mu = 49.5$	
$\mu = 50.0$	Accept H_0
$\mu = 50.5$	
$\mu = 51.0$	
$\mu = 51.5$	

9. Obtain a 95% confidence interval for the mean content volume from the bottle data in Beer.xls. How does your answer confirm your earlier conclusion regarding set-up in Exercise 6?

10. Suppose that a quality manager claims that 70% of units pass final inspection first time and that you check a sample of 40 units from the database and find that 23 of them passed final inspection first time.

    Calculate the percentage of the sample that passed final inspection first time and note your 'gut feeling' concerning the manager's claim. Use Minitab to obtain a 95% confidence interval for the proportion of the population of units which pass final inspection first time and state whether the result lends support to the manager's claim or otherwise. Was your gut feeling supported by the statistical analysis?

    Investigate the situation where checking a sample of 400 units revealed 230 failures.

11. The workbook Verify.xls contains information on whether or not a series of units passed verification first time. Are the data consistent with the manufacturing operation achieving a first-time pass rate of 80%?

12. A manufacturer of automatic teller machines introduced changes to the procedure for installing the printer in the carcass as part of a process improvement initiative. Random samples of installation times for a technician before and after the changes are tabulated in Table 7.22 and available in Printer.MTW.

    Investigate the evidence for a reduction in the mean installation time using both parametric and nonparametric tests. Whenever possible check any assumptions required.

    As an exercise, perform the two-sample t-tests using the data as presented in the worksheet, using the data in stacked form, and finally using the summarized data. These three methods of presenting the data to Minitab correspond to **Samples in one column, Samples in different columns** and **Summarized data** in the dialog box for the two-sample t-test.

13. A process improvement project on the manufacture of light bulbs was carried out in order to compare two different types of lead wire. Misfed lead wires require operator intervention. Data on the average hourly number of misfed leads for 12 production runs with the standard type of wire and for 12 production runs with a modified type of wire are given in Misfeeds.MTW. Investigate the evidence for a process improvement using both parametric and nonparametric tests. Whenever possible, check any assumptions required.

14. The PCS-12 is a generic measure of physical health status. The measure has been devised in such a way that in the general population of people in good health it has mean 50 and standard deviation 10. Table 7.23 gives PCS-12 scores for a random sample of patients who had hip joint replacement operations carried out, both

Table 7.22   Before and after samples of installation times.

Before	64	39	59	31	42	52	43
After	19	41	29	45	37	35	39

Table 7.23   Before and after PCS12 scores for a sample of 12 patients.

Patient	1	2	3	4	5	6	7	8	9	10	11	12
Pre	36	45	30	63	48	52	44	44	45	51	39	44
Post	39	42	33	70	53	51	48	51	51	51	42	50

immediately prior to the operation (Pre) and 6 months later (Post). The data are available in PCS12.MTW.

(a) Perform both parametric and nonparametric tests of hypotheses, investigating any assumptions required where possible, to investigate whether or not the data provide evidence of improvement in the physical health status of patients following the operation.

(b) Investigate whether or not the post-operative data are consistent with the mean for the general population.

15. Table 7.24 gives wear resistance data for four fabrics, obtained from a completely randomized single-factor experiment in which four samples of each one of a set of four fabrics of interest were tested. Wear was assessed by measuring the weight loss after a specific number of cycles in the wear-testing machine. The data, available in Fabrics. MTW, are from p.63 of *Fundamental Concepts in the Design of Experiments*, 5th edition, by Charles R. Hicks and Kenneth V. Turner, Jr, copyright © 1964, 1973, 1982, 1993, 1999 and used by permission of Oxford University Press, Inc., New York. Carry out a one-way ANOVA using Minitab, and perform diagnostic checks of assumptions and follow-up analysis if appropriate. Summarize your findings.

16. Sample sizes may be unequal in an experiment with a single factor. Box *et al.* (2005, p. 134) give an example on coagulation time for samples of blood from animals fed on a fixed set four diets A, B, C and D which were of interest. (Box *et al.*, 1978, p. 166). The data are available in Coagulation.MTW. (Reproduced by permission of John Wiley & Sons, Inc., New York.) Carry out an analysis of variance. The shorter the coagulation time is, the better from an animal health point of view. What recommendations would you make on the basis of the experiment?

17. Analyse the data in Exercises 15 and 16 using the Kruskal–Wallis procedure.

18. Analyse the data in Exercise 12 using ANOVA and verify that the P-value is exactly the same as that obtained from a two-sample t-test (assuming equal variances). In this situation the two tests are mathematically equivalent.

Table 7.24   Wear data for four fabrics.

	Fabric		
A	B	C	D
1.93	2.55	2.40	2.33
2.38	2.72	2.68	2.40
2.20	2.75	2.31	2.28
2.25	2.70	2.28	2.25

Table 7.25   Yield data for five batches.

Batch 1	Batch 2	Batch 3	Batch 4	Batch 5
74	68	75	72	79
76	71	77	74	81
75	72	77	73	79

19. A company supplies a customer with many batches of raw material in a year. The customer is interested in high yields of usable chemical in the batches. For quality control of incoming material purposes three sample determinations of yield are made for each batch.

    The data, available in Batches.MTW and displayed in Table 7.25, are from p.78 of *Fundamental Concepts in the Design of Experiments*, 5th edition, by Charles R. Hicks and Kenneth V. Turner, Jr, copyright © 1964, 1973, 1982, 1993, 1999 and used by permission of Oxford University Press, Inc., New York. Show that about 87% of the variation in yield is due to batch-to-batch variation with the remaining 13% of variation being due to variation within batches.

20. If four brands of car tyre A, B, C and D were to be tested using four tyres of each type and four cars, explain why design 1 displayed in Table 7.26 would be unsatisfactory.

    Table 7.27 gives the design and the results for the experiment actually carried out. State the type of design used. Set up the data in Minitab, analyse them and report your findings.

21. Table 7.28 gives weekly revenue (£000) for three city restaurants of the same size belonging to a restaurant chain. The weeks were a random sample of weeks during

Table 7.26   Design 1.

Design 1	Car			
	P	Q	R	S
	A	B	C	D
Tyre Brand	A	B	C	D
	A	B	C	D
	A	B	C	D

Table 7.27   Design 2.

Design 2	Car			
	P	Q	R	S
	B(1.9)	D(1.6)	A(1.8)	C(1.4)
Tyre Brand and	C(1.7)	C(1.7)	B(1.8)	D(1.4)
Wear (mm)	A(2.2)	B(1.9)	D(1.6)	B(1.4)
	D(1.8)	A(1.9)	C(1.5)	A(1.8)

Table 7.28   Weekly turnover data for three restaurants.

Week	Restaurant 1	2	3
1	8.3	7.4	9.2
2	10.7	10.0	12.8
3	9.5	8.5	14.6
4	3.2	3.9	7.2
5	12.7	12.6	13.2

2004. Analyse the data, stored in Restaurants.MTW, and report your findings. Is there evidence from the data of a clear winner of 'Restaurant of the Year' from the point of view of revenue?

22. Set up the data from Exercise 14 as data from a randomized complete block experiment with the patients as blocks. Carry out an ANOVA and verify that the P-value for testing the effect of the operation is the same as was obtained from the paired t-test. Paired data experiments are a special case of randomized complete block designs.

8

Process experimentation with two or more factors

Many important phenomena depend, not on the operation of a single factor, but on the bringing together of two or sometimes more factors at the appropriate levels. (Box, 1990, p. 365)

Overview

The design of experiments (DOE) is a vast topic – designed experiments are major tools for use in the improve phase of many Six Sigma projects. In the previous chapter experiments involving only a single factor of interest were considered. In this chapter experiments involving two or more factors of interest will be introduced. Much process improvement experimentation is done on a one-factor-at-a-time basis. Although such experimentation can lead to improvement, there is no doubt that multifactor experiments, in which factor levels are varied systematically according to a recognized design, can be much more informative. In particular, multifactor experiments can reveal the presence of important interactions between factors. Harnessing interaction effects can lead to dramatic process improvements.

Minitab can be of assistance both with the actual design of the experiment and with the display and analysis of the resulting data. Following a general introduction, in which the concept of interaction will be introduced, experiments in the 2^k series, with k factors, each with two levels, will be considered. Screening experiments and fractional factorial experiments in the 2^{k-p} series will be introduced, together with the concept of design resolution. The fundamentals of response surfaces will be described. Reference will be made to Taguchi experimental designs.

Six Sigma Quality Improvement with Minitab, Second Edition. G. Robin Henderson.
© 2011 John Wiley & Sons, Ltd. Published 2011 by John Wiley & Sons, Ltd.

8.1   General factorial experiments

8.1.1   Creation of a general factorial experimental design

As an introductory example, consider the process of making popcorn. Two factors are of interest – the first is the type of popper (X_1) and the second is the grade of corn used (X_2). The levels of interest for popper are air and oil, those for corn are budget, regular and luxury. The response of interest is the volume (ml) yield of popcorn (Y) from 250 ml of corn processed according to the instructions provided by the manufacturers of the machines. The design may be created using **Stat > DOE > Factorial > Create Factorial Design...** for which the initial dialog is shown in Figure 8.1.

The first step is to select **General full factorial design** with **Number of factors:** 2 specified. (With the generality of this type of design it is not possible to provide a catalogue of designs under **Display Available Designs. . ..**) The second step is to click on **Designs...** and engage in the subdialog as indicated in Figure 8.2.

The name of each factor may be entered together with the number of levels – in this case popper has two levels and corn has three. The default **Number of Replicates:** is 1, but in this experiment 4 was the number used. This means that the process was operated four times with each of the 2 (levels of popper) × 3 (levels of corn) = 6 factor-level combinations (FLCs) – we say that it was *replicated* four times. Thus the term 'replicate' is being used in a technical statistical sense in the context of designed experimentation. (When more than one replicate is employed the opportunity to block on replicates is made available but was not required in this case.) Having clicked **OK**, click on **Factors. . .** and complete the dialog shown in Figure 8.3. Here the categorical levels for the factors mean that **Type** should be selected as Text. (Numeric would be selected in the case of a factor such as temperature specified in degrees Celsius.)

Under **Options. . .** the default to **Randomize runs** is strongly advised as part of good experimental design practice. (If the user enters an integer under **Base for random data generator:** then subsequent use of **Randomize runs** with **Create Factorial Designs** and the

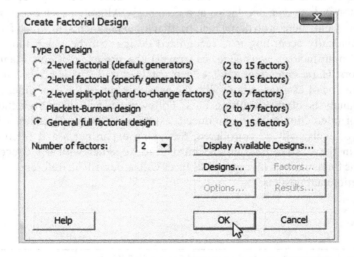

Figure 8.1   Initial dialog for general full factorial design.

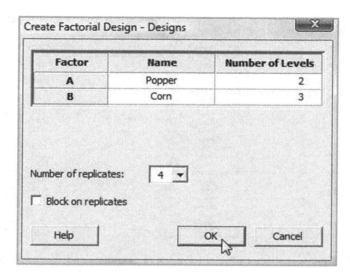

Figure 8.2 Specifying the factors.

same numbers of factors and levels and the same base leads to the same run order.) Opting to **Store design in worksheet**, which should of course be saved, is essential if data from the experiment are to be readily analysed later.

Having clicked **OK** and then **OK** again the design worksheet is created. The version obtained by the author is shown in Figure 8.4. It was augmented through the addition of a column named Yield for the recording of the yield obtained from each run and a column named Remarks in which those carrying out the experiment may note any unusual occurrences etc. It also has a column of residuals and a column of fitted values obtained during analysis of the data following completion of the experiment.

The columns created by Minitab are as follows:

- a column showing the standard order for each run – the reader is invited to create the worksheet again with the **Randomize runs** option unchecked and to scrutinize the systematic pattern in the standard order column;

- a column showing the actual run order created by the randomization procedure;

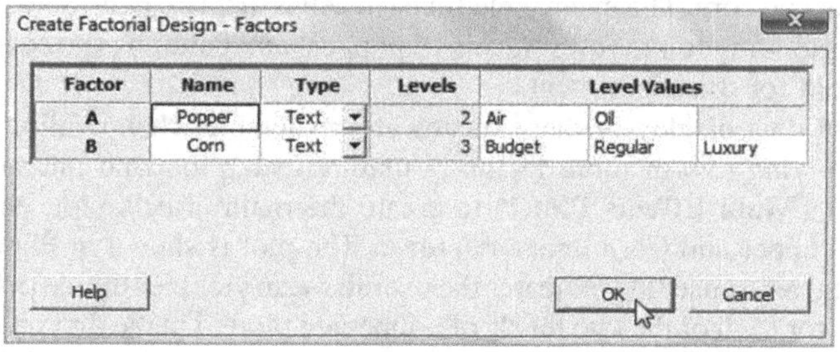

Factor	Name	Type	Levels		Level Values	
A	Popper	Text	2	Air	Oil	
B	Corn	Text	3	Budget	Regular	Luxury

Create Factorial Design - Factors

Help   OK   Cancel

Figure 8.3 Specifying the factors: name, type, number of levels and level values.

C1	C2	C3	C4	C5-T	C6-T	C7	C8-T	C9	C10
StdOrder	RunOrder	PtType	Blocks	Popper	Corn	Yield	Remarks	RESI1	FITS1
9	1	1	1	Air	Luxury	1360		-112.125	1472.12
12	2	1	1	Oil	Luxury	1349		52.625	1296.37
10	3	1	1	Oil	Budget	943	Much lower yield from Budget	-27.625	970.62
14	4	1	1	Air	Regular	1090		-199.250	1289.25
6	5	1	1	Oil	Luxury	1108		-188.375	1296.37
21	6	1	1	Air	Luxury	1518		45.875	1472.12
8	7	1	1	Air	Regular	1342		52.750	1289.25
16	8	1	1	Oil	Budget	948		-22.625	970.62
23	9	1	1	Oil	Regular	1289		175.500	1113.50
7	10	1	1	Air	Budget	1120		-26.375	1146.37
24	11	1	1	Oil	Luxury	1241		-55.375	1296.37
2	12	1	1	Air	Regular	1537		247.750	1289.25
13	13	1	1	Air	Budget	1106		-40.375	1146.37
11	14	1	1	Oil	Regular	1148		34.500	1113.50
17	15	1	1	Oil	Regular	1058		-55.500	1113.50
4	16	1	1	Oil	Budget	923		-47.625	970.62
22	17	1	1	Oil	Budget	1164		193.375	970.62
1	18	1	1	Air	Budget	1198		51.625	1146.37
19	19	1	1	Air	Budget	1066		-80.375	1146.37
5	20	1	1	Oil	Regular	887		-226.500	1113.50
3	21	1	1	Air	Luxury	1585		112.875	1472.12
15	22	1	1	Air	Luxury	1449		-23.125	1472.12
20	23	1	1	Air	Regular	1260		-29.250	1289.25
18	24	1	1	Oil	Luxury	1464	Phew!	167.625	1296.37

Figure 8.4 Design worksheet with yields, remarks, residuals and fits from additive model.

- a column indicating point type in what is known as the design space – an explanation will be given later in the chapter (here all points are of type 1);

- a column indicating block – when there is no actual blocking this column consists entirely of 1s.

Armed with a copy of the worksheet, with the additional columns for yield and remarks added, the experimenters could head for the kitchen. The first run would have been with the air popper using 250 ml of luxury grade corn and would have given a yield of 1360 ml of popped corn, the second run would have been with the oil popper using 250 ml of luxury grade corn and would have given a yield of 1349 ml of popped corn and so on. Once the experimental work is complete display and analysis of the data can begin. The first eight columns displayed in Figure 8.4 are provided in worksheet Popcorn.MTW. The columns of residuals and fitted values in Figure 8.4 will be referred to later in the chapter.

8.1.2   Display and analysis of data from a general factorial experiment

The data may be presented in the form of a two-way table (Table 8.1) with rows corresponding to the levels of the popper factor and with columns corresponding to the levels of the corn factor. Each cell in the table gives the four replicate yields obtained with one particular FLC for the factors popper and corn. The mean yield for each cell is displayed in Table 8.2 together with the means corresponding to the rows (levels of popper) and columns (levels of corn) and the overall mean yield for the experiment.

Two forms of data display of these means are invaluable when dealing with data from experiments involving two or more factors – main effects plots and interaction plots. Use **Stat > ANOVA > Main Effects Plot...** to create the main effects plot; enter Yield under **Reponses:** and Popper and Corn under **Factors:**. The plot is shown in Figure 8.5.

The horizontal reference line indicates the overall mean yield for the experiment of 1214.7. The mean yields for each of the two levels of popper are plotted in the first panel and the mean yields for each of the three levels of corn are plotted in the second. Two insights are obtained. The first is that, on average, switching from the air popper to the oil popper reduces yield by

Table 8.1   Raw data for popcorn experiment.

Yield	Budget				Regular				Luxury			
Air	1120				1090				1360			
		1066				1342				1518		
			1106				1537				1585	
				1198				1260				1449
Oil	943				1289				1349			
		948				887				1108		
			923				1148				1241	
				1164				1058				1464

Table 8.2   Means summary of yield.

Mean Yield	Budget	Regular	Luxury	Mean
Air	1122.5	1307.3	1478.0	1302.6
Oil	994.5	1095.5	1290.5	1126.8
Mean	1058.5	1201.4	1384.3	1214.7

about 200 ml. The second is that, on average, yield progressively increases as we switch from budget grade corn to regular to luxury.

Use **Stat > ANOVA > Interactions Plot...** to create the interactions plot; enter Yield under **Reponses:** and Popper and Corn under **Factors:**. The plot is shown in Figure 8.6.

In this plot the six means at the core (shaded) of Table 8.2, each of which corresponds to an FLC, are displayed. The levels of corn are indicated on the horizontal axis. The levels of popper (see the legend in the top right-hand corner of the display) are indicated through the connection by line segments of those points that have the same level of popper. The pairs of line segments

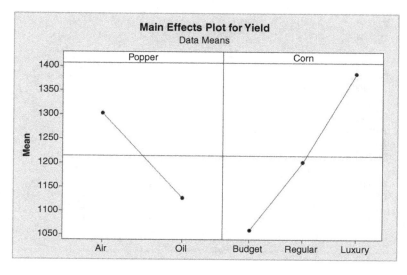

Figure 8.5   Main effects plot for popcorn experiment.

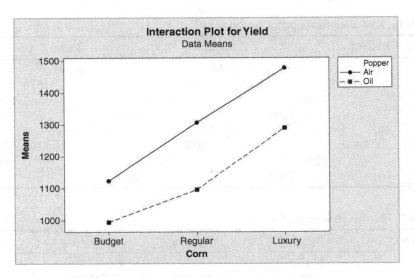

Figure 8.6   Interaction plot – first version.

lying vertically above each other are approximately parallel. This means that, when using one particular grade of corn, similar decreases in mean yield are experienced on changing from the level air to the level oil of popper. This is characteristic of a situation in which there is no interaction between factors.

To create an alternative version of the interactions plot enter Yield under **Reponses:** and Corn and Popper under **Factors:** i.e. reverse the order in which the factors are entered. The plot is shown in Figure 8.7. The levels of popper are indicated on the horizontal axis. The levels of the factor corn are indicated through connection by line segments of those points that have the same level of corn, as indicated by the legend. Again the segments lying vertically above and

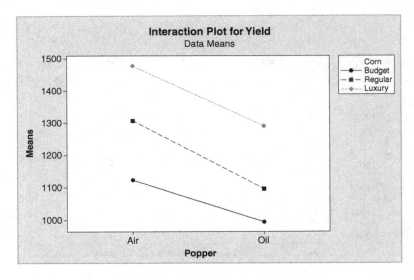

Figure 8.7   Interaction plot – second version.

GENERAL FACTORIAL EXPERIMENTS 309

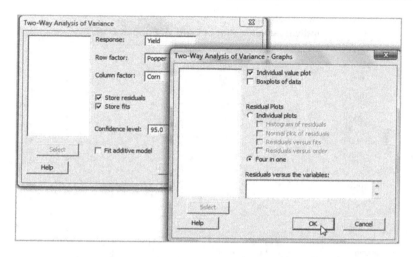

Figure 8.8   Dialog for two-way ANOVA.

below each other are approximately parallel. This means that, on changing the level of corn, similar changes in mean yield will be experienced, whether one is using an air popper or an oil popper.

If desired, both versions of the interaction plot may be created simultaneously by checking **Display full interaction plot matrix** in the dialog.

The data may be examined formally for evidence of interaction between the factors popper and corn by performing a two-way analysis of variance via **Stat > ANOVA > Two-Way...**. The terminology 'two-way' relates to the presentation of data from a general factorial experiment involving two factors in the form of a two-way table, as in Table 8.1. The dialog is shown in Figure 8.8.

Note carefully that the **Fit additive model** option must not be checked as we wish to formally assess the evidence for the presence of interaction. Under **Graphs...** it is strongly recommended that either the **Individual value plot** or **Boxplots of data** display be selected – when the number of replications is small the author recommends that the first of these plots be used. The **Four in one** option for **Residual Plots** is appropriate in this case since the observations of yield in the experiment are recorded in the worksheet in the time order in which they were obtained. The individual value plot is shown in Figure 8.9.

Note that the individual value plot in Figure 8.9 may be viewed as an 'exploded' version of the interactions plot in Figure 8.6 in which the individual yield values are plotted in addition to the yield means. The reader is invited to change the ordering of the factors in the dialog displayed on Figure 8.8 and to compare the individual value plot that is obtained with the interactions plot displayed in Figure 8.7.

As with the main effects plot in Figure 8.5 and the interaction plots in Figures 8.6 and 8.7, the plot in Figure 8.9 gives similar insights. It appears that:

- on average, switching from the air popper to the oil popper reduces yield by about 200 ml;

- on average, yield progressively increases as we switch from budget grade corn to regular to luxury.

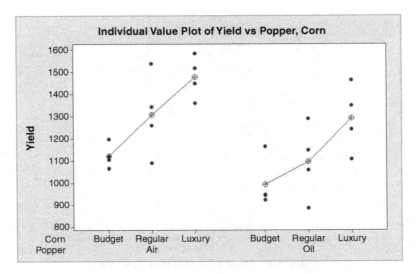

Figure 8.9   Individual value plot of yield data.

The standard residual plots are shown in Figure 8.10. The normal probability plot of the residuals is reasonably straight. Thus the assumption of a normal distribution of yield, for each combination of factor levels, is supported. Support for the assumption of a common variance for the distributions of yield for the six FLCs is provided by the similar vertical spread in all six groups of points in the plot of residuals versus fitted values. We can therefore proceed to interpret the ANOVA table with confidence that the assumptions underlying the valid application of the F-tests are reasonable.

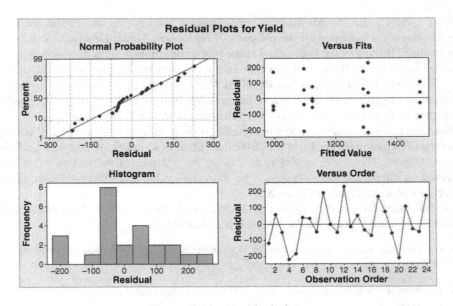

Figure 8.10   Residual plots.

```
Two-way ANOVA: Yield versus Corn, Popper

Source         DF       SS       MS       F      P
Corn            2   426586   213293   11.51  0.001
Popper          1   185328   185328   10.01  0.005
Interaction     2     7428     3714    0.20  0.820
Error          18   333423    18523
Total          23   952765

S = 136.1   R-Sq = 65.00%   R-Sq(adj) = 55.28%
```

Panel 8.1 ANOVA for model with interaction.

The ANOVA table is shown in Panel 8.1. In total we have a sample of 24 values of yield from the experiment so there are $24 - 1 = 23$ degrees of freedom (DF) in total. With 2 levels of popper there is $2 - 1 = 1$ degree of freedom for popper, and similarly $3 - 1 = 2$ degrees of freedom for corn. The number of degrees of freedom for the interaction between popper and corn is the product of the numbers of degrees of freedom for these two factors, i.e. $1 \times 2 = 2$. Thus the number of degrees of freedom for error is obtained by calculating $23 - (1 + 2 + 2) = 18$. The sums of squares (SS) are calculated using formulae that need not concern us; the mean squares (MS) are obtained by dividing each SS by its corresponding DF. Provided the assumptions referred to above are valid, the ratios of the MS values for popper, corn and interaction will have F-distributions. These enable the calculation of P-values.

The experiment provides no evidence of interaction between popper and corn since the P-value for interaction of 0.820 is well in excess of 0.05. This means that there is no evidence from the experiment that the effect on yield of changing the grade of corn used depends on the type of popper being used. When there is no evidence of interaction in a two-factor experiment of this nature one can proceed to fit an additive model. This may readily be achieved using **Edit Last Dialog** and checking **Fit additive model**. In addition **Store residuals** and **Store fits** were checked in order that fitted values and residuals were available for explanation that follows. These fitted values and residuals are displayed in the columns labelled RESI1 and FITS1 in Figure 8.4. The residual plots were again satisfactory (but are not displayed here) and the revised ANOVA table is shown Panel 8.2.

For popper the P-value is 0.004 so this means that the main effect of popper is significant at the 1% level. The main effect of corn is significant at the 0.1% level. Thus the experiment provides strong evidence that the factors of interest – type of popper (X_1) and grade of corn used (X_2) – influence the response of interest, i.e. the volume (ml) yield of popcorn (Y), from 250 ml of corn processed according to the instructions provided by the manufacturers of the machines.

We will now introduce and fit the formal additive model underlying the analysis. Reference to Table 8.2 enables the effects for popper to be calculated as indicated in Table 8.3. The overall

```
Two-way ANOVA: Yield versus Corn, Popper

Source  DF       SS       MS       F      P
Corn     2   426586   213293   12.52  0.000
Popper   1   185328   185328   10.87  0.004
Error   20   340851    17043
Total   23   952765

S = 130.5   R-Sq = 64.23%   R-Sq(adj) = 58.86%
```

Panel 8.2 ANOVA for additive model.

Table 8.3   Effects for popper.

Level	Mean yield	Effect
Air	1302.6	87.9
Oil	1126.8	− 87.9

mean yield for the experiment was 1214.7 ml. The effect corresponding to a particular level of a factor is the mean yield for that factor level minus the overall mean yield for the experiment. Hence the air popper effect is 1302.6 − 1214.7 = 87.9 and the oil popper effect is 1126.8 − 1214.7 = − 87.9. Thus we can interpret the effect of using the air popper as being to elevate yield from the overall mean of 1214.7 by 87.9 ml on average, and the effect of using the oil popper as being to depress (because of the negative effect) yield by 87.9 ml on average. Note that the two effects sum to 0.

Reference to Table 8.2 enables the effects for corn to be calculated as indicated in Table 8.4, and the reader is invited to confirm the calculations. Thus we can interpret the effect of using the budget grade corn as being to depress yield from the overall mean of 1214.7 by 156.2 ml on average, that of using the regular grade being to depress yield by 13.3 ml on average and that of using the luxury grade being to elevate yield by 169.6 ml on average. Note that the three effects sum to 0 (allowing for rounding of the means quoted in Table 8.2 to one decimal place).

The reader will recall from the previous chapter that

$$\text{Observed data value} = \text{Value fitted by model} + \text{Residual},$$

which we abbreviated as

$$\text{Data} = \text{Fit} + \text{Residual}.$$

In this case we take

$$\text{Fit} = \text{Overall mean} + \text{Popper effect} + \text{Corn effect}.$$

The first experimental run was carried out using an air popper with luxury grade corn and gave a yield of 1360 (see Figure 8.4). The fitted value is given by

$$\text{Fit} = \text{Overall mean} + \text{Air popper effect} + \text{Luxury grade of corn effect}$$
$$= 1214.7 + 87.9 + 169.6 = 1472.2.$$

Thus the corresponding residual may be obtained as

$$\text{Residual} = \text{Data} - \text{Fit} = 1360 - 1472.2 = - 112.2.$$

Table 8.4   Effects for corn.

	Corn		
Level	Budget	Regular	Luxury
Mean yield	1058.5	1201.4	1384.3
Effect	− 156.2	− 13.3	169.6

Scrutiny of the column of residuals computed by Minitab (see Figure 8.4) reveals a value of -112.125. Again the small discrepancy is due to rounding. The reader is invited to calculate the next few residuals in order to check his/her understanding.

8.1.3 The fixed effects model, comparisons

The *fixed* effects model for a general two-factor design, *with no interaction*, is specified in Box 8.1. The number of levels of the row factor is a, the number of levels of the column factor is b and the number of replications is n. For example, with $i = 2, j = 3$ and $k = 1$ we have the specific relationship

$$Y_{231} = \mu + \alpha_2 + \beta_3 + \varepsilon_{231}.$$

For the popcorn experiment this states that the yield from the first run ($k = 1$, indicating the first replicate) with the oil type of popper ($i = 2$, indicating the second level of the factor popper) and the luxury grade of corn ($j = 3$, indicating the third level of the factor corn) is made up of the overall mean plus the effect of oil type of popper plus the effect of luxury grade corn plus a random error. Earlier we estimated the overall mean, μ, as 1214.7, the effect of oil type popper, α_2, as -87.9 and the effect of luxury grade corn, β_3, as 169.6. Thus the fitted value for the combination of oil type of popper with luxury corn is the sum $1214.7 + (-87.9) + 169.6 = 1296.4$. The value $s = 130.5$, which is given beneath the ANOVA table and is the square root of the error MS in the ANOVA table in Panel 8.2, provides an estimate of the standard deviation σ of the random error component of the model. The model predicts that the population of yields obtained that would be obtained with the combination of oil type of popper and luxury grade corn would be normally distributed with mean 1296.4 and standard deviation 130.5. Similar statements may be made about the other FLCs.

Follow-up comparisons are available via **Stat > ANOVA > General Linear Model....** The dialog is shown in Figure 8.11. In **Model:** we are communicating to Minitab the nature of the model we are using, i.e. $Y_{ijk} = \mu + \alpha_i + \beta_j + \varepsilon_{ijk}$. Such models always include an overall mean and the random error term, so by entering Popper and Corn in **Model:** we are indicating the $\alpha_i + \beta_j$ terms in the core of the model. In the **Comparisons...** subdialog, **Pairwise comparisons** were selected, by the **Tukey** method. **Grouping information** and **Confidence interval** were both checked, with the default percentage **Confidence level:** 95.0 used. Part of the corresponding section of the Session window output is shown in Panel 8.3.

Observed data value = Overall mean + Row factor effect + Column factor effect
$\qquad\qquad$ + Random error

$Y_{ijk} = \mu + \alpha_i + \beta_j + \varepsilon_{ijk}, \quad i = 1, 2, \ldots, a, \quad j = 1, 2, \ldots, b, \quad k = 1, 2, \ldots, n$

$\displaystyle\sum_{i=1}^{a} \alpha_i = 0, \quad \sum_{j=1}^{b} \beta_j = 0, \quad \varepsilon_{ijk} \sim N(0, \sigma^2)$

Box 8.1 Additive fixed effects model for two factors.

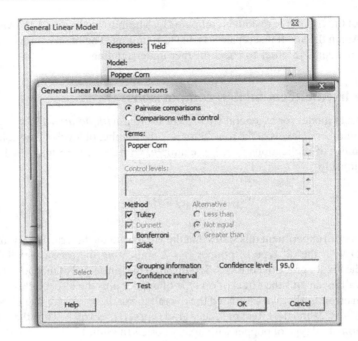

Figure 8.11 General Linear Model dialog.

The interpretation, with rounding of values, is as follows:

- On average the yield with the oil type popper is 176 ml lower than that with the air type popper, with confidence interval 65 ml to 287 ml (to the nearest integer).

- On average the yield with regular grade corn is 143 ml higher than with budget grade corn, with confidence interval − 22 to 308 ml (to the nearest integer).

- On average the yield with luxury grade corn is 326 ml higher than with budget grade corn, with confidence interval 161 to 491 ml (to the nearest integer).

- On average the yield with luxury grade corn is 183 ml higher than with regular grade corn, with confidence interval 18 to 348 ml (to the nearest integer).

Since the second confidence interval includes 0, we have no evidence that the mean yield with budget grade corn differs significantly from that with regular grade corn. The other confidence intervals provide evidence of the superiority in terms of yield of an air popper with luxury grade corn. The reader may, of course, arrive at the key conclusions by scrutinizing the grouping information provided in the Session window output prior to each set of confidence intervals.

If the aim is to maximize yield (Y) using a combination of the fixed set of levels of the factors popper (X_1) and corn grade (X_2) considered, the levels air and luxury should be selected. There are indications of a clear 'winning combination' in this scenario. Note that this combination corresponds to selection of the level for popper corresponding to the greatest mean for the levels of popper and to the selection of the level for corn corresponding to the

```
Grouping Information Using Tukey Method and 95.0% Confidence

Popper   N    Mean  Grouping
Air     12  1302.6  A
Oil     12  1126.8    B

Means that do not share a letter are significantly different.


Tukey 95.0% Simultaneous Confidence Intervals
Response Variable Yield
All Pairwise Comparisons among Levels of Popper
Popper = Air  subtracted from:

Popper  Lower  Center   Upper  ------+---------+---------+---------+
Oil    -286.9  -175.8  -64.58  (-------------*-------------)
                               ------+---------+---------+---------+
                              -240      -160      -80         0


Grouping Information Using Tukey Method and 95.0% Confidence

Corn      N    Mean  Grouping
Luxury    8  1384.2  A
Regular   8  1201.4    B
Budget    8  1058.5    B

Means that do not share a letter are significantly different.


Tukey 95.0% Simultaneous Confidence Intervals
Response Variable Yield
All Pairwise Comparisons among Levels of Corn
Corn = Budget  subtracted from:

Corn       Lower  Center  Upper  -+---------+---------+---------+-----
Regular   -22.36  142.9   308.1  (----------*----------)
Luxury    160.51  325.8   491.0          (----------*----------)
                                 -+---------+---------+---------+-----
                                  0       150       300       450


Corn = Regular  subtracted from:

Corn    Lower  Center  Upper  -+---------+---------+---------+-----
Luxury  17.64  182.9   348.1   (----------*----------)
                              -+---------+---------+---------+-----
                               0       150       300       450
```

Panel 8.3   Session window output of comparisons.

greatest mean for the levels of corn – see the main effects plot in Figure 8.5. However, it should be borne in mind that other responses might be used in decision-making, such as flavour and costs. It is also important to be aware that, when there are significant interaction effects, this pick-a-winner approach can lead to selection of an FLC that is not optimal.

As a second example, consider a hypothetical experiment performed by an internet retailer prior to the launch of new laptop computer on the market. The retailer manages its database of customers in homogeneous marketing groups of approximately 5000 customers. The factors of interest were price (X_1), with levels £499, £549 and £599, and offer (X_2), with levels software and wireless, and the response was number of sales (Y). Two replications were made of the full 3×2 factorial (the shorthand 3×2 indicating that the first factor has three levels and the second factor has two levels.) Thus each customer in two groups selected at random was sent an

Table 8.5   Data for marketing experiment.

Sales	Offer	
Price	Software	Wireless
£499	136	98
	140	110
£549	72	96
	74	114
£599	69	102
	61	92

e-mail inviting purchase of the new laptop for price £449 with the offer of free software. Customers in a second pair of groups selected at random received the offer of the new laptop for price £449 with the offer of a free wireless networking module, and so on. The data, displayed in Table 8.5, are available in the worksheet Marketing.MTW.

An individual values plot of the data is shown in Figure 8.12. The plot was obtained via **Stat > ANOVA > Two-Way...** and **Graphs...** with **Individual value plot** selected. **Fit additive model** was left unchecked and the ANOVA table shown in Panel 8.4 obtained.

There is evidence of interaction in this case, with P-value 0.001. In such cases emphasis is put on understanding the nature of the interaction rather than on interpretation of the main effects. Use was made of **Stat > ANOVA > Interactions Plot...** with the option **Display full interaction plot matrix** checked to create the interaction plots shown in Figure 8.13. Unlike the previous example the vertical bands of line segments are far from parallel in some cases. (In fact there is a crossing over in one case.) This is typical in situations where there is significant interaction between factors. There is evidence here that the effect on sales of changing the nature of the offer, from one of free software to one of a free wireless networking

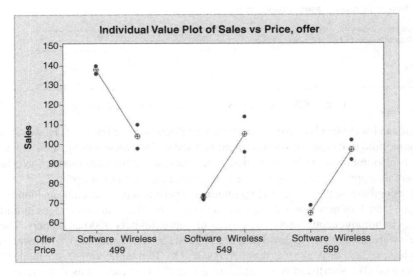

Figure 8.12   Individual value plot of marketing data.

```
Two-way ANOVA: Sales versus Price, Offer

Source         DF     SS        MS       F      P
Price           2   3584   1792.00   32.98   0.001
Offer           1    300    300.00    5.52   0.057
Interaction     2   2904   1452.00   26.72   0.001
Error           6    326     54.33
Total          11   7114

S = 7.371    R-Sq = 95.42%    R-Sq(adj) = 91.60%
```

Panel 8.4   ANOVA for marketing experiment.

module, depends on the level of price. For example, with price set at £499 there was a greater level of sales with the offer of software than with the offer of wireless, while at price £549 the opposite was the case. Having found evidence of a significant interaction effect the additive model fitted in the previous example will not be adequate.

Table 8.6 gives the mean sales corresponding to each of the $3 \times 2 = 6$ FLCs. The Overall Mean Sales for the experiment was 97 laptops per group.

Table 8.7 gives the main effects for the offer factor. Note that the effects sum to zero.

Table 8.8 gives the main effects for the price factor. Note that the effects sum to zero. As before, we have

$$\text{Data} = \text{Fit} + \text{Residual}.$$

In this case we take

$$\text{Fit} = \text{Overall mean} + \text{Price effect} + \text{Offer effect} + \text{Interaction effect}.$$

The interaction effect is obtained by computing, for each FLC, the value of Overall mean + Price effect + Offer effect and choosing the interaction effect to be such that the fit

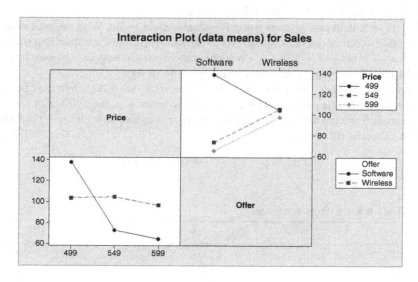

Figure 8.13   Interaction plots of marketing data.

Table 8.6   Mean sales for marketing experiment.

Price	Offer	
	Software	Wireless
£499	138	104
£549	73	105
£599	65	97

Table 8.7   Effects for offer.

Level	Mean sales	Effect
Software	92	− 5
Wireless	102	5

equals the mean sales in the experiment for the corresponding FLC. The detail is set out in Table 8.9. Note the mathematical symbols between the component sub-tables – the corresponding cell entries in the first four sub-tables are added together to give the corresponding cell entries in the final table.

For example, with level £499 for price and level software for offer, the sum of overall mean, price effect and offer effect equals $97 + 24 + (-5) = 116$. To obtain the mean sales figure of 138 for this FLC requires addition of a price–offer interaction effect of 22. With level £499 for price and level wireless for offer, the sum of overall mean, price effect and offer effect equals $97 + 24 + 5 = 126$. To obtain the mean sales figure of 104 for this combination of factor levels requires addition of a price–offer interaction effect of -22. The reader is invited to check the remaining entries in the fourth sub-table of Table 8.9. Note that *both* the rows *and* the columns of the matrix of interaction effects sum to 0.

Residuals are calculated as usual using Residual = Data − Fit. With price £499 and offer of software the two observed sales figures were 136 and 140, so the corresponding residuals are $136 - 138 = -2$ and $140 - 138 = 2$. (Whenever there are just two replications the residuals for each FLC will be equal in magnitude and opposite in sign.)

The *fixed* effects model for a general two-factor design *with interaction* may be specified as shown in Box 8.2. The number of levels of the row factor is a, the number of levels of the column factor is b and the number of replications is n. For example, with $i = 1, j = 2$ and $k = 3$ we have the specific relationship

$$Y_{123} = \mu + \alpha_1 + \beta_2 + (\alpha\beta)_{12} + \varepsilon_{123}.$$

Table 8.8   Effects for price.

Level	Mean sales	Effect
£499	121	24
£549	89	− 8
£599	81	− 16

Table 8.9 Fitting the model.

Overall mean	Offer	
Price	Software	Wireless
£499	97	97
£549	97	97
£599	97	97

+

Price effect	Offer	
Price	Software	Wireless
£499	24	24
£549	−8	−8
£599	−16	−16

+

Offer effect	Offer	
Price	Software	Wireless
£499	−5	5
£549	−5	5
£599	−5	5

+

Interaction effect	Offer	
Price	Software	Wireless
£499	22	−22
£549	−11	11
£599	−11	11

=

Cell mean/fitted value	Offer	
Price	Software	Wireless
£499	138	104
£549	73	105
£599	65	97

Observed data value = Overall mean + Row factor effect

$\qquad$ + Column factor effect + Interaction effect + Random error

$Y_{ijk} = \mu + \alpha_i + \beta_j + (\alpha\beta)_{ij} + \varepsilon_{ijk}, \quad i = 1, 2, \ldots, a, \quad j = 1, 2, \ldots, b, \quad k = 1, 2, \ldots, n$

$$\sum_{i=1}^{a} \alpha_i = 0, \quad \sum_{i=1}^{b} \beta_i = 0, \quad \sum_{j=1}^{b} (\alpha\beta)_{ij} = 0, \quad \sum_{i=1}^{a} (\alpha\beta)_{ij} = 0, \quad \varepsilon_{ijk} \sim N(0, \sigma^2)$$

Box 8.2 Fixed-effect model for two factors with interaction.

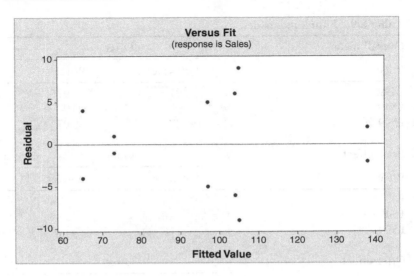

Figure 8.14   Plot of residuals against fits.

For the marketing experiment this states that the sales to the third group ($k = 3$, indicating the third replicate) with price £499 ($i = 1$, indicating the first level of the price factor) and with the offer of a wireless module ($j = 2$, indicating the second level of the offer factor) is made up of the overall mean plus the effect of price £499 plus the effect of the offer of wireless plus the interaction effect for level £499 of the price factor coupled with level wireless for the offer factor plus the random error component. Earlier we estimated the overall mean as 97, the effect of price £499 as 24, the effect of offer wireless as 5 and the interaction effect, $(\alpha\beta)_{12}$, for this combination of factor levels, as -22. Thus the fitted value for the combination of price £599 with offer wireless is the sum $97 + 24 + 5 + (-22) = 104$.

The value $s = 7.371$, which is given beneath the ANOVA table and is the square root of the error MS in the ANOVA table, provides an estimate of the standard deviation σ of the random error component of the model. The model predicts that the population of sales to groups obtained with the combination of price £499 with the wireless offer will be normally distributed with mean 104 and standard deviation 7.371. Similar statements may be made about the five other FLCs.

The response, sales, is a discrete random variable so, strictly speaking, cannot be normally distributed for a particular combination of factor levels. However, a normal probability plot of the residuals is satisfactory so one can be satisfied that the assumption of normality, underlying the valid use of the F-distribution to compute P-values, is approximately true. The plot of residuals against fits is shown in Figure 8.14.

The symmetry of the plot about the horizontal reference line, corresponding to residual value 0, stems from the fact noted earlier that, with two replications, the residuals occur in pairs of values with equal magnitude. The reader might feel that the wide variation in spread of the pairs of points might cast doubt on the model assumption of a random error with constant variance. However, use of **Stat > ANOVA > Test for Equal Variances. . .**with **Response:** Sales and **Factors:** Price Offer, yields a P-value of 0.668. Thus there is no evidence from Bartlett's test to cast doubt on a random error with constant variance. The main effects plot is shown in Figure 8.15.

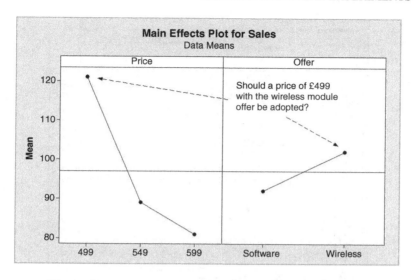

Figure 8.15 Main effects plot for marketing experiment.

In the case of the popcorn experiment, the best combination of factor levels, in terms of maximizing yield, could be identified by selecting the levels air and luxury on scrutiny of the main effects plot in Figure 8.5 for the levels of the factors corresponding to the greatest mean yields in each panel. In the case of the marketing experiment, scrutiny of the plot in Figure 8.15 might suggest the combination of price £499 with offer of wireless would be best, but, because of the interaction, it would appear that the combination of price £499 and offer of software is actually best – see Figure 8.12 and Table 8.6.

Grouping information from follow-up comparisons via the Tukey method with overall confidence level of 95% was obtained via **Stat > ANOVA > General Linear Model....** The dialog is shown in Figure 8.16. In **Model:** we are communicating to Minitab the nature of the model we are using, i.e. $Y_{ijk} = \mu + \alpha_i + \beta_j + (\alpha\beta)_{ij} + \varepsilon_{ijk}$. By entering Price, Offer and Price*Offer in **Model:** we are indicating the $\alpha_i + \beta_j + (\alpha\beta)_{ij}$ terms at the core of the model. In the **Comparisons** subdialog **Pairwise comparisons** were selected, by the **Tukey** method. **Grouping information**, with **Confidence level:** 95.0 was checked. In the **Terms:** window Price*Offer was entered as interest centres on sales from the different FLCs, i.e. price–offer combinations. Part of the corresponding section of the Session window output is shown Panel 8.5.

Note that the FLC of price £499 and software offer 'tops the league' with the highest observed mean sales of 138. This FLC comprises grouping A, and as this letter is not shared with any other grouping there is formal evidence of the superiority of the combination of price £499 and software offer over all other FLCs, in terms of the response, sales. Use of the grouping information option means that the Minitab user does not need to scrutinize and interpret a whole series of confidence intervals

8.1.4 The random effects model, components of variance

The *random* effects model for a general two-factor design *with interaction* may be specified as detailed in Box 8.3. The number of levels of the row factor is a, the number of levels of the

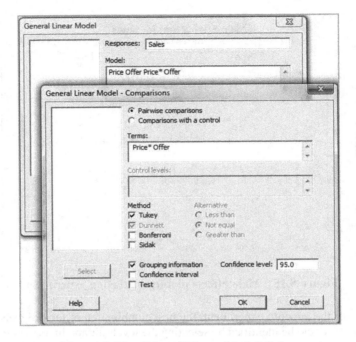

Figure 8.16   General Linear Model dialog.

column factor is b and the number of replications is n. In this model the effects are random variables – hence the terminology 'random effects model'. There are three null hypotheses that may be tested against alternatives:

$$H_0 : \sigma_a^2 = 0, \quad H_1 : \sigma_a^2 \neq 0;$$
$$H_0 : \sigma_b^2 = 0, \quad H_1 : \sigma_b^2 \neq 0;$$
$$H_0 : \sigma_{ab}^2 = 0, \quad H_1 : \sigma_{ab}^2 \neq 0.$$

In words, these null hypotheses state that there is no component of variation due to the row factor, the column factor, and the interaction, respectively.

```
Grouping Information Using Tukey Method and 95.0% Confidence

Price   Offer      N    Mean   Grouping
499     Software   2    138.0  A
549     Wireless   2    105.0      B
499     Wireless   2    104.0      B
599     Wireless   2     97.0      B C
549     Software   2     73.0        C D
599     Software   2     65.0          D

Means that do not share a letter are significantly different.
```

Panel 8.5   Session window output for follow-up comparisons.

Observed data value = Overall mean + Row factor effect + Column factor effect

+ Interaction effect + Random error

$Y_{ijk} = \mu + \alpha_i + \beta_j + (\alpha\beta)_{ij} + \varepsilon_{ijk}, \quad i = 1, 2, \ldots, a, \quad j = 1, 2, \ldots, b, \quad k = 1, 2, \ldots, n$

$\alpha_i \sim N(0, \sigma_a^2), \quad \beta_j \sim N(0, \sigma_b^2), \quad (\alpha\beta)_{ij} \sim N(0, \sigma_{ab}^2), \quad \varepsilon_{ijk} \sim N(0, \sigma^2)$

(The random variables α_i, β_j, $(\alpha\beta)_{ij}$ and ε_{ijk} are independent.)

Box 8.3   Random effects model.

As an example, consider an experiment in which three operators, selected at random from a pool of operators, each measured the height (mm) of each one of a random sample of ten bottles of a particular type on two occasions. On each occasion the bottles were presented to the operators in a random sequence and on the second occasion the operators were unaware of the results they had obtained on the first occasion. All measurements were taken with the same calibrated height gauge under similar environmental conditions. The data are available in Heights.xls and a segment of the data is shown in Figure 8.17. (Data reproduced by permission of Ardagh Glass Ltd., Barnsley.)

In this scenario the response, Y, is height and the factors are bottle (X_1) and operator (X_2). Both factors are random. Analysis via Minitab cannot be carried out using **Stat > ANOVA > Two-Way...** because it only deals with the case of *fixed* effects. However, the analysis may be carried out using **Stat > ANOVA > Balanced ANOVA....** Under **Results...** the option to **Display expected mean squares and variance components** was checked. Under **Graphs...** the options to create a normal probability plot of residuals and a plot of residuals versus fitted values were accepted. The remainder of the dialog, in which the model and the information that the bottle and operator factors are random are specified, is shown in Figure 8.18.

The ANOVA table, preceded by a list specifying the factors, types and levels from the Session window output, is displayed in Panel 8.6. The following conclusions may be reached concerning the hypotheses specified earlier:

- $H_0 : \sigma_a^2 = 0$ is rejected in favour of $H_1 : \sigma_a^2 \neq 0$ at the 0.1% level of significance (*P*-value given as 0.000 to three decimal places);

- $H_0 : \sigma_b^2 = 0$ is rejected in favour of $H_1 : \sigma_b^2 \neq 0$ at the 1% level of significance (*P*-value = 0.002);

- $H_0 : \sigma_{ab}^2 = 0$ cannot be rejected (*P*-value = 0.621).

The normal plot of the residuals and the plot of residuals versus fits were considered to be satisfactory and are not reproduced in this text.

As in the case of the random effects model for a single factor, expressions can be derived, in terms of the four variances in the model, for the expected mean squares. Use of these expressions gives rise to estimates of each of the four variances. The expressions and the estimates from the Session window output are shown in Panel 8.7. The estimate of σ_{ab}^2 is − 0.00002. A negative variance is impossible. Together with the *P*-value of 0.621 for interaction (Panel 8.6), the negative estimate provides a further indication that the interaction component should be dropped from the model. Thus the model was revised by removing the

↓	C1 Bottle	C2-T Operator	C3 Height
1	1	Neil	214.82
2	2	Neil	214.61
3	3	Neil	214.52
4	4	Neil	214.57
5	5	Neil	214.64
6	6	Neil	214.72
7	7	Neil	214.60
8	8	Neil	214.73
9	9	Neil	214.70
10	10	Neil	214.80
11	1	Neil	214.83
12	2	Neil	214.64
13	3	Neil	214.53
14	4	Neil	214.61
15	5	Neil	214.64
16	6	Neil	214.73
17	7	Neil	214.61
18	8	Neil	214.75
19	9	Neil	214.67
20	10	Neil	214.78

Figure 8.17   Segment of bottle height data.

bottle–operator interaction term from the model by specifying **Model:** Bottle Operator in the dialog, i.e. by deleting Bottle*Operator from the window. Revised Session window output is shown in Panel 8.8.

The final model is shown in Box 8.4. The variance components σ_a^2, σ_b^2 and σ^2, due to bottle, operator and random error respectively, are estimated as 0.009 34, 0.000 09 and 0.000 23 respectively. Since any observed data value involves a sum of independent random variables the result in Box 4.2 in Section 4.3.1 is applicable. The variance of Y_{ijk} is thus $0 + \sigma_a^2 + \sigma_b^2 + \sigma^2$ (μ, being a constant, has variance 0). Thus the estimated variance of Y_{ijk} is given by

$$0.009\,34 + 0.000\,09 + 0.000\,23 = 0.009\,66$$

and the estimated proportion of total variance accounted for by bottle is 0.009 34/ 0.009 66 = 96.7%. The fact that a relatively large proportion of the total variability is attributable to the product is desirable from the point of view of the performance of the measurement system. The topic of measurement system analysis will be considered in detail in Chapter 9.

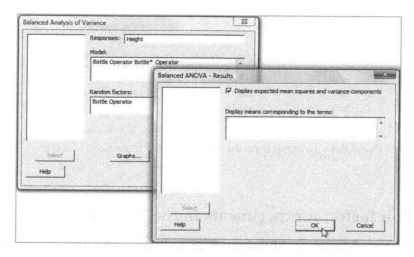

Figure 8.18   Dialog for balanced ANOVA.

```
ANOVA: Height versus Bottle, Operator

Factor      Type      Levels   Values
Bottle      random       10    1,  2,  3,  4,  5,  6,  7,  8,  9, 10
Operator    random        3    Lee, Neil, Paul


Analysis of Variance for Height

Source           DF        SS        MS       F       P
Bottle            9  0.506302  0.056256  273.43   0.000
Operator          2  0.003863  0.001932    9.39   0.002
Bottle*Operator  18  0.003703  0.000206    0.86   0.621
Error            30  0.007150  0.000238
Total            59  0.521018


S = 0.0154380   R-Sq = 98.63%   R-Sq(adj) = 97.30%
```

Panel 8.6   ANOVA table.

```
                                        Expected Mean Square
                         Variance  Error  for Each Term (using
     Source              component term   unrestricted model)
  1  Bottle               0.00934     3    (4) + 2 (3) + 6 (1)
  2  Operator             0.00009     3    (4) + 2 (3) + 20 (2)
  3  Bottle*Operator     -0.00002     4    (4) + 2 (3)
  4  Error                0.00024          (4)
```

Panel 8.7   Estimated variance components.

	Source	Variance component	Error term	Expected Mean Square for Each Term (using unrestricted model)
1	Bottle	0.00934	3	(3) + 6 (1)
2	Operator	0.00009	3	(3) + 20 (2)
3	Error	0.00023		(3)

Panel 8.8   Estimated variance components from revised model.

8.2   Full factorial experiments in the 2^k series

8.2.1 2^2 Factorial experimental designs, display and analysis of data

Factorial experiments in which all the factors of interest have just two levels are of particular importance in the improve phase of many Six Sigma projects. An experiment involving, for example, three factors, each with two levels, involves a total of $2 \times 2 \times 2$ or 2^3 FLCs. A 2^k factorial experiment involves k factors, each with two levels.

Consider the manufacture of a product, for use in the making of paint, in a batch process. Fixed amounts of raw material are heated under pressure in reactor 1 for a fixed period of time and the product is then recovered. Currently the process is operated at temperature 225 °C and pressure 4.5 bar. As part of a Six Sigma project, aimed at increasing product yield, a 2^2 factorial experiment with two replications was planned. Yields from the process with current temperature and pressure levels average 90 kg. It was decided after discussion amongst the project team to use the levels 200 °C and 250 °C for temperature and the levels 4.0 bar and 5.0 bar for pressure.

Use was then made of **Stat > DOE > Factorial > Create Factorial Design. . .** with the option **2-level factorial (default generators)** selected under **Type of Design**. Part of the dialog is shown in Figure 8.19. **Number of factors:** was specified as 2. Under **Designs . . .** the only available design is the full factorial design. The number of FLCs is $2^2 = 4$, so Minitab indicates this by listing the number of **Runs** required as 4. **Resolution**, Full in this case, will be discussed later in the chapter. Minitab represents 2^2 as 2**2. The **Number of replicates:** was specified as 2, the **Number of blocks:** as 1 and **Number of center points:** as 0. (Experimental designs involving centre points will be considered in Chapter 10.)

Observed data value $=$ Overall mean $+$ Bottle effect $+$ Operator effect

$+$ Random error

$$Y_{ijk} = \mu + \alpha_i + \beta_j + \varepsilon_{ijk}, \ i = 1, 2, \ldots, a, \ j = 1, 2, \ldots, b, \ k = 1, 2, \ldots, n$$

$$\alpha_i \sim N(0, \sigma_a^2), \ \beta_j \sim N(0, \sigma_b^2), \quad \varepsilon_{ijk} \sim N(0, \sigma^2)$$

(The random variables α_i, β_j and ε_{ijk} are independent.)

Box 8.4   Revised random effects model.

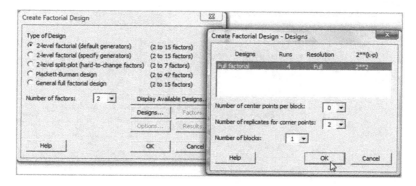

Figure 8.19   Dialog for design of a 2^2 experiment.

On clicking **OK** one may commence the remainder of the dialog. Under **Factors. . .** the factors temperature and pressure (both numeric in this case) and their levels were specified. The defaults were accepted under **Options. . .**, i.e. to **Randomize runs** and to **Store design in worksheet**. (No reference will be made in this book to **Fold Design** and **Fraction**.) Under **Results. . .** the defaults were also accepted. On clicking **OK, OK** the resulting worksheet should be augmented with a column in which the values of yield obtained can be recorded and a column in which those carrying out the experiment can note any unusual happening or information that might prove relevant to the analysis of the data from the experiment and to the project. Before commencing the experimentation the work to date should be stored in a Minitab project. Fictitious yield data are used in order to make the checking of key calculations straightforward for the reader.

Once the experiment has been completed one could carry out the display and analysis using facilities for plotting and analysis available under **Stat** > **ANOVA** > **. . .**, but Minitab has built-in facilities for the analysis of 2^k factorial experiments via **Stat** > **DOE** > **. . .**. For initial display of the data one may use **Stat** > **DOE** > **Factorial** > **Factorial Plots. . .**. The data and dialog involved in the creation of the three plots available are shown in Figure 8.20. The data are provided in Reactor1.MTW and the reader should use that worksheet should he/she wish to re-create the displays and analysis that follows as the worksheet, which was created using **Stat** > **DOE** > **Factorial** > **Create Factorial Design. . .**, includes hidden stored information on the experimental design.

Main Effects Plot, **Interaction Plot** and **Cube Plot** were all checked. Once a plot type has been checked one must then click on the corresponding **Setup. . .** button and enter the **Responses** to be plotted and **Factors to Include in Plots**. The arrow keys may be used to select or deselect highlighted factors from the list of **Available:** factors. Both temperature and pressure were selected for all three plots, the defaults. For all three plots the default option to display **Data Means** as **Type of Means to Use in Plots** was accepted. In the case of the interaction plot the option to **Draw full interaction plot matrix** was selected in order to obtain the display in Figure 8.22. (The reader should note that specification of a response is optional in the case of the cube plot. This allows experimenters, who wish to do so, to create a blank cube plot on which they can record data means for discussion prior to any formal analysis using software.)

The left-hand component of the annotated main effects plot in Figure 8.21 has the mouse pointer located at the point representing the mean yield of 88 kg from all the experimental runs

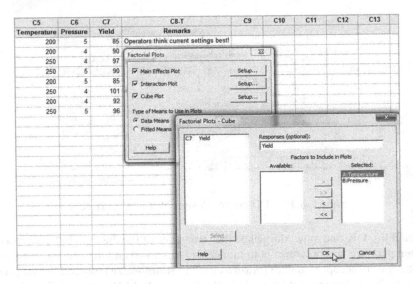

Figure 8.20   Dialog for creating factorial plots.

carried out with temperature 200 °C, as can readily be verified from the data in the worksheet in Figure 8.20. The mean yield from all the experimental runs carried out with temperature 250 °C was 96 kg. Hence, on average, increasing temperature from 200 °C to 250 °C increases yield of product by 8 kg. The reader is invited to confirm that the right-hand component of the plot indicates that, on average, increasing pressure from 4 bar to 5 bar decreases yield of product by 6 kg. The main effect of temperature is 8 kg and the main effect of pressure is − 6 kg.

Both versions of the interaction plot are shown in Figure 8.22. The mouse pointer is placed over the point in one version that corresponds to the FLC with temperature 250 °C and pressure

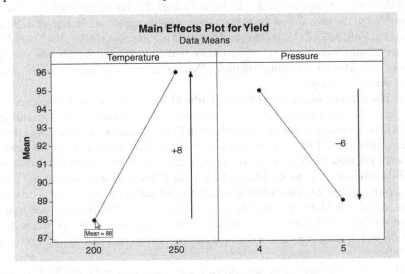

Figure 8.21   Main effects plot for reactor 1.

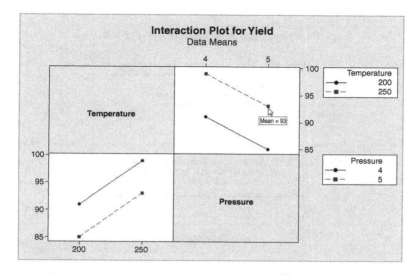

Figure 8.22 Interaction plots for reactor 1.

5 bar. Scrutiny of the data table visible in the worksheet in Figure 8.20 reveals yields of 90 and 96 kg with mean 93 kg. The parallel lines indicate no temperature–pressure interaction in this situation. The reader is encouraged to create the plot in Figure 8.22 for him/herself and, with a copy of the data at hand, to move the mouse pointer to each of the eight points in turn and confirm the mean yields displayed.

The mean yields for the four FLCs used in the experiment are plotted at the vertices of a square (a cube in two dimensions!) in the cube plot in Figure 8.23. The annotated version of the cube plot in Figure 8.24 indicates another way of determining the main effect of temperature. The annotated version of the cube plot in Figure 8.25 indicates another way of determining the main effect of pressure.

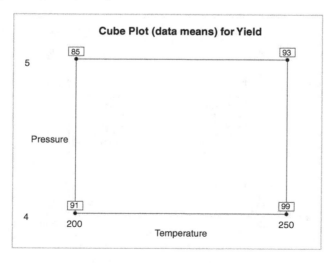

Figure 8.23 Cube plot for reactor 1.

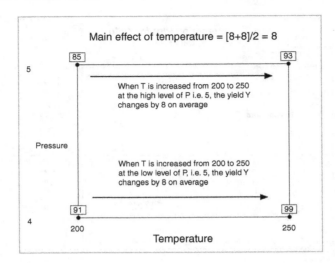

Figure 8.24 Annotated cube plot for main effect of temperature.

Having displayed the data in various ways, the next step is to perform the formal analysis using **Stat > DOE > Factorial > Analyze Factorial Design. . . .** (Note that the Minitab icon for **Analyze Factorial Design. . .** is based on the sort of cube plot we have been considering in Figure 8.23.) The dialog is shown in Figure 8.26.

It is necessary to specify Yield under **Response:** and to specify the model using **Terms. . . .** Here under **Terms:** A: Temperature, B: Pressure and AB, denoting the temperature–pressure interaction, are selected as the default. Part of the Session window output obtained is shown in Panel 8.9.

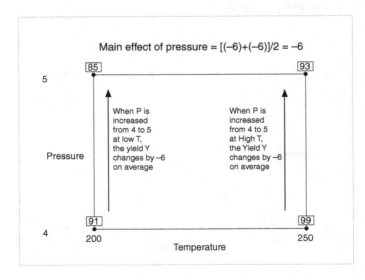

Figure 8.25 Annotated cube plot for main effect of pressure.

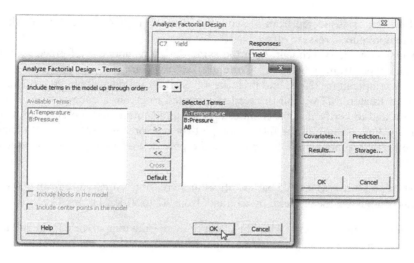

Figure 8.26 Analysing the factorial experiment.

Note that the temperature effect of 8 and the pressure effect of –6 estimated from the data earlier are given in the first column. The interaction plots indicated no temperature–pressure interaction and this corresponds to the interaction effect estimate of zero. (The redundant negative sign will have arisen due to the way that numbers are stored in the software.) The Coef (coefficient) column in the output has the value 92 as the first entry. This is the overall mean yield for the entire experiment and is referred to as the constant term in the model. The coefficients corresponding to temperature, pressure and temperature $\times$ pressure are simply half the corresponding effects. (Equations in which these coefficients are required will be discussed later in the chapter.) The heading SE Coef refers to the standard errors of the coefficients. For example, the constant term 92 is the mean of a sample of eight measurements. The standard error or standard deviation of a mean of eight values is the standard deviation for individual values divided by $\sqrt{8}$. The output gives the estimated standard deviation of individual yields obtained from the ANOVA to be $s = 2.64575$. Division by $\sqrt{8}$ yields 0.9354. The values of the coefficients divided by their standard errors give a Student's t-statistic for testing the null hypothesis that each coefficient is zero. The P-values for each are given in the final column. The fact that the P-values for both temperature and pressure are less than 0.05 may be taken as evidence that both temperature and pressure have a real effect on

```
Estimated Effects and Coefficients for Yield (coded units)

Term                     Effect     Coef   SE Coef       T       P
Constant                            92.000  0.9354    98.35   0.000
Temperature               8.000    4.000   0.9354     4.28   0.013
Pressure                 -6.000   -3.000   0.9354    -3.21   0.033
Temperature*Pressure     -0.000   -0.000   0.9354    -0.00   1.000


S = 2.64575     PRESS = 112
R-Sq = 87.72%   R-Sq(pred) = 50.88%   R-Sq(adj) = 78.51%
```

Panel 8.9 ANOVA for 2^2 factorial experiment on reactor 1.

yield. However, as suspected from viewing the interaction plots, there is no evidence of a nonzero temperature–pressure effect.

Consider now an experiment, with the same design as above, performed on reactor 2, a different type from reactor 1. Currently, as with reactor 1, the process is also operated at temperature 225 °C and pressure 4.5 bar, and yields average 90 kg. The data are available in Reaotr2.MTW, and the reader is invited to create a cube plot for this second experiment and to verify that the main effects for temperature and pressure are 8 and − 6, respectively. One version of the interaction plot for reactor 2 is displayed, with annotation, in Figure 8.27. The nonparallelism suggests the presence of a temperature–pressure interaction effect.

- When pressure was set at 5 bar the effect of increasing temperature from 200 to 250 °C was to increase mean yield, on average, from 87 to 91 kg, i.e. by 4 kg.

- When pressure was set at 4 bar the effect of increasing temperature from 200 to 250 °C was to increase mean yield, on average, from 89 to 101 kg, i.e. by 12 kg.

- The main effect of temperature is given by $(4 + 12)/2 = 8$.

- The temperature–pressure interaction effect is given by $(4 - 12)/2 = - 4$.

In the final calculation of the interaction effect it is important to note that the difference considered is that of the effect of temperature at the higher level of pressure less that of the effect at the lower level of pressure. (Had the changes in mean yield, indicated by the arrows in Figure 8.27, been equal then the line segments linking points with the same level of temperature in the interaction plot would have been parallel, as in the top right-hand panel in Figure 8.22. The difference between the changes would be zero and the interaction effect would be zero.) The second version of the interaction plot is shown in Figure 8.28.

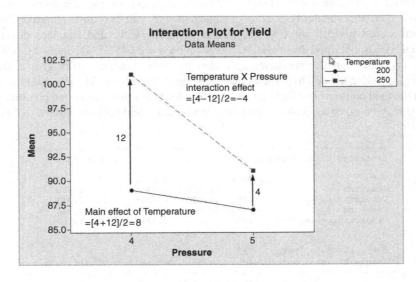

Figure 8.27   First version of interaction plot for reactor 2.

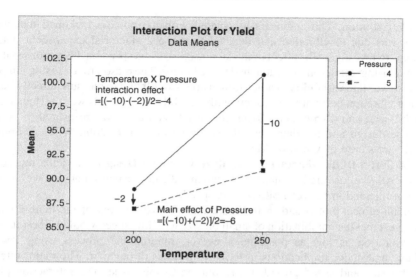

Figure 8.28   Second version of interaction plot for reactor 2.

- When temperature was set at 250 °C the effect of increasing pressure from 4 to 5 bar was to change mean yield, on average, from 101 to 91 kg, i.e. by -10 kg.

- When temperature was set at 200 °C the effect of increasing pressure from 4 to 5 bar was to change mean yield, on average, from 89 to 87 kg, i.e. by -2 kg.

- The main effect of pressure is given by $[(-10) + (-2)]/2 = -6$.

- The temperature-pressure interaction effect is given by $[(-10) - (-2)]/2 = -4$.

The reader is also invited to verify using **Stat > DOE > Factorial > Analyze Factorial Design...** that the output in Panel 8.10 is obtained. Note that the effects calculated earlier are confirmed and also that the interaction effect differs significantly from zero at the 10% level of significance. Thus there is slight evidence of a real temperature–pressure interaction in the case of reactor 2.

Montgomery (2009, p. 566) gives an example from the electronics industry. It concerned registration notches cut in printed circuit boards using a router. Although the process was stable, producing a satisfactory mean notch size, variability was too high. High variability leads to problems in assembly, when components are being placed on the boards, due to

Factorial Fit: Yield versus Temperature, Pressure

Estimated Effects and Coefficients for Yield (coded units)

Term	Effect	Coef	SE Coef	T	P
Constant		92.000	0.7906	116.37	0.000
Temperature	8.000	4.000	0.7906	5.06	0.007
Pressure	−6.000	−3.000	0.7906	−3.79	0.019
Temperature*Pressure	−4.000	−2.000	0.7906	−2.53	0.065

Panel 8.10   ANOVA for 2^2 factorial experiment on reactor 2.

improper registration. The engineers associated with the process reckoned that the high variability was due to vibration and decided to use a 2^2 factorial experiment, with *four* replications, to investigate the effect of the factors bit size and speed on the response vibration, as measured using accelerometers attached to the boards. The reason for choosing vibration as the response was that variability in notch dimension was difficult to measure directly but it was known that vibration correlated positively with variability. The levels were 1/16 inch and 1/8 inch for bit size, and 40 rpm and 80 rpm for speed. Vibration was measured in cycles per second. The design and the data, reproduced by permission of John Wiley & Sons, Inc., New York, are stored in Vibration.MTW.

Use of **Stat** > **DOE** > **Factorial** > **Analyze Factorial Design. . .** yields the output in Panel 8.11. All *P*-values are less than 0.001, which indicates, in particular, that we have strong evidence of interaction between bit size and speed.

The main effects plot is shown in Figure 8.29. Naïve scrutiny of the main effects plot would suggest that the combination of small bit size with low speed would be best in terms of keeping vibration as low as possible. However, running the process using low speed would have had major implications in terms of process throughput. The major interaction between bit size and speed provided a resolution to this issue. The interaction plots are presented in Figure 8.30. These plots reveal that, with bit size 1/16 inch, vibration is low at both speeds 40 rpm and 80 rpm. Thus it was decided to operate the process with bit size 1/16 inch and speed 80 rpm. This led to a major reduction in the variability of the size of the registration notches.

The residual plots are shown in Figure 8.31. They were created using **Graphs. . .** in **Analyze Factorial Design. . .** with the **Four in one** option selected. Although the variability of the set of residuals for the lowest of the four fitted values appears relatively low, use of **Stat** > **ANOVA** > **Test for Equal Variances. . .** with **Response:** Vibration and **Factors:** Bit Size Speed yields no formal evidence of nonconstant variability. The plots appear to be generally satisfactory.

8.2.2 Models and associated displays

In order to introduce further important ideas some notation is required. Were one to design a 2^2 experiment with no centre points, one replication, no blocking, no randomization of the run order and using the factor names X1 and X2, with the default levels, then the worksheet displayed in Figure 8.32 would be obtained. The columns named X1 and X2 in the worksheet contain the essence of the design. The X1 × X2 interaction effect may be calculated by using an additional column X1X2. This additional X1X2 column may be obtained from the X1 and X2 columns by simply multiplying corresponding row entries together, hence the X1 × X2 notation.

```
Factorial Fit: Vibration versus Bit Size, Speed

Estimated Effects and Coefficients for Vibration (coded units)

Term             Effect    Coef   SE Coef      T       P
Constant                  23.831   0.6112    38.99   0.000
Bit Size         16.638    8.319   0.6112    13.61   0.000
Speed             7.538    3.769   0.6112     6.17   0.000
Bit Size*Speed    8.713    4.356   0.6112     7.13   0.000
```

Panel 8.11 ANOVA for 2^2 factorial router experiment.

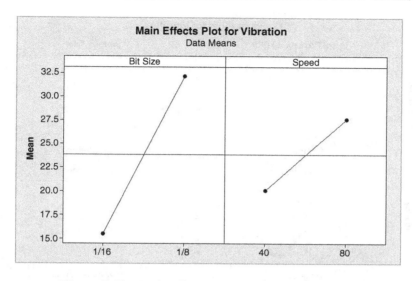

Figure 8.29 Main effects plot for router experiment.

In terms of the reactors referred to earlier, X1 and X2 may be considered as coded variables representing temperature and pressure, respectively. Recall that current operating conditions were temperature 225 °C and pressure 4.5 bar. These would have coded values X1 = 0 and X2 = 0, respectively. The high level of temperature used in the experiment, i.e. 250 °C, would be coded as X1 = 1 and the low level, 200 °C, as − 1. Similarly, the low and high levels of pressure used, 4 bar and 5 bar, would be coded as X2 = − 1 and X2 = 1, respectively. The author finds it helpful to think in terms of a reactor control panel as shown in Figure 8.33. The low temperature of 200 °C is 'one notch down' (X1 = − 1) from the current

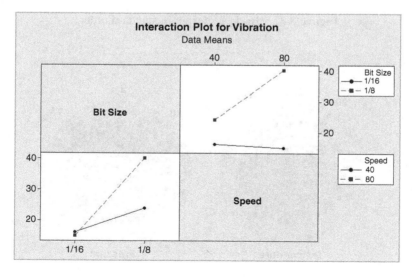

Figure 8.30 Interaction plots for router experiment.

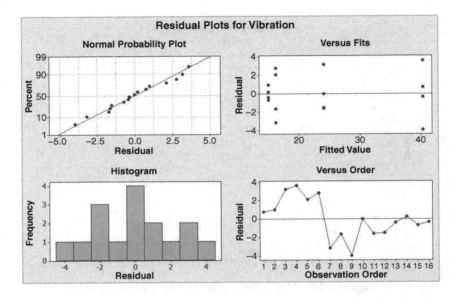

Figure 8.31 Residual plots for router experiment.

StdOrder	RunOrder	CenterPt	Blocks	X1	X2
3	1	1	1	-1	1
1	2	1	1	-1	-1
2	3	1	1	1	-1
4	4	1	1	1	1

Figure 8.32 Basic 2^2 design in standard order.

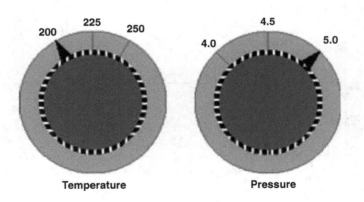

Figure 8.33 Reactor control panel.

Table 8.10 Reactor 1 data.

FLC	X1	X2	X1X2	Mean Y
1	− 1	− 1	1	91
2	1	− 1	− 1	99
3	− 1	1	− 1	85
4	1	1	1	93

operational setting and the high pressure of 5 bar is 'one notch up' ($X2 = 1$) from the current operational setting.

Table 8.10 lists the four FLCs used in the experiment with reactor 1. The means in the final column are the means of the pairs of yields obtained with each FLC. The overall mean and effects may be calculated as follows. The overall mean is simply $(91 + 99 + 85 + 93)/4 = 92$. The main effect of temperature is given by

$$[\text{Mean yield with } X1 = 1] - [\text{Mean yield with } X1 = -1]$$

$$= \frac{99 + 93}{2} - \frac{91 + 85}{2} = 96 - 88 = 8.$$

The main effect of pressure is

$$[\text{Mean yield with } X1 = 1] - [\text{Mean yield with } X2 = -1]$$

$$= \frac{85 + 93}{2} - \frac{91 + 99}{2} = 89 - 95 = -6.$$

Finally the temperature–pressure interaction effect is given by

$$[\text{Mean yield with } X1X2 = 1] - [\text{Mean yield with } X1X2 = -1]$$

$$= \frac{91 + 93}{2} - \frac{99 + 85}{2} = 92 - 92 = 0.$$

These calculations are summarized in Table 8.11. Each coefficient is half the corresponding effect.

Table 8.12 lists the four FLCs used in the experiment with reactor 2. The means in the final column are the means of the pairs of yields obtained with each FLC. The reader is invited to check the calculation of the three effects and corresponding coefficients displayed in Table 8.13 and to confirm that the overall mean is 92.

Table 8.11 Reactor 1 effects.

	Mean response at level		Effect	Coefficient
	−	+		
X1	88	96	8	4
X2	95	89	−6	−3
X1X2	92	92	0	0

Table 8.12 Reactor 2 data.

Reactor 2	Coded Factor Levels		X1X2	Mean Y
FLC	X1	X2		
1	-1	-1	1	89
2	1	-1	-1	101
3	-1	1	-1	87
4	1	1	1	91

The constant (overall mean) and the coefficients generated by Minitab provide statistical models for the expected response of the form

$$Y = \text{Overall mean} + \text{Coeff. of X1} \times \text{X1} + \text{Coeff. of X2} \times \text{X2} + \text{Coefficient of X1X2} \times \text{X1X2}.$$

For reactor1 we obtain

$$Y = 92 + 4X1 - 3X2 + 0X1X2 = 92 + 4X1 - 3X2,$$

and for reactor 2 we have

$$Y = 92 + 4X1 - 3X2 - 2X1X2$$

The models are of the form $Y = f(\mathbf{X})$ referred to in Section 1.4. Here the bold X represents the pair of factors X1 and X2. The model may be used to predict the expected yield for any FLC. (The models may also be obtained using multiple regression methods that will be referred to in Chapter 10.) As an example, consider reactor 1 operating at temperature 250 °C (X1 = 1) and pressure 4 bar (X2 = −1). Substitution of these values into the reactor 1 model equation gives expected yield

$$Y = 92 + 4 \times 1 - 3 \times (-1) = 92 + 4 + 3 = 99.$$

Of course this is the mean yield obtained for reactor 1 with temperature 250 °C and pressure 4 bar in the experiment and is the fitted value from the model for the particular FLC of temperature 250 °C and pressure 4 bar.

As a second example, consider reactor 2 operating at temperature 200 °C (X1 = −1) and pressure 5 bar (X2 = 1). Substitution of these values into the reactor 2 model equation gives expected yield

$$Y = 92 + 4 \times (-1) - 3 \times 1 - 2 \times (-1) \times 1 = 92 - 4 - 3 + 2 = 87.$$

Table 8.13 Reactor 2 effects.

	Mean response at level		Effect	Coefficient
	$-$	$+$		
X1	88	96	8	4
X2	95	89	-6	-3
X1X2	94	90	-4	-2

This is the mean yield obtained for reactor 2 with temperature 200 °C and pressure 5 bar in the experiment and is the fitted value from the model for that particular FLC.

In terms of optimizing the performance of the reactors by achieving as great a yield as possible, it would appear that both operate best, in the ranges of temperature and pressure considered in the experiments, at temperature 250 °C and pressure 4 bar. The ranges of factor levels considered in an experiment constitute the design space. With the reactor models above it is a fairly simple matter to substitute coded values of temperature and pressure into the model equations to obtain predicted responses.

Minitab provides a facility for investigation of process optimization using the models fitted to data from 2^k factorial experiments. With worksheet Reactor2.MTW open, for example, this facility is available via **Stat > DOE > Factorial > Response Optimizer. . ..** Recall that the reactors have been achieving yields averaging 90 kg with the current operating settings of 225 °C and 4.5 bar. Suppose that the project team had decided that it was desirable to achieve yields as high as possible, with target 105 kg on average and no worse than 95 kg on average. The Response Optimizer may be used to explore model predictions in a visual way and without having to perform calculations directly from the model equations. Part of the necessary dialog is shown in Figure 8.34.

It is necessary here to have **Selected:** Yield as the (only) response and under **Setup. . .** to set **Goal** to Maximize, **Lower** to 95 and **Target** to 105. No value is entered for **Upper** when seeking to maximize a response. **Weight** and **Importance** need not concern us when dealing with a single response. Under **Options. . .** check only **Optimization Plot**; there is no need to enter any **Starting value** for the factors.

The outcome from implementation is an interactive screen, part of which has been captured in Figure 8.35. The reader is advised to have this screen on display on his/her computer for the

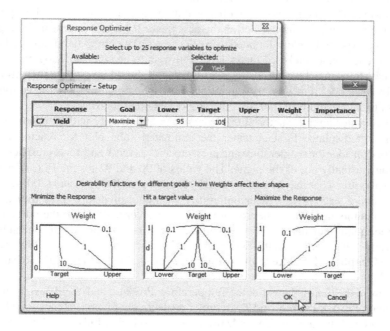

Figure 8.34   Dialog for Response Optimizer.

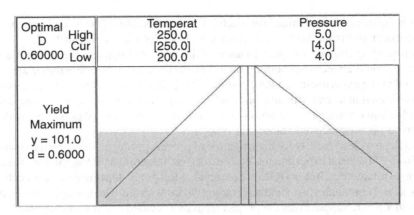

Figure 8.35 Response Optimizer.

discussion that follows. Under the headings Temperat(ure) and Pressure the High and Low values specify the design space for the experiment. The Cur values in square brackets of 250.0 and 4.0 indicate that the maximum mean yield that can be achieved with reactor 2, based on the model fitted to the data from the experiment, will be obtained with temperature 250 °C and pressure 4 bar. The predicted maximum yield with this combination of factor levels is stated as $y = 101$. (Had the desired target of 105 been achieved then the desirability score, d, would have been 1.0. Since 101 lies 0.6 of the way between the lower value of 95 and the target value of 105 the desirability score is $d = 0.6$.)

One can explore predictions by changing the Cur(rent) temperature and pressure levels in one of two ways. The first method is to click and drag the red vertical lines that appear in the diagrams. On releasing the mouse button the new factor level will appear in red together the corresponding expected yield calculated using the new set of current factor levels. The second method is to click on a bracketed factor level you wish to change. A dialog box appears via which the new level may be entered. As an exercise you should investigate the optimum yield predicted by the model if new safety regulations were to restrict the operating temperature for the process to a maximum of 240 °C – the author obtained 98.6 kg.

Before considering a 2^3 factorial experiment the opportunity will be taken to introduce further relevant displays. First consider the annotated cube plot for reactor 1 shown in Figure 8.36. Consider the temperature and pressure axes as the X and Y axes respectively, and a Z axis rising vertically out of the plane. The yields of 90 and 92 for the FLC of temperature 200 °C and pressure 4 bar may be represented by the points (200, 4, 90) and (200, 4, 92) relative to the X, Y and Z axes. (The reader unfamiliar with coordinates in three dimensions may imagine two points in space at the tips of rods that rise vertically from the bottom left-hand corner of the square. The first would be at height 90 above the square, the second at height 92.) All eight yield values plotted in this way provide a three-dimensional scatterplot, available in Minitab via **Graph > 3D Scatterplot > Simple. . ..** The dialog for the creation of such a plot is shown in Figure 8.37.

Note how in the dialog the variables Temperature, Pressure and Yield have been specified as the **X variable:**, **Y variable:** and **Z variable:** respectively for consistency with Figure 8.36 and the associated discussion. Under **Data View. . .** both **Symbols** and **Project lines** were checked. The plot is shown in Figure 8.38. The reader is advised to scrutinize the data for

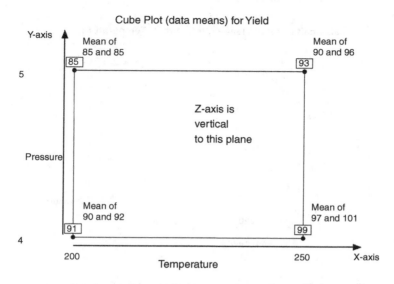

Figure 8.36   Annotated cube plot for reactor 1.

reactor 1 in worksheet Reactor1.MTW and to relate the yields to the points in the display – for example, with temperature 200 °C and pressure 5 bar, both the observed yields were 85 kg, hence the single point at the tip of the rod at the back left of the cuboid.

The fitted model equation may be used to compute an expected yield for any temperature–pressure combination within the design space. One may imagine a forest of vertical rods, like the four in the scatterplot, with points at their tips corresponding to the calculated expected yield. For example, with temperature 225 °C and pressure 4.5 bar, i.e. with $X1 = 0$ and $X2 = 0$, the predicted yield is 92 kg. The points representing the yields form the response surface predicted by the model. Minitab enables this surface to be viewed in two ways.

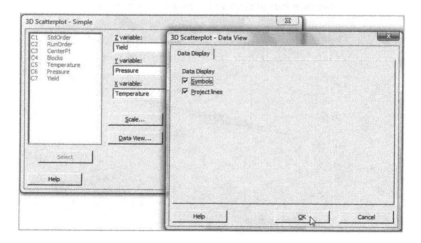

Figure 8.37   Dialog for 3D Scatterplot.

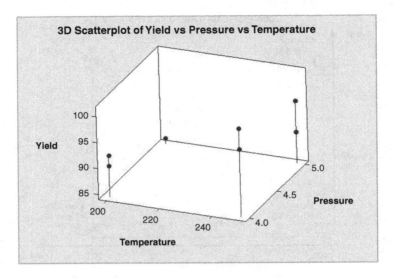

Figure 8.38   Three-dimensional scatterplot of reactor 1 data.

One may use **Stat > DOE > Factorial > Contour/Surface Plots. . .**. In order to produce the displays in Figures 8.39 and 8.40 check both **Contour plot** and **Surface plot**. Click on **Setup. . .** for both and ensure that the column containing the response Yield is selected in both cases. The surface plot created using the default ordering of the factors is shown in Figure 8.39. The reader is invited to view the plot obtained by reversing the order of the two factors.

The response surface in the case of rector 1, where there is no interaction term, is a plane passing through the points corresponding to the mean yields at the four corners of the rectangle in Figure 8.36. The contour plot, with anotation, is shown in Figure 8.40. The contour lines in

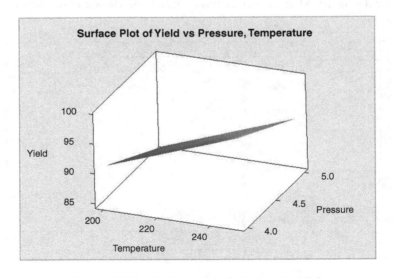

Figure 8.39   Surface plot of reactor 1 model.

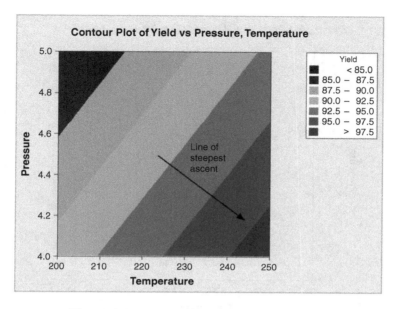

Figure 8.40   Contour plot of reactor 1 model.

the case of a plane are parallel straight lines. The line of steepest ascent is at right angles to the contour lines. The model indicates that it might be worth carrying out further experimentation in the direction of the line of steepest ascent, i.e. at temperatures above 250 °C and pressures below 4 bar, given that the aim of the project was to increase yield. The model for reactor 2 includes a nonzero interaction term. The response surface contour plot, obtained using **Contour lines** instead of **Area** via the **Contours...** button during setup, is shown in Figure 8.41. The response surface is not a plane since the contour lines are curved.

8.2.3   Examples of 2^3 and 2^4 experiments, the use of Pareto and normal probability plots of effects

As a first example of a 2^3 experiment we will consider data given by Iman and Conover (1989, p. 676) from an experiment, carried out by a large department store, as part of a project to improve on the collection of delinquent accounts. The store selected, at random, 12 delinquent accounts that had previously been in arrears and 12 delinquent accounts that had not previously been in arrears. Half of each of these 12 sets was sent only a second notice; the other half was sent a second notice accompanied by a strongly worded letter. These billings were divided again so that half contained a return envelope and half contained a prepaid return envelope. The store recorded the percentage paid on each of these 24 accounts over the 30-day period following sending out of the second notices. The data, reproduced by permission of the authors, are displayed in Table 8.14 and stored, with the design, in the worksheet Accounts.MTW.

Use of **Stat > DOE > Factorial > Analyze Factorial Design. . .** yields the Session window output shown in Panel 8.12. With the three factors there are three main effects, three first-order interactions (interactions between pairs of factors) and one second-order interaction (an interaction between three factors). In terms of the P-values given, none of the

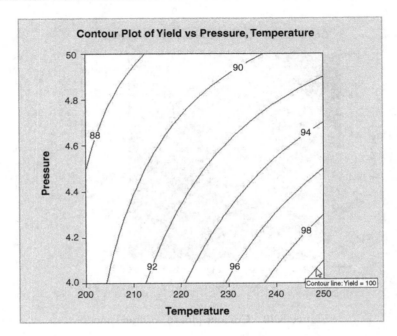

Figure 8.41   Contour plot of reactor 2 model.

interaction effects differs significantly from zero. The main effect of billing and the main effect of previous arrears are significant at the 5% level; the main effect of postage is bordering on being significant at this level. The main effects and interaction plots are displayed in Figures 8.42 and 8.43, respectively.

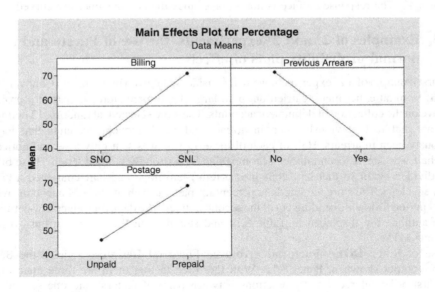

Figure 8.42   Main effects plot for accounts experiment.

Table 8.14 Data for accounts experiment.

		Account previously in arrears			
		No		Yes	
Percentage of outstanding arrears paid off within 30 days of second billing notice being sent out		Postage		Postage	
		Unpaid	Prepaid	Unpaid	Prepaid
Billing	Second	80	100	5	10
	Notice	20	40	0	60
	Only (SNO)	50	60	25	80
	Second	100	100	25	50
	Notice with	50	100	40	100
	Letter (SNL)	80	75	80	50

As there are no significant interactions the main effects plot may be interpreted at face value. Enclosure of a strongly worded letter had a beneficial effect on the response – the higher the proportion of a delinquent account paid off the better from the store's point of view. Customers with a record of previous arrears paid off a lower percentage on average than those with no such record. There is slight evidence that the provision of a prepaid envelope leads to the payment of a higher proportion of the arrears.

The nonsignificant interactions are reflected in the pairs of near parallel lines in the interaction plots displayed in Figure 8.43. The central plot in the top row of the matrix has been annotated with two vertical arrows. These indicate the similar change in the average percentage of arrears paid off, attributable to the inclusion of a strongly worded letter with the second notice, for both customers with no previous history of arrears and customers with such a history.

The next example to be considered is to be found in Box *et al.* (2005, p. 183). Three factors of interest for a pilot chemical plant were temperature, with levels 160 °C and 180 °C, concentration, with levels 20% and 40% and catalyst supplier, with levels A and B. The

```
Factorial Fit: Percentage versus Billing, Previous Arrears, Postage

Estimated Effects and Coefficients for Percentage (coded units)

Term                                Effect    Coef   SE Coef      T       P
Constant                                     57.50     5.492   10.47   0.000
Billing                              26.67   13.33     5.492    2.43   0.027
Previous Arrears                    -27.50  -13.75     5.492   -2.50   0.024
Postage                              22.50   11.25     5.492    2.05   0.057
Billing*Previous Arrears              0.83    0.42     5.492    0.08   0.940
Billing*Postage                      -5.83   -2.92     5.492   -0.53   0.603
Previous Arrears*Postage              6.67    3.33     5.492    0.61   0.552
Billing*Previous Arrears*Postage     -5.00   -2.50     5.492   -0.46   0.655


S = 26.9065      PRESS = 26062.5
R-Sq = 51.84%    R-Sq(pred) = 0.00%    R-Sq(adj) = 30.76%
```

Panel 8.12 ANOVA for 2^3 billing factorial experiment.

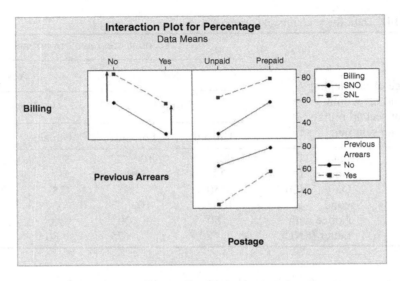

Figure 8.43   Interaction plots for accounts experiment.

response considered was yield (g) and two replications were used. We will work through the steps involved in designing the experiment and analysing the data had it been done using Minitab.

Use of **Stat > DOE > Factorial > Create Factorial Design. . .** offers the option to **Display Available Designs. . ..** With the default **Type of Design** 2-level factorial (default generators) selected, clicking on the button yields the display in Figure 8.44. For three factors, two designs are listed. The full 2^3 factorial experiment requires $2^3 = 8$ runs. A run is simply a factor–level combination. (The reader may find the term 'run' a little confusing. For example,

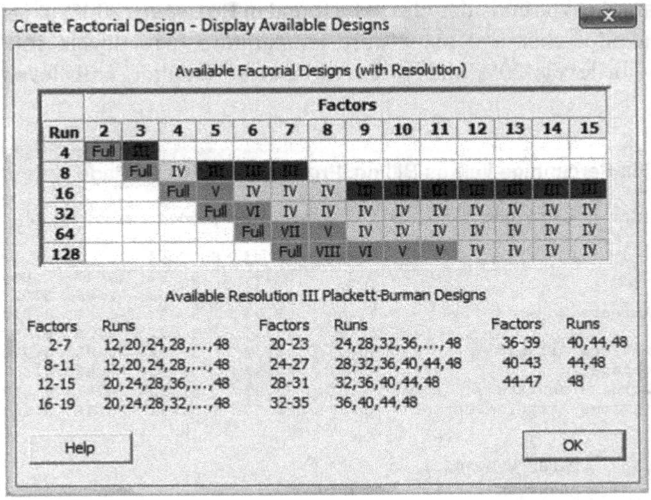

Figure 8.44   Designs available in Minitab for factors each with two levels.

in the 2^2 factorial reactor experiments considered earlier, eight yields were obtained in each. Many would say that eight runs were performed with each reactor during the experimentation. Minitab refers to four runs being replicated twice in the two experiments.) The design involving a half fraction of the full 2^3 factorial experiment requires 4 runs. The roman numbers on rectangles give the resolutions of the corresponding experimental designs. Fractional factorial experiments and the concept of resolution will be introduced later in the chapter. Note that up to 15 factors may be used. (Plackett–Burman experimental designs will not be considered in this book.)

Having clicked on **OK**, begin the dialog by setting **Number of factors:** to 3. Then click on **Designs. . .**, highlight **Full factorial**, set **Number of replicates for coner points:** to 2 and accept the defaults of no centre points and a single block. Next the names and levels of the factors must be specified under **Factors. . .**, noting that temperature and concentration should be specified as numeric while catalyst supplier should be specified as text. The worksheet Pilot. MTW contains the data, reproduced by permission of John Wiley & Sons, Inc., New York.

The main effects plot shown in Figure 8.45 suggests that, of the three factors involved in the experiment, temperature is the major player in terms of its leverage in influencing yield. However, one must beware the temptation to claim that concentration and catalyst supplier are unimportant factors without considering interaction effects.

The major feature of the interaction plots shown in Figure 8.46 is the apparent temperature–catalyst interaction. The two versions of this specific interaction plot appear in the top right and bottom left of the matrix of plots – in the first the line segments are markedly nonparallel and in the second the line segments actually intersect.

The cube plot (a 'real' cube in the case of three factors!) is shown in Figure 8.47. One can relate features of the cube to the two earlier plots. The means of the two yields obtained for each FLC on the left face of the cube (all at temperature 160 °C) are generally much lower than the means of the two yields obtained for each FLC on the right face of the cube (all at

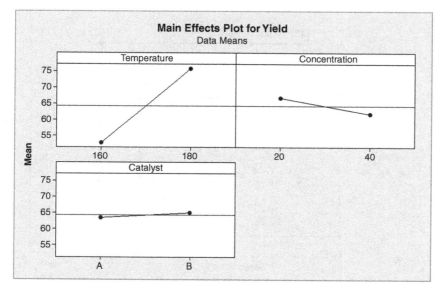

Figure 8.45 Main effects plot for pilot chemical plant experiment.

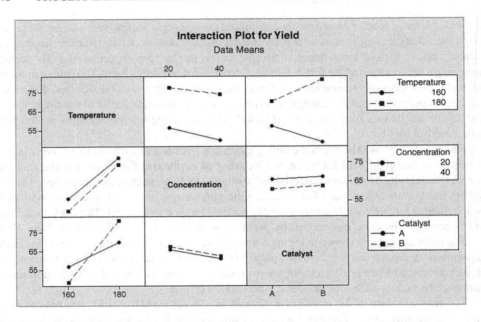

Figure 8.46   Interaction plots for pilot chemical plant experiment.

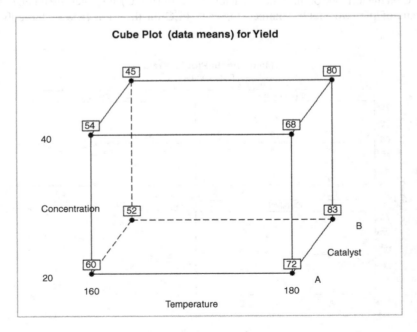

Figure 8.47   Cube plot for pilot chemical plant experiment.

```
Factorial Fit: Yield versus Temperature, Concentration, Catalyst

Estimated Effects and Coefficients for Yield (coded units)

Term                                Effect    Coef  SE Coef       T      P
Constant                                    64.250   0.7071   90.86  0.000
Temperature                         23.000  11.500   0.7071   16.26  0.000
Concentration                       -5.000  -2.500   0.7071   -3.54  0.008
Catalyst                             1.500   0.750   0.7071    1.06  0.320
Temperature*Concentration            1.500   0.750   0.7071    1.06  0.320
Temperature*Catalyst                10.000   5.000   0.7071    7.07  0.000
Concentration*Catalyst              -0.000  -0.000   0.7071   -0.00  1.000
Temperature*Concentration*Catalyst   0.500   0.250   0.7071    0.35  0.733


S = 2.82843   R-Sq = 97.63%   R-Sq(adj) = 95.55%
```

Panel 8.13 Session window output for 2^3 pilot chemical plant experiment.

temperature 180 °C). This corresponds to the apparent major temperature main effect. The apparent temperature–catalyst interaction is indicated by the fact that changing catalyst from A to B (moving from front to back of cube) at the lower temperature leads to *decreased* yields, while on changing catalyst from A to B at the higher temperature leads to *increased* yields.

Part of the Session window output from using **Stat > DOE > Factorial > Analyze Factorial Design. . .** is shown in Panel 8.13. Scrutiny of the *P*-values provides evidence of a temperature–catalyst interaction and of a concentration main effect. The temperature main effect is also flagged as significant, but Box *et al.* (2005, p. 185) comment that 'the main effect of a factor should be *individually* interpreted only if there is no evidence that the factor interacts with other factors'. The main conclusions that may be drawn are as follows:

- The estimated effect of changing concentration from 20% to 40% is to *reduce* yield by 5 g on average (concentration effect is -5 from the Session window output in Panel 8.13), irrespective of the levels of the other factors, since there is no evidence of any interaction between concentration and either of the other two factors.

- The effects of temperature and catalyst cannot be interpreted separately because of the significant interaction between them. The interaction represents the different response to change in the level of temperature with the two catalysts.

This interaction may be highlighted in a cube plot that does not include the factor concentration. This is readily achieved by using **Setup. . .** for the cube plot and the arrow key for factor removal as shown in Figure 8.48. Note the right-hand arrow for selection of individual factors and the double-headed arrows for both selection and removal of all factors simultaneously. Having highlighted Concentration in the **Selected:** list clicking on the arrow pointing to the left removes the factor from the list. Clicking **OK**, removing the checks for the plots already done and clicking **OK** yields the 'collapsed' cube plot in Figure 8.49. Arrows and comments have been added to the plot.

Some final comments from the experts are relevant:

> A result of great practical interest was the very different behaviors of the two 'catalyst types' in response to temperature. The effect was unexpected, for although obtained from two different suppliers, the catalysts were supposedly identical. Also, the yield from catalyst B at 180 °C was the highest that had been

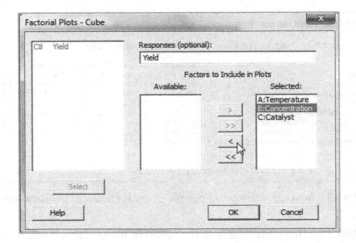

Figure 8.48   Removal and selection of factors.

seen up to that time. This finding led to a very careful study of the catalysts and catalyst suppliers in a later investigation. (Box *et al.*, 2005, p. 186)

The residual plots are satisfactory, so one can have confidence in the conclusions based on the *P*-values given in the Session window output. As an alternative to the direct use of these *P*-values to detect significant effects in the case of a 2^k factorial experiment with replication

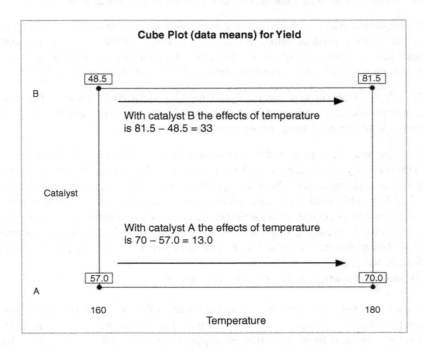

Figure 8.49   Cube plot to highlight temperature–catalyst interaction.

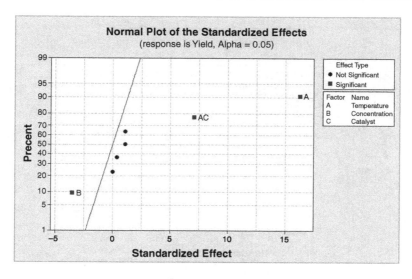

Figure 8.50   Normal probability plot of effects.

one may use a normal probability plot, a half-normal plot or a Pareto plot of the standardized effects. The standardized effects, obtained by dividing the coefficients (recall that the coefficients are half the corresponding effects) by their estimated standard errors, are in fact the *t* values given in Panel 8.13.

These plots are available under the **Graphs. . .** options in **Stat > DOE > Factorial > Analyze Factorial Design. . .**. Under **Effects Plots** one checks **Normal**, **Half Normal** or **Pareto** (or any combination) as desired. (The conclusions that may be reached from the plots are identical, so one would normally create only one of them.) Note that the default significance level **Alpha:** is 0.05. The basis of the normal plot of the effects is that, were all main effects and interaction effects in reality zero, all the estimated effects calculated from the data would constitute a random sample from a normal distribution with mean zero. Points in a normal probability plot of the effects which 'stand out from the crowd' may be taken as being associated with 'real' nonzero effects. The normal probability plot of the seven effects for the pilot chemical plant experiment is shown in Figure 8.50.

Effects that differ significantly from zero, at the chosen level of significance (5% in this case and indicated by Alpha = 0.05 in the subheading in Figure 8.50), are labelled. Use of the keys in the text boxes to the right of the plot enables the same conclusions to be reached as via the *P*-values in Panel 8.13.

The Pareto plot of the effects is shown in Figure 8.51. Bars that protrude through the vertical reference line correspond to effects that are significant at the chosen level. (The interested reader might check that the critical value 2.31 is the critical value for a two-tailed test using Student's *t* with 8 degrees of freedom.) Again the conclusions are exactly the same as before. One of the advantages of the Pareto chart is that standardized effects are ranked by absolute magnitude. The half normal plot also displays the absolute values of the standardized effects.

Table 8.15 was created by using Minitab by designing a full 2^3 factorial experiment, with one replication, no blocking, no centre points and no randomization of the run order.

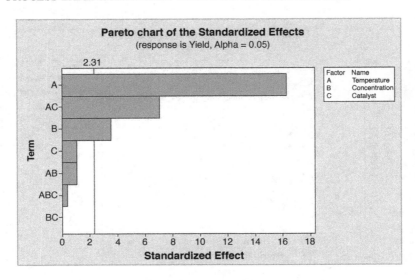

Figure 8.51   Pareto plot of effects.

The default factor names A, B and C were used together with the default levels -1 and $+1$ to represent all eight FLCs. The shaded columns give the full 2^3 factorial design in standard order. The AB, AC and BC columns were calculated by direct multiplication of the entries in the appropriate columns. The ABC column may be calculated in a variety of ways, e.g. by multiplication of the entries in the AB column by those in the C column. The I column consists entirely of 1s. The final column contains the mean response for the two replicate yields obtained for each FLC. For example, with temperature 160 °C ($A = -1$), concentration 20% ($B = -1$) and catalyst A ($C = -1$) the two yields obtained were 59 g and 61 g with mean 60 g.

The A, B and C columns may be used to calculate the main effects. The AB, BC and AC columns may be used to calculate the first-order interaction effects and the ABC column may be used to calculate the second-order interaction effect. The calculations are set out in Table 8.16. The reader is invited to check some of the calculations and confirm that the results agree with the Session window output in Panel 8.13.

Table 8.15   Pilot plant data.

Pilot plant	Temp.			Conc.		Catalyst			Mean response
FLC	I	A	B	AB	C	AC	BC	ABC	Mean Y
1	1	-1	-1	1	-1	1	1	-1	60
2	1	1	-1	-1	-1	-1	1	1	72
3	1	-1	1	-1	-1	1	-1	1	54
4	1	1	1	1	-1	-1	-1	-1	68
5	1	-1	-1	1	1	-1	-1	1	52
6	1	1	-1	-1	1	1	-1	-1	83
7	1	-1	1	-1	1	-1	1	-1	45
8	1	1	1	1	1	1	1	1	80

Table 8.16 Effects for pilot plant experiment.

	Mean response at level		Effect	Coefficient
	−	+		
A	52.75	75.75	23	11.5
B	66.75	61.75	−5	−2.5
AB	63.5	65.0	1.5	0.75
C	63.5	65.0	1.5	0.75
AC	59.25	69.25	10	5
BC	64.25	64.25	0	0
ABC	64.0	64.5	0.5	0.25

We can consider the 2^3 experiment as two 2^2 experiments – one carried out by varying temperature and concentration with catalyst A only, and the other carried out by varying temperature and concentration with catalyst B only. The data for the first of these 2^2 experiments are presented in Table 8.17. The temperature–concentration interaction effect with catalyst A is given by the AB interaction from Table 8.17:

$$[\text{Mean yield with } AB = 1] - [\text{Mean yield with } AB = -1]$$

$$= \frac{60 + 68}{2} - \frac{72 + 54}{2} = 64 - 63 = 1.$$

The data for the second of these 2^2 experiments is presented in Table 8.18. The temperature–concentration interaction effect with catalyst B is given by AB interaction from Table 8.18:

Table 8.17 Pilot plant data with catalyst A.

FLC	Temp. A	Conc. B	AB	Mean response Mean Y
1	− 1	− 1	1	60
2	1	− 1	− 1	72
3	− 1	1	− 1	54
4	1	1	1	68

Table 8.18 Pilot plant data with catalyst B.

FLC	Temp. A	Conc. B	AB	Mean response Mean Y
5	−1	−1	1	52
6	1	−1	−1	83
7	−1	1	−1	45
8	1	1	1	80

$$[\text{Mean yield with } AB = 1] - [\text{Mean yield with } AB = -1]$$

$$= \frac{52 + 80}{2} - \frac{83 + 45}{2} = 66 - 64 = 2.$$

We can sum up by stating that:

- the AB interaction for $C = -1$ is 1, i.e. the temperature–concentration interaction effect with catalyst A is 1;

- the AB interaction for $C = 1$ is 2, i.e. the temperature–concentration interaction effect with catalyst B is 2.

The AB interaction effect for the full 2^3 experiment is the mean of the AB interaction for $C = -1$ and the AB interaction for $C = 1$, i.e. $(2 + 1)/2 = 1.5$. The ABC interaction effect measures the extent to which the AB interaction effects differ for the two levels of factor C. It is given by half the difference between the two AB interaction effects, i.e. by $(2 - 1)/2 = 0.5$.

A final comment before leaving this example is that the coefficient in the Session window output labelled Constant is the overall mean for the experiment. One can think of it as being generated by the column labelled I in Table 8.15. There are no negative values in the column so it corresponds to finding the mean of all 8 values in the mean response column. Familiarity with the notation introduced in Table 8.15 is vital for understanding of topics in the next section.

It is also worth examining the remainder of the Session window output. The ANOVA table is shown in the Panel 8.14. The P-value of 0.000 for main effects indicates that there is very strong evidence that the null hypothesis that all main effects are zero would be rejected. Similarly, the P-value of 0.001 for two-way (first order) interactions indicates that there is strong evidence that the null hypothesis that all interactions involving two factors are zero would be rejected. Finally, the P-value of 0.733 for three-way (second order) interactions means that there is no evidence that the ABC interaction differs from zero. The ANOVA provides a global test of significance for the various categories of effect. The P-values in the previous section of the output discussed earlier enable one to identify the specific important main effects and interaction effects.

The coefficients in the first section of the Session window output in Panel 8.13, which are half of the effects, can be used to create a model with coded variables A, B and C representing the factors and Y representing yield as follows:

$$Y = 64.25 + 11.50A - 2.50B + 0.75C + 0.75AB + 5.00AC + 0.25ABC.$$

With the process run at temperature $160\,°C$ ($A = -1$), concentration 20% ($B = -1$) and with catalyst A ($C = -1$) the model predicts yield

```
Analysis of Variance for Yield (coded units)

Source                DF    Seq SS    Adj SS    Adj MS      F       P
Main Effects           3   2225.00   2225.00   741.667   92.71   0.000
2-Way Interactions     3    409.00    409.00   136.333   17.04   0.001
3-Way Interactions     1      1.00      1.00     1.000    0.12   0.733
Residual Error         8     64.00     64.00     8.000
  Pure Error           8     64.00     64.00     8.000
Total                 15   2699.00
```

Panel 8.14   ANOVA for 2^3 pilot chemical plant experiment.

$$Y = 64.25 + 11.50(-1) - 2.50(-1) + 0.75(-1)$$
$$+ 0.75(-1)(-1) + 5.00(-1)(-1) + 0.25(-1)(-1)(-1)$$
$$= 64.25 - 11.5 + 2.5 - 0.75 + 0.75 + 5.00 - 0.25 = 60.$$

This predicted yield is the fit and is simply the mean yield obtained in the experiment with the FLC temperature 160 °C, concentration 20% and catalyst A.

A second set of coefficients is provided beneath the ANOVA table in the Session window output. These coefficients enable the model equation to be written down in terms of the uncoded variables temperature, concentration and catalyst.

The final section of the Session window output is shown in Panel 8.15. It lists the symbol I, corresponding to estimation of the overall mean, the main effects and the interaction effects. The fact that each is listed in a separate row is indicative of a full factorial experiment and that each effect may be estimated from the data. Alias structure is important in both the discussion of fractional factorial experiments and the concept of resolution in the next section.

A final example is given of a full 2^4 factorial experiment. It is taken from p. 283 of *Experimental Design with Applications in Management, Engineering, and the Sciences,* 1st edition, by Berger and Maurer, © 2002, and reprinted with permission of Brooks/Cole, a division of Thomson Learning: www.thomsonrights.com. As the title indicates, it includes nonmanufacturing applications. Table 8.19 gives numbers of coupons redeemed out of 1000 issued to each one of 16 groups of 1000 customers. The effects on this response of the factors customer willingness to use (CWTU), ease of use, value and product type were of interest. The data are available in Coupons.MTW.

With two factors, the cube plot was two-dimensional, as was the design space; with three factors, the cube plot was three-dimensional, as was the design space. The cube plot of the data in Table 8.19 is displayed in Figure 8.52. Minitab displays two cubes – one for the low level of

```
Alias Structure
I
Temperature
Concentration
Catalyst
Temperature*Concentration
Temperature*Catalyst
Concentration*Catalyst
Temperature*Concentration*Catalyst
```

Panel 8.15   Alias structure for pilot plant experiment.

Table 8.19   Data for 2^4 factorial experiment.

CWTU	Ease of Use	Value			
		Low		High	
		Product type		Product type	
		Food	Paper	Food	Paper
Low	Low	4	2	8	6
Low	High	4	4	8	8
High	Low	4	5	9	9
High	High	7	6	8	8

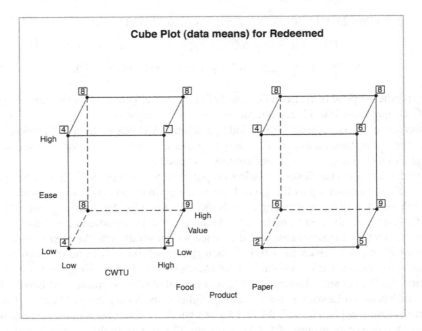

Figure 8.52   Cube plot for 2^4 factorial experiment.

product type and the other for the high level (the hypercube in four dimensions has been represented as two cubes in three dimensions!).

Part of the Session window output from **Analyze Factorial Design. . .** is displayed in Panel 8.16. The missing-value symbols indicate that no P-values can be calculated in this situation. The half normal probability plot of the standardized effects, with the default significance level $\alpha = 0.10$ selected is displayed in Figure 8.53. Thus only the main effect

```
Analysis of Variance for Redeemed (coded units)

Source                      DF   Seq SS    Adj SS    Adj MS   F   P
Main Effects                 4  61.2500   61.2500   15.3125  *   *
  CWTU                       1   9.0000    9.0000    9.0000  *   *
  Ease                       1   2.2500    2.2500    2.2500  *   *
  Value                      1  49.0000   49.0000   49.0000  *   *
  Product                    1   1.0000    1.0000    1.0000  *   *
2-Way Interactions           6   4.7500    4.7500    0.7917  *   *
  CWTU*Ease                  1   0.2500    0.2500    0.2500  *   *
  CWTU*Value                 1   1.0000    1.0000    1.0000  *   *
  CWTU*Product               1   1.0000    1.0000    1.0000  *   *
  Ease*Value                 1   2.2500    2.2500    2.2500  *   *
  Ease*Product               1   0.2500    0.2500    0.2500  *   *
  Value*Product              1   0.0000    0.0000    0.0000  *   *
3-Way Interactions           4   4.7500    4.7500    1.1875  *   *
  CWTU*Ease*Value            1   2.2500    2.2500    2.2500  *   *
  CWTU*Ease*Product          1   2.2500    2.2500    2.2500  *   *
  CWTU*Value*Product         1   0.0000    0.0000    0.0000  *   *
  Ease*Value*Product         1   0.2500    0.2500    0.2500  *   *
4-Way Interactions           1   0.2500    0.2500    0.2500  *   *
  CWTU*Ease*Value*Product    1   0.2500    0.2500    0.2500  *   *
Residual Error               0        *         *         *
Total                       15  71.0000
```

Panel 8.16   ANOVA for coupon experiment.

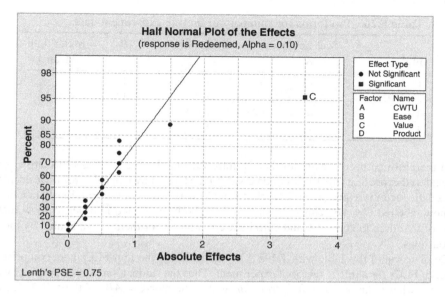

Figure 8.53   Normal probability plot of effects.

of the factor labelled C, i.e. of the value factor, appears to be significantly different from zero. (Lenth's PSE, given in the bottom corner of the plot, refers to the technical detail of the method used to determine significant effects when there is no replication – see Lenth, 1989.) The reader is invited to create a main effects plot and to interpret the main effect of value.

8.3   Fractional factorial experiments in the 2^{k-p} series

8.3.1   Introduction to fractional factorial experiments, confounding and resolution

In order to introduce this topic part of a data set given in Box *et al.* (2005, p. 216) will be used. The data are reproduced by permission of John Wiley & Sons, Inc., New York. However, the context presented here was developed by the author for use in introducing the topic of fractional factorial experimental designs on training courses.

Consider a student of DOE who was asked by his tutor to design and execute a factorial experiment. The student decided to investigate the effect of the factors seat height (A, levels 26 and 30 inches) and generator state (B, levels off and on) on the response, time (Y, seconds) to travel a fixed distance up an inclined section of a road. The design he chose was a 2^2 factorial with two replications.

As no software was to be involved in the data analysis the student had prepared Table 8.20 in advance in order to display a summary of the results and as an aid to calculate the effects. The shaded columns indicate the coded levels of the factors; the four shaded rows indicating the four factor–level combinations to be used (replicated twice) in the experiment.

On showing the table to his tutor prior to carrying out the experiment, the tutor remarked that, from experience, he thought it unlikely that there would be an interaction effect between seat height and generator state. The response from the student was to ask if the AB column

Table 8.20 Pro forma for summary of bicycle experiment data.

FLC	Seat height A	Generator state B	AB	Mean time Mean Y
1	−1	−1	1	
2	1	−1	−1	
3	−1	1	−1	
4	1	1	1	

could therefore be used for the levels of a third factor. The answer was affirmative, so the student left the meeting with the revised version of the table shown in Table 8.21. He decided to allocate the factor tyre pressure (C, levels 40 and 55 psi) to the column previously labelled AB and now labelled $C = AB$. This column has now also been shaded in order to show that the levels of the three factors A, B and C to be used in the experiment are indicated by the four shaded rows.

Comparison of this table with Table 8.15 indicates that the four FLCs here comprise *half* the set of FLCs for a full 2^3 factorial experiment. Thus the student's modified proposal was to carry out what is known as a *half fraction*, with generator $C = AB$, of the full 2^3 factorial experiment. Generators are used in the creation of fractional factorial designs and also to determine their statistical properties. The interested reader will find further details in Montgomery (2009, pp. 587–594).

Half of 2^3 is $2^3/2 = 2^3/2^1 = 2^{3-1}$. Thus the design is referred to as the 2^{3-1} fractional factorial experiment with generator $C = AB$. The family of such fractional factorial designs is designated by 2^{k-p}, so, for example, a 2^{5-2} fractional factorial experiment would be a one-quarter fraction of a full 2^5 experiment since $2^{5-2} = 2^5/2^2 = 2^5/4$. A 2^{5-2} fractional factorial experiment would involve $2^5/2^2 = 32/4 = 8$ FLCs. Let us now turn to Minitab for the design and analysis of the student's bicycle experiment.

Once again use of **Stat > DOE > Factorial > Create Factorial Design. . .**, with the option **2-level factorial (default generators)** selected, is required. **Number of factors:** is specified as 3. Under **Designs. . .** the 1/2 fraction is selected with **Number of replicates for corner points:** 2 and with the defaults of no centre points and a single block The number of Runs or FLCs is four; note that Minitab represents 2^{3-1} as 2**(3-1). Under **Factors. . .** the factors (two Numeric and one Text in this case) and their levels are specified. The defaults were accepted under both **Options. . .** (i.e to **Randomize runs** and to **Store design in worksheet**) and **Results. . ..** The resultant worksheet should be augmented with a column showing the response values, i.e. the times, and a column in which the student could note any unusual happening or information that might prove relevant to the analysis of the data from the

Table 8.21 Revised pro forma for bicycle experiment data.

FLC	Seat height A	Generator state B	Tyre pressure C = AB	Mean time Mean Y
1	−1	−1	1	
2	1	−1	−1	
3	−1	1	−1	
4	1	1	1	

C1	C2	C3	C4	C5	C6-T	C7	C8	C9-T
StdOrder	RunOrder	CenterPt	Blocks	Seat Height	Generator State	Tyre Pressure	Time	Remarks
6	1	1	1	30	Off	40	41	Snow on road
8	2	1	1	30	On	55	41	
5	3	1	1	26	Off	55	50	
3	4	1	1	26	On	40	54	
2	5	1	1	30	Off	40	43	Melting snow
4	6	1	1	30	On	55	44	
1	7	1	1	26	Off	55	48	
7	8	1	1	26	On	40	60	Snow

Figure 8.54 Worksheet for bicycle experiment.

experiment and to the project. As ever, the worksheet should be saved both for storage of the data and to facilitate analysis once the experimental work has been completed (see Figure 8.54, where the response data and comments have been added; the data are provided in Bicycle.MTW).

The information that appears in the Session window, on creation of the experimental design, is displayed in Panel 8.17. Base Design: 3, 4 indicates that three factors are involved and that four FLCs are employed. The design has resolution III, which means that 'no main effects are aliased with any other main effect, but main effects are aliased with two-factor interactions and some two-factor interactions may be aliased with each other' (Montgomery, 2009, p. 589). The student will make a total of eight experimental runs. (Minitab is inconsistent in its use of the term 'runs' – in the dialog for the creation of the design, 'runs' refers to the number of FLCs.) Two runs will be made with each FLC. A half fraction of the full factorial is being used and the eight runs will constitute a single block. No centre points are to be used.

The price paid for using this fractional factorial is that some main effects are confounded with two-way interactions (A with BC, B with AC, and C with AB – in each case the main effect and the interaction are aliases of each other), which means that if an interaction between a pair of factors exists then we will be unable to 'disentangle' it from a main effect in our analysis. This type of confounding is a feature of resolution III designs. Note also that the overall mean (I) is confounded with the ABC interaction.

The equation $C = AB$ indicates how the design may be generated. Note that this was how a factor was allocated to the third column in the discussion between student and tutor.

```
Fractional Factorial Design

Factors:   3    Base Design:        3, 4    Resolution:  III
Runs:      8    Replicates:            2    Fraction:    1/2
Blocks:    1    Center pts (total):    0

* NOTE * Some main effects are confounded with two-way interactions.

Design Generators: C = AB


Alias Structure

I + ABC

A + BC
B + AC
C + AB
```

Panel 8.17 Specification of 2^{3-1} design for bicycle experiment.

As stated above, the design considered here is of resolution III. With resolution IV designs no main effect is aliased with a two-factor interaction, but at least one two-factor interaction is aliased with another two-factor interaction. With resolution V designs no main effect or two-factor interaction is aliased with either a main effect or a two-factor interaction. The higher the resolution of an experimental design, the greater its potential for the unambiguous identification of important main effects and interaction effects. Full factorial design experiments involving replication have the capability to provide information, without any confounding, on all main effects and all interactions and may be said to have 'infinite' resolution. Resolution III designs are often used for screening experiments, carried out with the aim of identifying potentially important factors from a set of factors identified as being worth investigation. 'For most practical purposes, a resolution 5 design is excellent and a resolution 4 design may be adequate' (NIST/SEMATECH, 2005, Section 5.7).

The cube plot of the data is shown in Figure 8.55. Note the four vertices corresponding to the four FLCs used in the experiment. The mouse pointer is shown located at the vertex corresponding to the FLC with seat height 30, generator on and tyre pressure 55. The two times achieved were 41 and 44, giving the mean of 42.5 displayed. (The other vertices provide an alternative half fraction of the full 2^3 design.)

The main effects plot is shown in Figure 8.56. The major effect would appear to be that of seat height, with runs done with the seat in the high position taking around 11 seconds less on average than those runs done with the seat in the low position. Runs with the generator on took around 4 seconds longer than those with the generator off. This is not surprising as, with the generator on, some of the cyclist's energy will be required to drive the generator. Runs with the tyre pressure high took about 4 seconds less on average than those with tyre pressure low. Thus a pick-a-winner approach on the basis of this plot would suggest seat height set high, generator turned off and tyre pressure set high to achieve the shortest possible time. However, we must bear in mind that this could be totally in error if there are significant interaction effects.

Proceeding to formal analysis via **Analyze Factorial Design. . .**, it is informative to examine the default offered under **Terms. . .** as displayed in Figure 8.57. The default is to **Include terms in the model up through order:** 1, i.e. to include only terms corresponding

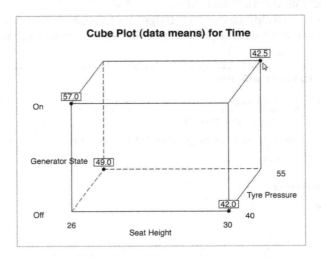

Figure 8.55   Cube plot for bicycle experiment.

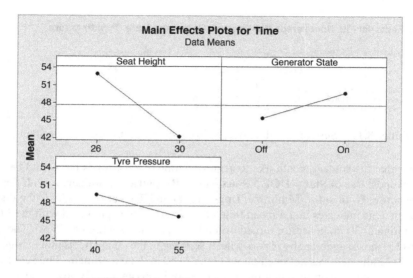

Figure 8.56 Main effects plot for bicycle experiment.

to main effects. These are listed under **Selected Terms:**. The arrow keys may be used to change the selection of terms.

Part of the Session window output from the analysis is shown in Panel 8.18. If we assume that there are no significant interactions, then the analysis provides strong evidence of a seat height effect, slight evidence of a generator state effect and marginal evidence of a tyre pressure effect (at the 10% significance level). Let us decide to keep all three terms in the model and let

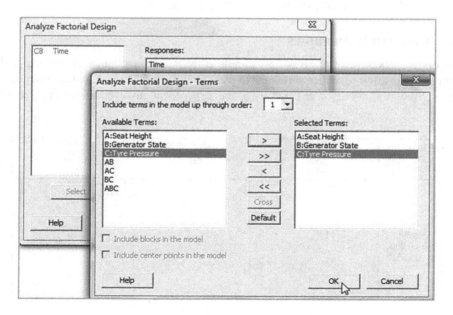

Figure 8.57 Default terms for model for bicycle experiment.

Factorial Fit: Time versus Seat Height, Generator State, Tyre Pressure

Estimated Effects and Coefficients for Time (coded units)

Term	Effect	Coef	SE Coef	T	P
Constant		47.625	0.9100	52.33	0.000
Seat Height	-10.750	-5.375	0.9100	-5.91	0.004
Generator State	4.250	2.125	0.9100	2.34	0.080
Tyre Pressure	-3.750	-1.875	0.9100	-2.06	0.108

Panel 8.18   Session window output from analysis of bicycle experiment.

us suppose that the student would like to achieve a time of 35 seconds but would be satisfied with 40 seconds. Use of **Stat > DOE > Factorial > Response Optimizer. . .** with time as the (only) response, **Goal set** to Minimize, **Upper** to 40 and **Target** to 35 yields the display in Figure 8.58. This indicates that a mean time of 38.25 seconds is predicted with the FLC seat height set high at 30 in, generator turned off and tyre pressure set high at 55 psi. These levels correspond to those suggested by pick-a-winner scrutiny of the main effects plot in Figure 8.56.

The FLC indicated was not one that was employed in the experiment. Thus confirmation trials with the predicted optimal combination of factor levels are clearly advisable. (In fact the student actually carried out the full 2^3 factorial. Both times achieved with the seat height set high at 30 in, generator turned off and tyre pressure set high at 55psi were 39 seconds. Thus the prediction from the fractional factorial was good. The analysis of the full data set provided no evidence of any significant interaction effects.) The user can, of course, vary the factor levels with the response optimizer, as indicated in the previous section, in the case of seat height and tyre pressure. However, with the factor generator state being of text type, no intermediate values are possible. If the user tries to drag the red line to an intermediate position, release of the mouse button causes the factor level to 'jump' to the other of the two discrete levels. The two solid circles indicate the nature of this particular factor.

8.3.2 Case study examples

Montgomery (2009, p. 594) gives an example of a 2^{7-3} fractional factorial experiment conducted in an effort to reduce shrinkage of parts manufactured using an injection moulding process. Since $2^{7-3} = 2^7/2^3 = 128/8 = 16$, the experiment involved 16 of the 128 FLCs

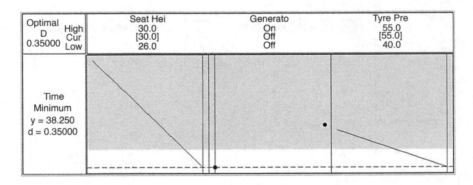

Figure 8.58   Response Optimizer for bicycle experiment.

```
Alias Structure

I + ABCE + ABFG + ACDG + ADEF + BCDF + BDEG + CEFG

A + BCE + BFG + CDG + DEF + ABCDF + ABDEG + ACEFG
B + ACE + AFG + CDF + DEG + ABCDG + ABDEF + BCEFG
C + ABE + ADG + BDF + EFG + ABCFG + ACDEF + BCDEG
D + ACG + AEF + BCF + BEG + ABCDE + ABDFG + CDEFG
E + ABC + ADF + BDG + CFG + ABEFG + ACDEG + BCDEF
F + ABG + ADE + BCD + CEG + ABCEF + ACDFG + BDEFG
G + ABF + ACD + BDE + CEF + ABCEG + ADEFG + BCDFG
AB + CE + FG + ACDF + ADEG + BCDG + BDEF + ABCEFG
AC + BE + DG + ABDF + AEFG + BCFG + CDEF + ABCDEG
AD + CG + EF + ABCF + ABEG + BCDE + BDFG + ACDEFG
AE + BC + DF + ABDG + ACFG + BEFG + CDEG + ABCDEF
AF + BG + DE + ABCD + ACEG + BCEF + CDFG + ABDEFG
AG + BF + CD + ABDE + ACEF + BCEG + DEFG + ABCDFG
BD + CF + EG + ABCG + ABEF + ACDE + ADFG + BCDEFG
ABD + ACF + AEG + BCG + BEF + CDE + DFG + ABCDEFG
```

Panel 8.19 Alias structure for a 2^{7-3} fractional factorial design.

required for a full 2^7 factorial. The experiment was a 1/8 fraction of the full factorial. The factors were mould temperature, A; screw speed, B; holding time, C; cycle time, D; moisture content, E; gate size, F; and holding pressure G. The response, Y, was shrinkage measured as a percentage, and a single replication was used. The data are available in Shrinkage.MTW and are reproduced by permission of John Wiley & Sons, Inc., New York. The alias structure of this design is displayed in Panel 8.19. The reader is invited to check this using **Stat > DOE > Factorial > Create Factorial Design. . .** with the option **2-level factorial (default generators)** selected and **Number of factors: 7** specified. Under **Designs. . .** the 1/8 fraction is selected with defaults otherwise The number of runs or FLCs is 16; note that Minitab represents 2^{7-3} as 2**(7-3). The alias structure in Panel 8.19 then appears in the Session window.

The design is of resolution IV. Main effects and interactions between sets of three factors are confounded with each other (and with higher-order interactions). Interactions between pairs of factors are confounded with each other (and with higher-order interactions). Resolution IV designs are such that 'no main effect is aliased with any other main effect *or* with any two-factor interaction, but two-factor interactions are aliased with each other' (Montgomery, 2009, p. 592).

On the basis of the annotated normal probability plot of the effects displayed in Figure 8.59, the engineers involved with the process made the tentative conclusion that the important factors were likely to be mould temperature A and screw speed B. The level of significance 0.001 was used in creating the plot. The conclusion has to be a tentative one since, on the assumption that all interactions involving three or more factors are zero, the effect labelled AB in the plot could be important because any of the AB, CE and FG interactions could be important or indeed any combination involving two or more of them. This can be confirmed by checking the alias structure in Panel 8.19 (tenth line).

The engineers agreed to set temperature at the low level and screw speed at the low level on the strength of scrutiny of the interaction plot displayed in Figure 8.60. Clearly at the low screw speed the effect of temperature appears to be negligible – in such situations the choice of level of temperature to use can be based on cost considerations.

The project team fitted a model involving only the mould temperature, the screw speed and the interaction between them. They plotted the residuals from the fitted model against the

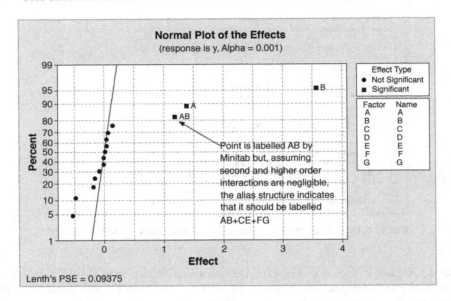

Figure 8.59 Normal probability plot of effects for injection moulding experiment.

levels of the other factors not in the model. This may be readily achieved in Minitab using **Residuals versus variables:** under **Graphs. . .** in **Analyze Factorial Design. . .** and selecting the five factors not in the model. Of particular interest was the plot of residuals against holding time, shown in Figure 8.61, as it suggests that at the low level of holding time (factor C) there would be less variability in shrinkage.

This led the project team to investigate running the process with low mould temperature, low screw speed and low holding time. The aim was to ascertain whether or not it could be

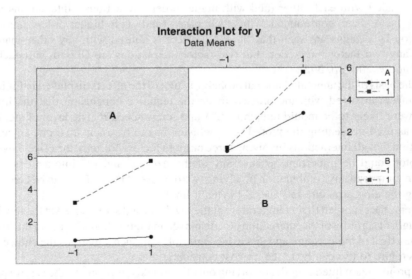

Figure 8.60 *AB* interaction plot for the injection moulding experiment.

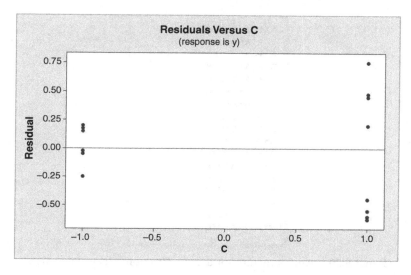

Figure 8.61 Plot of residuals versus holding time.

confirmed that factors A and B were indeed influencing the response in terms of its location and that factor C was influencing the response in terms of its variability.

An advertising agency conducted an experiment in which various forms of newspaper advertisement for a new leisure product were created. The factors and levels were: size of advertisement (A, with levels small and large); type of advertisement (B, simple and complex); colours (C, two and four); setting (D, sole and group); and newspaper style (E, compact and broadsheet). The response was the proportion (expressed as a percentage) of a sample of people who were able to remember reading the advertisement, having read through a mock-up newspaper containing it. The experiment was a 2^{5-1} design, i.e. it was a half fraction of the full 2^5 factorial and therefore involved 16 FLCs. A single replication was used. The data are available in Advertising.MTW.

On use of **Analyze Factorial Design...** to perform an analysis of the data the alias structure for the design that was obtained, which has resolution V, is shown in Panel 8.20. Main effects are confounded with third-order interactions (i.e. interactions involving four factors), and first-order interactions (i.e. interactions involving pairs of factors) are confounded with second-order interactions (i.e. interactions involving sets of three factors). This alias structure highlights the important property of resolution V designs, namely that 'no main effect or two-factor interaction is aliased with any other main effect or two-factor interaction, but two-factor interactions are aliased with three-factor interactions' (Montgomery, 2009, p. 592) Thus, if second- and higher-order interactions are negligible, as is frequently the case, then main effects and two-factor interactions can be separately estimated. The normal probability plot of the effects is shown in Figure 8.62.

The main effects of factors B and E are flagged as being potentially important, together with the first-order interaction effect CD. The nature of these effects my be gleaned from scrutiny of the main effects plot in Figure 8.63 and the interaction plot in Figure 8.64. The main effect of B may be interpreted as a reduction in recall of around 7% on changing from simple to complex type. The main effect of E may be interpreted as an increase in recall of around 10% on changing from compact to broadsheet style of newspaper.

```
Alias Structure
I + A*B*C*D*E
A + B*C*D*E
B + A*C*D*E
C + A*B*D*E
D + A*B*C*E
E + A*B*C*D
A*B + C*D*E
A*C + B*D*E
A*D + B*C*E
A*E + B*C*D
B*C + A*D*E
B*D + A*C*E
B*E + A*C*D
C*D + A*B*E
C*E + A*B*D
D*E + A*B*C
```

Panel 8.20   Alias structure for a 2^{5-1} fractional factorial design.

The *CD* interaction may be interpreted as follows: changing from two to four colours (i.e. from $C = -1$ to $C = 1$) when the advertisement is the only one on the page ($D = -1$) leads to a decrease in the recall rate, whereas changing from two to four colours (i.e. from $C = -1$ to $C = 1$) when the advertisement is one of a group on the page ($D = 1$) leads to an increase in the recall rate.

8.4   Taguchi experimental designs

Figure 8.65 shows the two possible half fractions of the full 2^3 factorial experiment. The circular symbols represent the principal half fraction of the full 2^3 factorial that was introduced

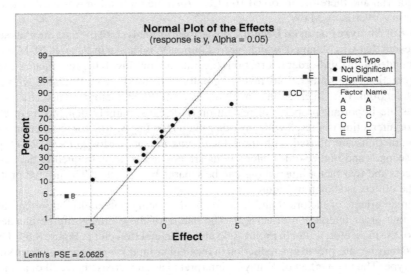

Figure 8.62   Normal probability plot of effects for advertising experiment.

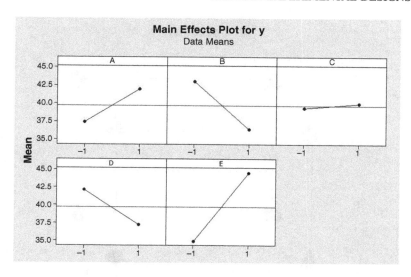

Figure 8.63   Main effects plot for advertising experiment.

earlier for the bicycle experiment – see Table 8.21 and Figure 8.55. The generator of this fraction is $C = AB$. An alternative half fraction is represented by the triangular symbols in Figure 8.65. The generator of this fraction is $C = -AB$, as the reader can readily verify from scrutiny of Table 8.23.

The numbers in the A, B and C columns define three vectors in a space of four dimensions. The inner product of two vectors is calculated by obtaining the sum of the products of corresponding components of the two vectors. For example, the inner product of A and B is

$$(-1) \times (-1) + 1 \times (-1) + (-1) \times 1 + 1 \times 1 = 1 - 1 - 1 + 1 = 0.$$

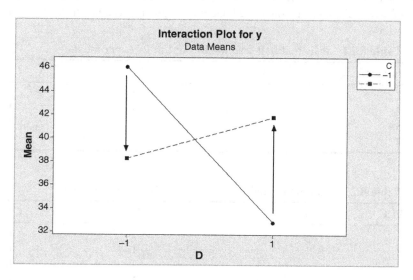

Figure 8.64   CD interaction plot for advertising experiment.

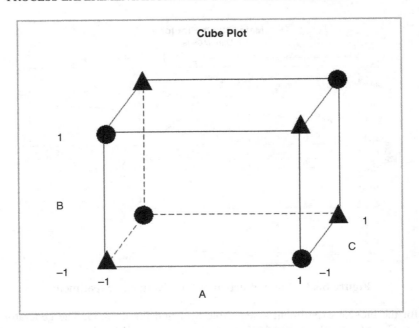

Figure 8.65   The two possible half fractions of the full 2^3 factorial.

The reader is invited to verify that the inner product of A and C and the inner product of vectors B and C are also zero. Vectors with inner product of zero are said to be *orthogonal*. Since the A, B and C columns of factor levels for the experimental design in Table 8.23 are made up of orthogonal vectors the design constitutes an orthogonal array.

If the factor levels in this design are denoted by 1 and 2 rather than -1 and 1 then, with some rearrangement of the rows, we have the experimental design,designated *Taguchi's $L_4(2^3)$ orthogonal array*, displayed in Table 8.24. Genichi Taguchi was a Japanese engineer who did

Table 8.22   Principal half fraction of the full 2^3 experiment.

FLC	A	B	C
1	-1	-1	1
2	1	-1	-1
3	-1	1	-1
4	1	1	1

Table 8.23   Alternative half fraction of the full 2^3 experiment.

FLC	A	B	C
1	-1	-1	-1
2	1	-1	1
3	-1	1	1
4	1	1	-1

Table 8.24 Taguchi's $L_4(2^3)$ orthogonal array.

FLC	A	B	C
1	1	1	1
2	1	2	2
3	2	1	2
4	2	2	1

Figure 8.66 Linear graph for the $L_4(2^3)$ orthogonal array.

much to promote the use of designed experiments in industry and made other significant contributions to quality improvement thinking and methodology. Taguchi created linear graphs associated with each orthogonal array; that for the $L_4(2^3)$ array is displayed in Figure 8.66. The graph, reproduced by permission of the American Supplier Institute Inc., Dearborn, indicates that, if factors A, B and C are respectively assigned to columns 1, 2 and 3 of the orthogonal array, the interaction between the factors A (1) and B (2) will be aliased or confounded with the main effect of C (3). Both Kolarik (1995, pp. 892–901) and Hicks and Turner (1999, pp. 370–398) give comprehensive lists of Taguchi orthogonal arrays and their associated linear graphs. In Minitab the available Taguchi experimental designs may be accessed via **Stat** > **DOE** > **Taguchi** > **Create Taguchi Design. . ..**

Taguchi advocated the use of signal to noise ratios in the analysis of data from designed experiments. The three principal ratios are given in Box 8.5. Each corresponds to a different

1. Target (nominal is best)

$$SN_T = 10 \log_{10} \left(\frac{\bar{y}^2}{s^2} \right) \quad \text{or} \quad SN_T = -10 \log_{10} s^2$$

2. Maximisation (larger is better)

$$SN_L = -10 \log_{10} \left(\frac{1}{n} \sum_{i=1}^{n} \frac{1}{y_i^2} \right)$$

3. Minimisation (smaller is better)

$$SN_S = -10 \log_{10} \left(\frac{1}{n} \sum_{i=1}^{n} y_i^2 \right)$$

Box 8.5 Taguchi signal to noise ratios.

objective. The first may be employed when the objective is to achieve a target value for the response, the second when it is desired to maximize a response, and the third when minimization of a response is required. These signal to noise ratios were developed in such a way that in all cases the larger the value of the ratio the better in terms of achieving objectives. At a symposium on Taguchi methods in 1985, J. Quinlan and colleagues from Flex Products reported on an experiment that they had carried out as part of a quality improvement project with the objective of minimization of the shrinkage of extruded thermoplastic speedometer cable casing (Quinlan, 1985, pp. 257–266). In their introduction they refer to the fact that the previous manufacturer had 'conducted much one factor at a time experimentation with high costs and disappointing results'. The 15 factors of interest in the experiment are listed in Table 8.25 together with the two levels for each that were selected by the experimenters.

In order to design a Taguchi experiment one may use **Stat > DOE > Taguchi > Create Taguchi Design. . ..** On selecting **Number of factors:** 15, Minitab offers the $L_{16}(2^{15})$ and $L_{32}(2^{15})$ orthogonal arrays as shown in Figure 8.67. The researchers chose the former design, which involves 16 FLCs. For each of these combinations 3000 feet of the casing was produced. From each length four sections were removed and the percentage shrinkage measured. The data are provided in Casing.MTW, displayed in Table 8.26 and reproduced by permission of the American Supplier Institute Inc., Dearborn, MI. With this type of experimental design Minitab offers no randomisation procedure. However, in the conduct of the experiment the various FLCs were randomized 'as much as possible'.

Before discussing the analysis of the actual experimental data we will consider hypothetical data shown in Table 8.27 for three FLCs. It would appear that FLCs II and III are equally good in that both give mean percentage shrinkage of 0.15, compared with 0.50 for combination I. For FLC I the appropriate 'smaller is better' signal to noise ratio is calculated as

Table 8.25 4 Factors and levels for speedometer cable casing experiment.

Factor	Name	Level 1	Level 2
A	Liner OD	Existing	Changed
B	Liner die	Existing	Changed
C	Liner material	Existing	Changed
D	Liner line speed	Existing	80% of existing
E	Wire braid type	Existing	Changed
F	Braiding tension	Existing	Changed
G	Wire diameter	Smaller	Existing
H	Liner tension	Existing	More
J	Liner temperature	Ambient	Preheated
K	Costing material	Existing	Changed
L	Coating die type	Existing	Changed
M	Melt temperature	Existing	Cooler
N	Screen pack	Existing	Denser
O	Cooling method	Existing	Changed
P	Line speed	Existing	70% of existing

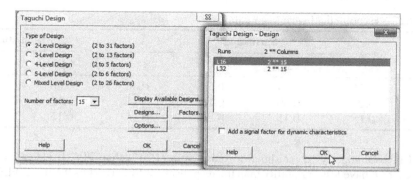

Figure 8.67   Selecting the $L_{16}(2^{15})$ orthogonal array.

$$SN_S = -10\log_{10}\left(\frac{1}{n}\sum_{i=1}^{n}y_i^2\right)$$

$$= -10\log_{10}\left(\frac{1}{4}(0.56^2 + 0.44^2 + 0.54^2 + 0.46^2)\right)$$

$$= -10\log_{10}\left(\frac{1.0104}{4}\right)$$

$$= -10\log_{10}0.2526$$

$$= -10 \times (-0.598) = 5.98.$$

Table 8.26   Data for speedometer cable casing experiment.

A	B	C	D	E	F	G	H	J	K	L	M	N	O	P	y1	y2	y3	y4
1	1	1	1	1	1	1	1	1	1	1	1	1	1	1	0.49	0.54	0.46	0.45
1	1	1	1	1	1	1	2	2	2	2	2	2	2	2	0.55	0.60	0.57	0.58
1	1	1	2	2	2	2	1	1	1	1	2	2	2	2	0.07	0.09	0.11	0.08
1	1	1	2	2	2	2	2	2	2	2	1	1	1	1	0.16	0.16	0.19	0.19
1	2	2	1	1	2	2	1	1	2	2	1	1	2	2	0.13	0.22	0.20	0.23
1	2	2	1	1	2	2	2	2	1	1	2	2	1	1	0.16	0.17	0.13	0.12
1	2	2	2	2	1	1	1	1	2	2	2	2	1	1	0.24	0.22	0.19	0.25
1	2	2	2	2	1	1	2	2	1	1	1	1	2	2	0.13	0.19	0.19	0.19
2	1	2	1	2	1	2	1	2	1	2	1	2	1	2	0.08	0.10	0.14	0.18
2	1	2	1	2	1	2	2	1	2	1	2	1	2	1	0.07	0.04	0.19	0.18
2	1	2	2	1	2	1	1	2	1	2	2	1	2	1	0.48	0.49	0.44	0.41
2	1	2	2	1	2	1	2	1	2	1	1	2	1	2	0.54	0.53	0.53	0.54
2	2	1	1	2	2	1	1	2	2	1	1	2	2	1	0.13	0.17	0.21	0.17
2	2	1	1	2	2	1	2	1	1	2	2	1	1	2	0.28	0.26	0.26	0.30
2	2	1	2	1	1	2	1	2	2	1	2	1	1	2	0.34	0.32	0.30	0.41
2	2	1	2	1	1	2	2	1	1	2	1	2	2	1	0.58	0.62	0.59	0.54

Table 8.27   Hypothetical shrinkage data.

FLC	Shrinkage				Statistics	
	y1	y2	y3	y4	Mean	SN$_S$
I	0.56	0.44	0.54	0.46	0.50	5.98
II	0.21	0.09	0.19	0.11	0.15	16.00
III	0.15	0.15	0.16	0.14	0.15	16.47

FLC III gives a slightly larger signal to noise ratio (16.47) than FLC II (16.00) since the responses for it have less variability than the responses for FLC II. Thus the 'best' FLC yields the highest signal to noise ratio.

In order to analyse the data from the actual experiment use may be made of **Stat > DOE > Taguchi > Analyze Taguchi Design. . ..** The dialog is displayed in Figure 8.68. **Response data are in:** y1-y4 indicates the location of the response values in the worksheet. (The reader should note that the layout of the data differs from that in the case of the factorial experiments considered earlier, where replicate responses appeared in different rows of the same column of the Minitab worksheet for analysis.) Under **Options. . .** a signal to noise ratio appropriate to the context is selected – here **Smaller is better** is appropriate. Under **Graphs. . .** and **Analysis. . .** the defaults were accepted. **Storage. . .** was used to specify that columns containing the means and signal to noise ratios should be computed for each FLC.

The main effects plot for both the mean response and the signal to noise ratio are displayed in Figures 8.69 and 8.70. The 'main players' appear to be factors E and G. However, Minitab has provided no information on the aliasing involved in this design. In fact the design is of resolution III, so main effects are confounded with first-order interactions between pairs of factors. Thus any major apparent observed effect might be due to an important interaction.

Assuming that there are no important interactions, and bearing in mind that the objective was to reduce shrinkage as much as possible, the main effects plot of means for factor E (wire braid type) indicates that level 2 (changed) should be used since it was desirable that shrinkage be as low as possible.

In scrutinizing the corresponding main effects plot of the signal to noise ratios there is no conflict. Level 2 for factor E (wire braid type) is again indicated since the signal to noise ratio is defined in such a way that we always seek to achieve the largest possible values. Similarly, both

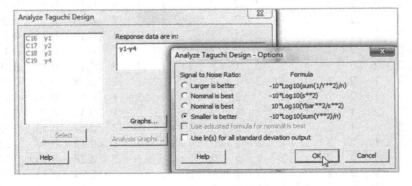

Figure 8.68   Dialog for analysis of a Taguchi design experiment.

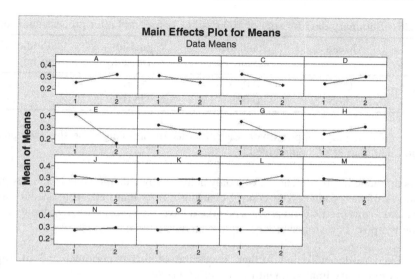

Figure 8.69 Main effects plot for shrinkage.

plots indicate that level 2 (existing) should be used for factor G (wire diameter). Factors such as O and P appear to have no impact on shrinkage.

Quinlan *et al.* used ANOVA on the signal to noise ratios and concluded that the eight factors E, G, K, A, C, F, D and H had significant effects. Table 8.28 gives a summary of data for 100 casings produced with the existing choice of factor levels prior to the experimentation and for 100 casings produced with the choice of factor levels indicated by the experiment as being optimal.

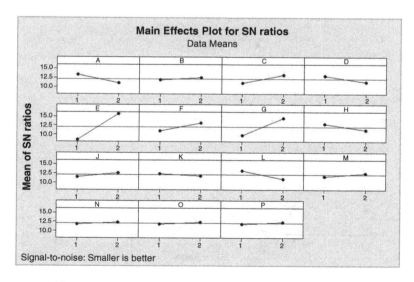

Figure 8.70 Main effects plot for signal to noise ratio.

Table 8.28   Summary data for sets of 100 casings before and after process changes.

Levels	Mean	Standard deviation	Signal to noise ratio	Predicted signal to noise ratio
Existing	0.26	0.050	11.6	12.6
Optimal	0.05	0.025	25.0	24.3

Figure 8.71 displays approximate distributions of shrinkage for the two phases of production. Quinlan *et al.* commented:

> this dramatic improvement . . . was only achieved by changing one of the design criteria of the product. The control charting efforts that had been assiduously applied to this product and process could not have been successful in reducing the average post extrusion shrinkage by the amount shown.

The interested reader will find further discussion of this experiment and of Taguchi's contribution to quality improvement in Box *et al.* (1988).

In the previous example each column of the orthogonal array had a factor allocated to it. In the next example, reported by Pignatiello and Ramberg (1985, pp. 198–206), the $L_8(2^7)$ array, which may be used to investigate seven factors, was used to investigate four factors. The process that they investigated involved the manufacture of vehicle leaf springs. The spring assembly passes through a high-temperature furnace and is then transferred to a press where the curvature is induced. Finally, the assembly is quenched in oil. The process should yield springs with a free height of 8 inches. The four controllable factors were:

- *A*, furnace temperature (°F) with levels 1880 (1) and 1840 (2);

- *B*, heating time (s) with levels 23 (1) and 25 (2);

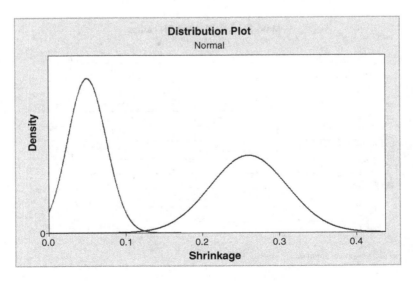

Figure 8.71   Approximate distributions of shrinkage before and after process changes.

- C, transfer time (s) with levels 10 (1) and 12 (2);

- D, hold-down time (s) with levels 3 (1) and 2 (2).

The reader should note that the experimenters did not always use the lowest value of a factor setting as level 1.

The linear graphs associated with the Taguchi $L_8(2^7)$ array are displayed in Figure 8.72 and are reproduced by permission of the American Supplier Institute Inc., Dearborn, MI.

Engineers involved with the process suspected that interactions were likely between all possible pairs of the factors A, B and C but that factor D was unlikely to interact with the other three. Thus they used Graph 1 to make the allocation of factors A, B, C and D to columns 1, 2, 4 and 7 respectively as indicated by the shading in Table 8.29. These four columns give the eight FLCs to be used in the experiment. The line segment labelled 3 in Graph 1, which joins the triangle vertices labelled 1 (allocated to factor A) and 2 (allocated to factor B), indicates that column 3 corresponds to the AB interaction. The reader is invited to verify from Graph 1 that columns 5 and 6 correspond to the AC and BC interactions respectively. Columns 3, 5 and 6 have no bearing on the actual running of the experiment but may be used in subsequent analysis of the experimental data to investigate AB, AC and BC interaction effects. This array was what is referred to in the Taguchi methodology as the inner array for the experiment.

Table 8.29 The $L_8(2^7)$ orthogonal array.

A	B	AB	C	AC	BC	D
1	1	1	1	1	1	1
1	1	1	2	2	2	2
1	2	2	1	1	2	2
1	2	2	2	2	1	1
2	1	2	1	2	1	2
2	1	2	2	1	2	1
2	2	1	1	2	2	1
2	2	1	2	1	1	2

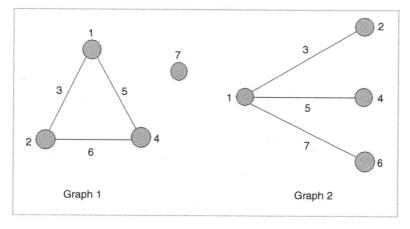

Figure 8.72 Linear graphs for the $L_8(2^7)$ orthogonal array.

Table 8.30   Data for spring free height experiment.

Inner array				Outer array O					
Factor				1				2	
A	B	C	D						
1	1	1	1	7.56	7.81	7.69	7.81	7.50	7.79
1	1	2	2	7.59	7.56	7.75	7.63	7.75	7.56
1	2	1	2	7.69	8.09	8.06	7.56	7.69	7.62
1	2	2	1	8.15	8.18	7.88	7.88	7.88	7.44
2	1	1	2	7.56	7.62	7.44	7.18	7.18	7.25
2	1	2	1	7.50	7.56	7.50	7.50	7.56	7.50
2	2	1	1	7.94	8.00	7.88	7.32	7.44	7.44
2	2	2	2	7.78	7.78	7.81	7.50	7.25	7.12

The temperature of the oil used for quenching the springs once formed was difficult to control so it was treated as a noise factor. For each FLC, three springs were produced with factor O, oil quench temperature (°F), at level 1 (130–150°F) and three were produced with O at level 2 (150–170°F). Factor O with its two levels provides what is referred to in the Taguchi methodology as the outer array for the experiment. The design and data are given in Table 8.30.

We will now work through the design and analysis of the experiment, using Minitab. The first step is to create the design using **Stat > DOE > Taguchi > Create Taguchi Design**. On specifying 2-Level Design under **Type of Design** and **Number of factors: 4**, clicking on **Designs. . .** reveals four available designs – $L_8(2^7)$, $L_{12}(2^7)$, $L_{16}(2^7)$ and $L_{32}(2^7)$. In this case L_8 was used. **Factors. . .** may be used to name the factors and specify the levels as before, as shown in Figure 8.73.

Abbreviated names were used. Note that the default allocation of the factors A, B, C and D is to columns 1, 2, 4 and 7 which is what is required in this case. (Were the design associated

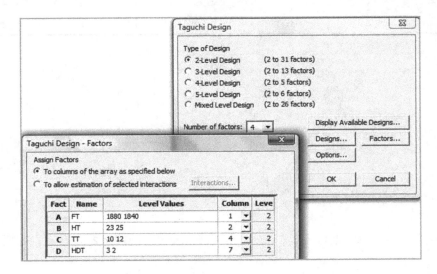

Figure 8.73   Specifying the factors and levels.

FT	HT	TT	HDT	h1 (O Low)	h2 (O Low)	h3 (O Low)	h4 (O High)	h5 (O High)	h6 (O High)
1880	23	10	3	7.56	7.81	7.69	7.81	7.50	7.79
1880	23	12	2	7.59	7.56	7.75	7.63	7.75	7.56
1880	25	10	2	7.69	8.09	8.06	7.56	7.69	7.62
1880	25	12	3	8.15	8.18	7.88	7.88	7.88	7.44
1840	23	10	2	7.56	7.62	7.44	7.18	7.18	7.25
1840	23	12	3	7.50	7.56	7.50	7.50	7.56	7.50
1840	25	10	3	7.94	8.00	7.88	7.32	7.44	7.44
1840	25	12	2	7.78	7.78	7.81	7.50	7.25	7.12

Figure 8.74 Design and data.

with the other linear graph in Figure 8.72 required, then the drop-down menu for the **Column** associated with factor D could be used to change the 7 to a 6.) The reader should note that under **Options. . .** there is no scope for randomization. The data are shown, added to the design worksheet, generated by Minitab, in Figure 8.74. This worksheet is supplied as Springs. MTW, and the data are reproduced with permission from the *Journal of Quality Technology* (© 1985 American Society for Quality).

Columns C5, C6 and C7 contain the heights of springs obtained with the eight controllable FLCs for the Taguchi design and with low oil quench temperature. Columns C8, C9 and C10 contain the heights of springs obtained with the eight controllable FLCs for the Taguchi design and with high oil quench temperature.

Central to Taguchi's methodology is the determination of levels of the controllable factors which maximize signal to noise ratio and then using factors which do not influence the signal to noise ratio to adjust the actual response, i.e. in this case the spring free height, to the desired target value.

The analysis may be performed using **Stat > DOE > Taguchi > Analyze Taguchi Design. . ..** Under **Response data are in:** the six columns containing the data were selected. The nominal is best signal to noise ratio $SN_T = 10 \log_{10}(\bar{y}^2/s^2)$ was selected under **Options. . ..** Under **Graphs. . .** the options to **Generate plots of main effects and interactions in the model for** both signal to noise ratios and means were checked. For the **Interaction plots** only **Display interaction plot matrix** was checked. **Terms . . .** was used to select the four main effects of interest (A, B, C and D) plus the three two-factor (first-order) interactions of interest (AB, AC and BC). Under **Analysis. . . Signal to Noise ratios** and **Means** were checked, both under **Display response tables for** and under **Fit linear model for**. No use was made of **Analysis Graphs. . .** or **Storage. . ..**

The main effects plot for the signal to noise ratios is displayed in Figure 8.75. The most important effect appears to be that of B, heating time. Bearing in mind that signal to noise ratios were developed in such a way that the higher the value the better the indication is that a heating time of 23 seconds appears advisable.

The interaction plot matrix for the signal to noise ratios is shown in Figure 8.76. The experimenters were led to believe that the interaction between heating time and transfer time (BC) was most important. Note that the greatest mean signal to noise ratio is associated with the FLC of heating time 23 seconds and transfer time 12 seconds. The experimenters used ANOVA to provide formal evidence that the main effect of heating time and the interaction effect between heating time and transfer time effects were 'active'.

The main effects plot for means of the spring heights is displayed in Figure 8.77. This plot suggested to the experimenters that furnace temperature and hold-down time could be used to adjust spring height to the target value of 8 inches. Thus the levels for furnace temperature of 1880°F and for hold-down time of 3 seconds are indicated.

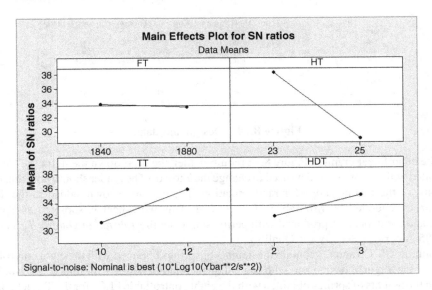

Figure 8.75   Main effects plot for signal to noise ratios.

Finally, the experimenters determined that the mean free height obtained with low oil quench temperatures was 7.77 inches and that the mean free height obtained with high oil quench temperatures was 7.51 inches. This led to the decision to attempt to control the oil quench temperature in the range 130–150°F.

Changes made to the process operating standards on the basis of the experimental findings led to a 60% reduction in the process variability and to a reduction in the deviation of mean free height from target from 0.026 inches to 0.014 inches on average. In conclusion the authors

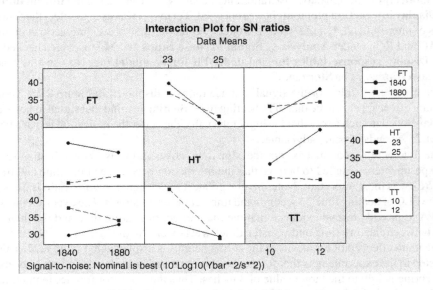

Figure 8.76   Interaction plots for signal to noise ratios.

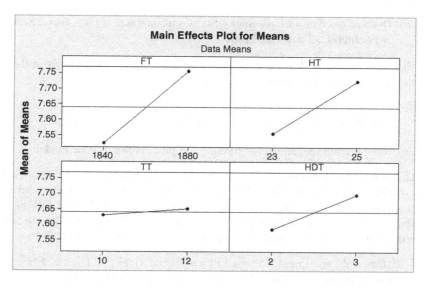

Figure 8.77   Main effects plot for means.

commented: 'Engineers without extensive background in experimental design were able to use the Taguchi method to guide them through a major reduction in process variability and an assessment of control factors'.

Hicks and Turner (1999, p. 398) make a number of observations on the use of Taguchi experimental designs. Three are quoted below:

- 'Taguchi's arrays, when used on the maximum number of factors, are all of resolution III. This means that all main effects are independent of other main effects but are confounded with two-way interactions. Such designs are seldom recommended because of their low resolution number.'

- 'Taguchi-type designs should be considered only as a screening type of experiment.'

- 'To really find out what factors and interactions are important, one should run either a complete factorial or a fractional factorial of resolution V or higher.'

8.5   Exercises and follow-up activities

1. Chemical process engineers conducted a designed experiment in order to investigate the influence of the factors cure temperature (°C) and amount of filler (kg) on the response hardness of a rubber compound. The levels of temperature selected were 120, 160 and 200, and the levels of amount selected were 10, 15 and 20.

   (i) Given that it was decided to have two replications, use **Stat > DOE > Factorial > Create Factorial Design** to create a pro forma that the experimenters could have used to conduct the experiment and to record the results. The actual design used and the results of the experiment are provided in worksheet Hardness.MTW.

(ii) Display the data in both main effects and interaction plots using **Stat** > **DOE** > **Factorial** > **Factorial Plots. . ..**

(iii) State the factor–level combination that appears best on the basis of these plots, given that high hardness is desirable.

(iv) Use **Stat** > **DOE** > **Factorial** > **Analyze Factorial Design. . .** to carry out a formal analysis, including examination of plots of the residuals.

(v) State recommendations for running the process that can be made.

2. The marketing department of a large company carried out a two-factor designed experiment in order to assess the potential impact of both the design and size of a newspaper advertisement on sales. The design had levels 1, 2 and 3 and the size had levels 1 and 2, with 2 corresponding to the larger size. The response was sales projection. Display and analyse the data, available in Adverts.MTW.

3. A 2^2 factorial experiment was used to investigate yield (g), Y, from a batch culture process for yeast. The factors of interest were the amount of sugar used $(A, g\,l^{-1})$, and air flow $(B, l\,h^{-1})$. The levels for A were 40 and 60 and the levels for B were 2 and 3. Use **Stat** > **DOE** > **Factorial** > **Create Factorial Design** with the option **2-level factorial (default generators)** to set up a pro forma for the experiment, with two replications. Create and print two copies of a 'blank' cube plot that could be used to record and display mean yield for each factor–level combination.

Given the data obtained in Table 8.31 for scenario 1,

(i) calculate the mean yield for each factor–level combination and add the means to one of your blank cube plots;

(ii) with the aid of your cube plot, calculate the main effects of sugar and air flow and their interaction effect.

Given further that the design and data are available in Yeast1.MTW,

(iii) use Minitab to display main effects and interaction plots;

(iv) use **Stat** > **DOE** > **Factorial** > **Analyze Factorial Design. . .** to carry out a formal analysis and to create residual plots – check your calculated effects against those in the Session window output and state which effects differ significantly from zero at the 5% level of significance;

Table 8.31   Data for yield – scenario 1.

Yield		Air flow		
		2		3
Sugar	40	9.4		4.7
			12.2		5.6
	60	17.5		10.1
			21.1		16.2

Table 8.32   Data for yield – scenario 2.

	Yield	Air flow 2		Air flow 3	
Sugar	40	7.5		5.2	
			10.4		9.0
	60	21.8		9.8	
			20.1		12.1

(v) create both a surface plot and a contour plot of yield versus sugar and air flow and, given that the higher the yield the better, state the nature of further experimentation that might prove informative.

Repeat the above five steps for the data in Table 8.32 for scenario 2; the design and alternative set of data are provided in Yeast2.MTW. (Note that in this exercise the run order obtained when you create the design is unlikely to match that in the supplied worksheets.)

4. Hellstrand (1989) reported on a 2^3 experiment carried out by SKF, a manufacturer of deep groove bearings. The factors were inner ring heat treatment, outer ring osculation and cage design, and the response was life (hours) of the bearings determined in an accelerated life test procedure. For each factor the levels were standard and modified, with standard representing the current level for production of the type of bearing. The data are displayed in Table 8.33 and are reproduced by permission of The Royal Society, London.

Use Minitab to display the data in the form of a cube plot and to create an interaction plot for the heat–osculation interaction effect. (In setting up the design using Minitab, uncheck **Randomize Runs** under **Options...**, so that the factor–level combinations will be in standard order as in Table 8.33.) Box (1990, p. 367) commented that the experiment 'led to the extraordinary discovery that, in this particular application, the life of a bearing can be increased fivefold if the two factors outer ring osculation and inner ring heat treatments are increased together'.

5. In Exercise 8 of Chapter 3 you set up a data set on battery life based on work by Wasiloff and Hargitt (1999). Open the worksheet that you saved (or, if you did not save it, refer to Table 3.4), analyse the data and comment on the theory referred to in that exercise.

Table 8.33   Data for bearing life experiment.

Heat	Osculation	Cage	Life
Standard	Standard	Standard	17
Modified	Standard	Standard	26
Standard	Modified	Standard	25
Modified	Modified	Standard	85
Standard	Standard	Modified	19
Modified	Standard	Modified	16
Standard	Modified	Modified	21
Modified	Modified	Modified	128

6. Box *et al.* (2005, p. 236) give an example of a 2^{4-1} fractional factorial experiment with a single replication. An industrial chemist performed the experiment during the formulation of a household liquid product. The factors of interest were acid concentration, catalyst concentration, temperature and monomer concentration. The response of interest was stability, with values in excess of 25 being desirable. The data are provided in Stability.MTW and reproduced by permission of John Wiley & Sons, Inc., New York.

(i) Before opening the worksheet containing the data, create a 2^{4-1} design with a single replication using all the defaults and observe the output in Panel 8.21 that appears in the Session window. Observe that the design has resolution IV and that first-order (two-factor) interactions are confounded in pairs.

(ii) On opening Stability.MTW, the design involved in the experiment may be checked using the Show Design icon ⊞. On clicking the icon the information in Figure 8.78 is displayed. The final line lists the terms as A, B, C, D, AB, AC and AD. This indicates that, if it is assumed that the three-factor interactions and the four-factor interaction are negligible, the experiment will provide estimates of the four main effects. The alias structure displayed in Panel 8.21 reveals that the estimate corresponding to the term AB will reflect both the AB and CD interactions if neither is negligible.

(iii) Verify from a normal probability plot of the effects obtained via **Graphs. . .** under **Stat** > **DOE** > **Factorial** > **Analyze Factorial Design. . .** that only the main effects of acid concentration and catalyst concentration are significant at the 10% level. (Note that the default list of terms, displayed under **Terms. . .**, is A, B, C, D, AB, AC and AD, as specified in the Show Design information in Figure 8.78.)

(iv) The experiment may now be regarded as a 2^2 factorial in the factors A and B, with two replications. Reanalyse the data using **Analyze Factorial Design. . .** with the **Terms. . .** list changed to A, B and AB. There are now sufficient degrees of

```
Fractional Factorial Design

Factors:    4    Base Design:         4, 8    Resolution:    IV
Runs:       8    Replicates:             1    Fraction:     1/2
Blocks:     1    Center pts (total):     0


Design Generators: D = ABC


Alias Structure

I + ABCD

A + BCD
B + ACD
C + ABD
D + ABC
AB + CD
AC + BD
AD + BC
```

Panel 8.21   Design details for stability experiment.

```
Factors:        4    Base Design:          4,  8
Runs:           8    Replicates:                1
Blocks:      none    Center pts (total):        0

Display Order:  Run Order
Display Units:  Uncoded


Factors and Their Uncoded Levels

Factor  Name            Low   High
A       Acid             20     30
B       Catalyst          1      2
C       Temperature     100    150
D       Monomer          25     50


Responses and Models

Response:   Stability
Terms:      A B C D AB AC AD
```

Figure 8.78   Details of the experimental design.

freedom to enable *P*-values to be computed, and you should find the information in Panel 8.22 displayed in the Session window.

Thus we have formal evidence, at the 1% and 5% levels of significance respectively, of the influence of both acid and catalyst concentrations on stability. Note that clicking the Show Design icon now provides the same information as in Figure 8.78 except that list of terms has been revised in view of the changes made to the fitted model.

(v) Use **Stat > DOE > Factorial > Contour/Surface Plots...** to confirm the plot in Figure 8.79.

(vi) State recommendations for further experimentation.

The authors commented that 'although none of the tested conditions produced the desired level of stability, the experiments did show the direction in which such conditions might be found. A few exploratory runs performed in this direction produced for the first time a product with stability greater than the goal of 25.'

```
Estimated Effects and Coefficients for Stability (coded
units)
```

Term	Effect	Coef	SE Coef	T	P
Constant		14.625	0.4146	35.28	0.000
Acid	-5.750	-2.875	0.4146	-6.93	0.002
Catalyst	-3.750	-1.875	0.4146	-4.52	0.011
Acid*Catalyst	0.250	0.125	0.4146	0.30	0.778

Panel 8.22   Output from analysis of Stability experiment.

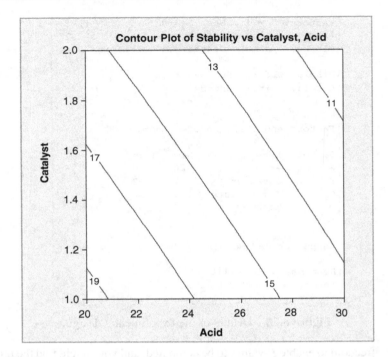

Figure 8.79   Contour plot of Stability versus Acid and Catalyst concentrations.

7. Designing, executing and analysing a factorial experiment provides valuable learning opportunities. One of the most widely used sources of data for such experiments are helicopters made according to the template in Figure 8.80.

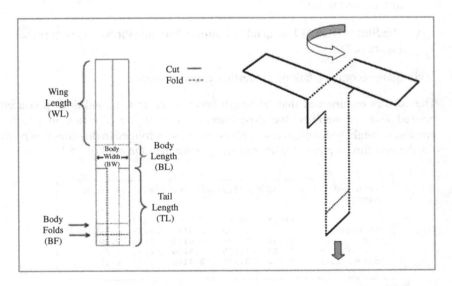

Figure 8.80   Helicopter template and sketch of helicopter in flight.

Table 8.34 Factors for helicopter design.

Factor	Level for prototype (mm)
Wing length WL	75
Body width BW	30
Body length BL	25
Tail length TL	75
Body folds BF	10

As a prototype it is suggested that a helicopter be constructed from paper with density $80\,\text{g/m}^2$ with the dimensions given in Table 8.34. The response of interest is the flight time (seconds) when the helicopter is released from a fixed height and allowed to descend to the floor. Flight time can be measured using an electronic stopwatch. The project team should make several flights with the prototype in order to establish standard operating procedures for launching and timing.

Given that the longer the flight time the better, the project team should then design, execute and analyse a screening experiment in the form of a fractional factorial, of at least resolution IV, that involves at least five factors and at least two replications. (Not all factors in Table 8.34 need be used. For example, paper density, with levels $80\,\text{g/m}^2$ and $100\,\text{g/m}^2$, could be used as a factor; the presence or absence of adhesive tape to maintain right angles between the wings and the body could be used as a factor.) Finally, recommendations for further experimentation should be made. Interested readers will find useful further information on helicopter experiments and on the use of experimental design to improve products and services in Box (1999) and Box and Liu (1999).

8. Two former students of the author at Edinburgh Napier University carried out a designed experiment as part of their course on experimental design taught by Dr Jeff Dodgson. The factors and levels considered are listed in Table 8.35. The design and data are available in Golf.MTW and in Dodgson (2003) and are reproduced with permission from the *Journal of Quality Technology* (© 2003 American Society for Quality).

(i) State the resolution of the design.

(ii) Demonstrate that there is evidence that the effects A, D, E and AE are significant.

(iii) Give an interpretation of the significant main effect of D.

(iv) Give an interpretation of the significant AE interaction effect.

Table 8.35 Factors and levels for golf experiment.

Factor	Level	
	-1	$+1$
A Ability (handicap)	8	4
B Ball type	Balata	Two piece
C Club type	Wood	Metal
D Ground condition	Soft	Hard
E Teeing	No tee	Tee

Table 8.36 Factors and levels for paint experiment.

Factor	Description	Levels	
		−	+
A	Component I	0.01	0.03
B	Component II	0.003	0.015
C	Component III	0.008	0.016
D	Component IV	0.01	0.02
E	Component V	0.57	0.77
F	Component VI	0.20	0.40
G	Substratum	Rigid	Flexible
H	Primer	Without	With
I	Oven Temperature	110 °C	130 °C
J	Time in Oven	20 min	30 min

9. Teresa López-Alvarez and Aguirre-Torres (1997) employed a fractional factorial experiment in the development of a new paint product for use in the automotive industry. The experiment was a 2^{10-5} design of resolution IV. The factors and levels are listed in Table 8.36.

   The response of interest was a measure of yellowing of the paint on a car bumper component that had been subjected to an accelerated ageing process. The data for this example, available in the file Yellow.MTW, are reproduced with permission from *Quality Engineering* (© 1997 American Society for Quality).

   Use Minitab to confirm the conclusions reached by the experimenters:

   • 'Flexible substratum promotes the appearance of yellowing.'

   • 'Control of factors C, D and F provides the largest opportunities for improvement.'

   • 'Increasing factor E does not result in any significant improvement.'

   The authors report that the conclusion that factor E did not have a significant effect was a surprise to research personnel. Make suggestions for further experimentation.

10. It is possible to incorporate blocking into factorial and fractional factorial experiments. Consider, for example, a full 2^3 factorial with factors A, B and C in standard order as displayed in Table 8.37.

    Imagine that the planned experiment was to involve a single replication and required the use of a raw material supplied in batches. Given that a batch of raw material is sufficient for four experimental runs of the process and that it is known that there is significant variation between batches, one possibility would be to carry out the first four runs with one batch of raw material and the remaining four with a second batch, as indicated by the final column in Table 8.37. However, scrutiny of the final two columns of Table 8.37 indicates that any block effect would be confounded with the main effect of factor C.

    The correct procedure, for the design of a full 2^3 factorial experiment in two blocks, is based on the knowledge that three-factor (second-order) interactions are often

Table 8.37   Naïve blocking of a full 2^3 experiment.

A	B	C	Batch
− 1	− 1	− 1	1
1	− 1	− 1	1
− 1	1	− 1	1
1	1	− 1	1
− 1	− 1	1	2
1	− 1	1	2
− 1	1	1	2
1	1	1	2

negligible. The reader is invited to complete the calculation of the *ABC* column in Table 8.38 and the allocation of factor–level combinations for which *ABC* is −1 to batch 1 and those for which ABC is 1 to batch 2. The design is shown in standard order in Table 8.39.

Finally, the reader is invited to use **Stat > DOE > Factorial > Create Factorial Design. . .** and to select, under **Design. . .**, the full 2^3 design with **Number of center points:** 0, **Number of replicates:** 1 and **Number of blocks:** 2. The author obtained the design in Table 8.40 (randomization was used).

If one thinks of the two batches of available raw material labelled 1 and 2 then the first four experimental runs would be carried out using the second batch – think of tossing a coin

Table 8.38   Blocking of a full 2^3 experiment.

A	B	C	ABC	Batch
− 1	− 1	− 1	− 1	1
1	− 1	− 1	1	2
− 1	1	− 1	1	2
1	1	− 1	− 1	1
− 1	− 1	1		
1	− 1	1		
− 1	1	1		
1	1	1		

Table 8.39   Blocking of a full 2^3 experiment – standard order.

Standard Order	A	B	C	Batch
1	− 1	− 1	− 1	1
2	1	1	− 1	1
3	1	− 1	1	1
4	− 1	1	1	1
5	1	− 1	− 1	2
6	− 1	1	− 1	2
7	− 1	− 1	1	2
8	1	1	1	2

Table 8.40 Blocking of a full 2^3 experiment with randomization.

StdOrder	RunOrder	CenterPt	Blocks	A	B	C
6	1	1	2	-1	1	-1
5	2	1	2	1	-1	-1
7	3	1	2	-1	-1	1
8	4	1	2	1	1	1
4	5	1	1	-1	1	1
1	6	1	1	-1	-1	-1
2	7	1	1	1	1	-1
3	8	1	1	1	-1	1

to decide which batch is to be used first. Table 8.40 indicates that the corresponding factor–level combinations in standard order are numbers 5,6,7 and 8. The randomization implemented by Minitab dictates that these combinations should be run in the order 6, 5, 7 and 8 in the experiment. Similarly,with the first batch of raw material the factor–level combinations 1, 2, 3 and 4 in standard order should be run in the order 4, 1, 2 and 3.

11. An experiment was designed during the design of a chiller for pasteurized milk. The chiller consists of refrigerated plates over which the milk flows. The factors of interest were spacing of the plates, temperature of the plates and milk flow rate. The response was the score awarded to milk chilled using each factor–level combination by a panel of expert tasters. As only four experimental runs could be carried out with one tanker of milk, it was decided to run a single replicate of a full 2^3 factorial experiment in two blocks. The data are available in Milk.MTW. Create a normal probability of the effects and interpret it.

12. John Douglass and Shirley Coleman presented a case study of the application of a Taguchi experimental design at *Industrial Statistics in Action 2000* (Douglass and Coleman, 2000, pp. 11–18). The designed experiment was employed in a quality improvement project that had the aim of improving the yield at final test of breathing apparatus demand valves manufactured by Draeger Safety UK Ltd., Northumberland, UK.

    One response of interest was static pressure and the factors (and levels) were as follows:

    1. demand valve type (standard/US version);

    2. lever spring tension (standard/modified);

    3. seal diameter (9.51 mm/9.63 mm);

    4. bore diameter (9.16 mm/9.18 mm);

    5. lever height setting (low/high);

    6. diaphragm resistance (low/high).

    The specification tolerance for static pressure was 2.7 to 3.5 bar and at times reject levels approached 20% for static pressure, with the 'great majority of these rejects exceeding the upper specification limit'.

The design used was the Taguchi $L_{32}(2^{31})$ orthogonal array with five replications. The data are available in the Excel spreadsheet Valves.xls and are reproduced by permission of Draeger Safety UK Ltd. On opening the spreadsheet as a Minitab worksheet and selecting **Stat > DOE > Taguchi > Analyze Taguchi Design. . .**, a message is displayed requesting information in order to enable the software to perform the analysis. On clicking on **Yes**, a **Define Custom Taguchi Design** dialog box appears in which the six factors must then be selected. On clicking on **OK** one can then proceed to carry out the analysis.

(i) By referring to the mean response, explain why it would be desirable to operate the process using modified tension in the lever springs and high lever height setting.

(ii) With Spring and Lever as **Terms** use **Stat > DOE > Taguchi > Predict Taguchi Results. . .** to verify that a mean static pressure of 3.01 is predicted. (Under **Levels**, check Select levels from a list and specify level 2 for each of the factors of interest using the drop-down menus that appear.)

Following the introduction of these levels in routine production the reject rate was reduced to below 3%.

9

Evaluation of measurement processes

Determining the capability of a measurement system is an important aspect of most process and quality improvement efforts. (Burdick *et al.*, 2003, p. 342)

Overview

The measure phase in a Six Sigma process improvement project is of crucial importance. It is natural therefore that teams working on projects need to be confident that the measurement processes they employ are sound and effective. Discussion of measurement processes has been postponed until this stage in the book so that use can be made of concepts from statistical models, of knowledge of designed experiments, and of components of variance in particular.

First the measurement of continuous variables will be considered. Following the introduction of the concepts of bias, linearity, repeatability and reproducibility, reference will be made to the problem of inadequate measurement units and how it manifests itself in control charts for variability. Gauge repeatability and reproducibility studies will be described and associated indices of measurement system performance introduced. Finally some reference will be made to scenarios involving attribute data.

9.1 Measurement process concepts

9.1.1 Bias, linearity, repeatability and reproducibility

A measurement system may be defined as 'the collection of instruments or gages, standards, operations, methods, fixtures, software, personnel, environment and assumptions used to quantify a unit of measure or fix assessment to the feature characteristic being measured; the

Six Sigma Quality Improvement with Minitab, Second Edition. G. Robin Henderson.
© 2011 John Wiley & Sons, Ltd. Published 2011 by John Wiley & Sons, Ltd.

complete process used to obtain the measurement' (Automotive Industry Action Group (AIAG), 2002, p. 5).

Imagine that we wish to determine the concentration of mercury (Hg) in seawater in order to monitor effluent from a chemical manufacturing plant. Suppose, too, that we have available a batch of seawater with known mercury concentration of 30 ng/ml. There are three measurement systems available – Acme, Brill and Carat. Fifty determinations of mercury concentration were made with each system. Figure 9.1 displays the data in the form of fitted normal curves. The mean result from Acme was 30.10, which is close to the true value of 30. The mean from Brill was 29.89, which is also close to the true value. However, Carat gave a mean of 28.92: a good deal further from the true value than the means for the other two measurement systems.

The Carat result appears to show *bias*. Bias 'is the difference between the true value (reference value) and the observed average of measurements on the same characteristic on the same part' (AIAG, 2002, p. 49). Thus the bias for Carat is given by:

$$\text{Bias} = \text{Observed average value} - \text{Reference value} = 28.92 - 30.00 = -1.08.$$

This indicates that Carat yields measurements that, on average, are 1.08 ng/ml on the low side when the measurement system is used on seawater with mercury concentration of 30 ng/ml. The reader is invited to confirm that the bias for Acme and Brill is 0.10 and −0.11, respectively.

One way to deal with bias is by *calibration*, which may involve the adjustment of a gauge to account for the difference between the observed average value for a standard and the true reference value for that standard – which we have just defined as bias.

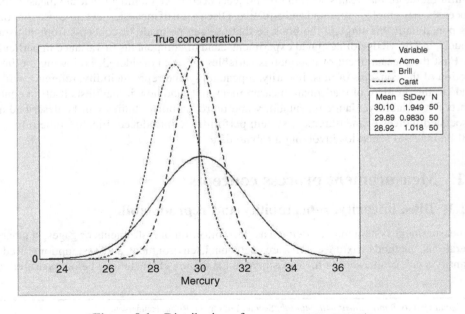

Figure 9.1   Distribution of mercury measurements.

One desirable property of gauges is good *repeatability*. Repeatability is defined as 'the variation in measurements obtained with *one measurement instrument* when used several times by *one appraiser* while measuring the identical characteristic on the *same part*' (AIAG, 2002, p. 52).

One could quote the standard deviation of such a set of measurements as a measure of variation. Historically, 5.15 times the standard deviation was used as an index of repeatability (AIAG, 2002, p. vi). The reason for this is that the interval from $\mu - 2.576\sigma$ to $\mu + 2.576\sigma$, with a range of 5.15σ, accounts for 99.00% of a normal distribution. The interval from $\mu - 3\sigma$ to $\mu + 3\sigma$, with a range of 6σ, accounts for 99.73% of a normal distribution. Minitab uses 6 times the standard deviation as the default.

Thus, adopting the Minitab default, Acme, Brill and Carat have estimated repeatability of $6 \times 1.949, 6 \times 0.983$ and 6×1.018, i.e. 11.7, 5.9 and 6.1, respectively. Thus, if the bias of Carat could be removed by calibration, then it would be on a par with Brill in terms of repeatability.

Another desirable property of gauges is good *reproducibility*. Reproducibility is 'the variation in the average of the measurements made by *different appraisers* using the *same measuring instrument* when measuring the identical characteristic on the *same part*' (AIAG 2002, p. 53).

The upper panel in Figure 9.2 shows the distributions of measurements of the standard mercury solution where reproducibility is relatively good. All three operators, Edward, Fiona and George, employed the Brill system to make 50 measurements of the standard – the display shows normal distributions fitted to the data – and their means, indicated by fulcrums, were 29.8, 30.2 and 30.0 respectively with range 0.4. The lower panel in Figure 9.2 shows a situation where the reproducibility is relatively poor. All three operators, Una, Veronica and Walter, also employed the Brill system to make 50 measurements of the standard – the display shows

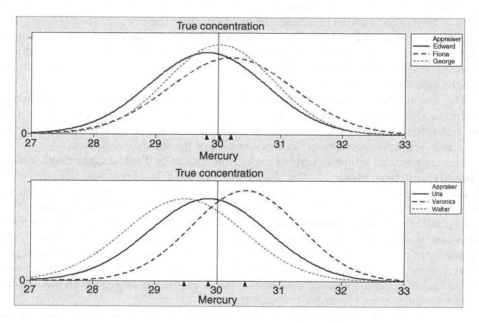

Figure 9.2   Two groups of operators with different levels of reproducibility.

normal distributions fitted to the data – and their means, indicated by fulcrums, were 29.8, 30.4 and 29.5 respectively with range 0.9.

A measurement system needs to have *stability*. 'Stability (or drift) is the total variation in the measurements obtained with a measurement system *on the same master or parts* when measuring *a single characteristic over an extended time period*' (AIAG 2002, p. 50).

Control charts may be used to monitor the stability of a measurement system. If an individual measurement of a standard or master part is made at regular intervals then an individuals or X chart may be used, with the centre line set at the reference value for the standard. If samples of measurements of a standard or master part are made at regular intervals then a mean or Xbar chart may be used with the centre line set at the reference value for the standard. A signal of special cause variation on the charts could indicate the need for calibration of the measurement system.

A final useful property of measurement systems is linearity. Linearity is an indication that 'gauge response increases in equal increments to equal increments of stimulus, or, if the gauge is biased, that the bias remains constant throughout the course of the measurement process' (NIST/SEMATECH, 2005). Readers should note that definition of linearity given by AIAG is strictly a definition of nonlinearity. Minitab provides a procedure for formally assessing linearity of a measurement system.

Suppose that the measurement systems Brill and Carat could be used with seawater samples containing up to 50 ng/ml of mercury. Standard solutions with known concentrations 10, 20, 30, 40 and 50 ng/ml are available. Five determinations of mercury concentration are made using each of the two systems. The data are tabulated in Table 9.1 and available in Mercury.MTW.

Figure 9.3 displays the data in the form of a scatterplot. The mean determination for each standard solution from each of the two systems is plotted against the true value. With no bias the means would equal the corresponding true values so the line with equation $y = x$ has been added to the diagram in order to indicate the ideal situation. The circular symbols for the Brill system are all close to the ideal line, suggesting no major bias. On the other hand, the triangular symbols for the Carat system diverge further from the ideal line as mercury concentration increases. This indicates that the Carat system is not exhibiting linearity, i.e. with Carat the bias changes with the true (reference) value.

The data in Table 9.1 are summarized in Table 9.2. The bias for each of the measurement systems is also given. The reader is invited to verify some of the bias values. There is no apparent pattern in the Brill bias values, but the Carat bias values are all negative and their magnitude generally increases as the concentration of the standard increases.

Minitab provides a formal analysis under **Stat > Quality Tools > Gage Study > Gage Linearity and Bias Study....** The dialog box is shown in Figure 9.4. The use of the phrase 'Part numbers' reflects the widespread use of measurement process evaluation in the automotive industry. We enter **Part numbers:** 'Standard No.', **Reference values:** 'True Concentration' and **Measurement data:** 'Brill Estimate'. No entry was made in the **Process variation:** window, an option that will not be considered in this book. Information on the measurement system used etc. may be entered under **Gage Info...**, and under **Options...** one may enter a title and choose to estimate the repeatability standard deviation either using sample ranges or sample standard deviations. The default sample range method was selected, and the resulting output is shown in Figure 9.5.

For the standard solution with concentration 10 ng/ml the Brill system gave the five estimates 10.00, 10.52, 11.11, 10.30 and 11.53. The corresponding deviations from the

Table 9.1 Data from linearity investigation.

Standard no.	True concentration	Brill estimate	Carat estimate
1	10	10.00	9.04
1	10	10.52	9.16
1	10	11.11	10.59
1	10	10.30	8.85
1	10	11.53	10.58
2	20	20.76	19.45
2	20	21.61	17.92
2	20	18.22	18.51
2	20	20.44	20.97
2	20	20.46	20.48
3	30	28.95	28.42
3	30	29.46	28.36
3	30	30.52	30.46
3	30	30.69	26.76
3	30	30.05	30.30
4	40	40.70	37.54
4	40	39.67	40.02
4	40	38.35	38.37
4	40	39.39	37.48
4	40	40.51	37.86
5	50	50.37	50.00
5	50	50.69	48.24
5	50	51.31	47.73
5	50	50.67	48.79
5	50	47.61	47.29

true value of 10 are 0.00, 0.52, 1.11, 0.30 and 1.53. The mean of these gives the bias of 0.692. The five deviations (solid circles) and their mean (solid square) are plotted against the reference value of 10. Similar plotting has been done for the other four standard solutions. Thus the solid square symbols constitute the scatterplot of bias versus reference value.

The five values of bias, together with their overall mean, are shown in the bottom right hand corner of the display together with corresponding P-values. (The P-values arise from t-tests being performed on each sample of bias values, with null hypothesis that the mean is zero and alternative hypothesis that the mean is nonzero.) None of these are less than 0.05, so there is no evidence of bias for the Brill measurement system. Further discussion of the output in this case is unnecessary.

The corresponding output for the Carat system is shown in Figure 9.6. Here we have evidence of bias since three of the P-values are less than 0.05. The regression line fitted to the scatterplot of bias versus reference value has a slope that differs significantly from zero at the 5% level of significance (P-value of 0.029 quoted in the top right of the display). Thus we have evidence of nonlinearity here, i.e. that for the Carat system the bias is not constant but is related to the reference value. In summary, Acme is inferior to both Brill and Carat because of its inferior repeatability (see Figure 9.1). However, Carat is inferior to Brill because of its bias and nonlinearity.

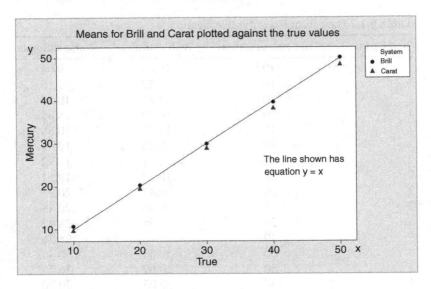

Figure 9.3   Scatterplot of means versus true values.

Table 9.2   Summarized data with bias values.

True	Brill mean	Brill bias	Carat mean	Carat bias
10	10.692	0.692	9.644	−0.356
20	20.298	0.298	19.466	−0.534
30	29.934	−0.066	28.860	−1.140
40	39.724	−0.276	38.254	−1.746
50	50.130	0.130	48.410	−1.590

9.1.2 Inadequate measurement units

Consider now some length (mm) data collected for random samples of four rods taken at 15-minute intervals from a production line. The measurement system used gives a digital display of length to two decimal places. Xbar and R charts of the data are displayed in Figure 9.7. There is no evidence from the charts of any special causes affecting the process – it appears to be behaving in a stable and predictable manner. The same data, rounded to one decimal place, give the Xbar and R charts shown in Figure 9.8. With the rounded data there are a number of signals of special cause variation on the charts. The data for both sets of charts are available in Inadequate.MTW.

Wheeler and Lyday (1989, pp. 3–9) refer to the problem of inadequate measurement units or inadequate discrimination due to a measurement unit that is too large. They state that the problem of inadequate discrimination 'begins to affect the control chart when the measurement unit exceeds the process standard deviation'. For the first pair of charts above the measurement unit was 0.01 mm whereas for the second the measurement unit was 0.1 mm.

C1	C2	C3	C4	C5
Standard No.	True Concentration	Brill Estimate	Carat Estimate	
1	10	10.00	9.04	
1	10	10.52	9.16	
1	10	11.11	10.59	
1	10	10.30	8.85	
1	10	11.53	10.58	
2	20	20.76	19.45	
2	20	21.61	17.92	

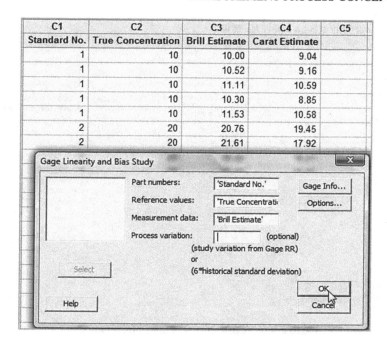

Figure 9.4   Dialog for gage linearity and bias analysis of the Brill system.

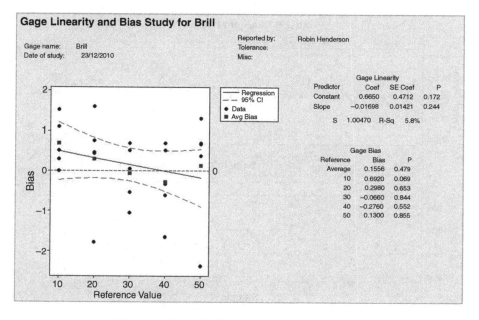

Figure 9.5   Brill linearity and bias analysis.

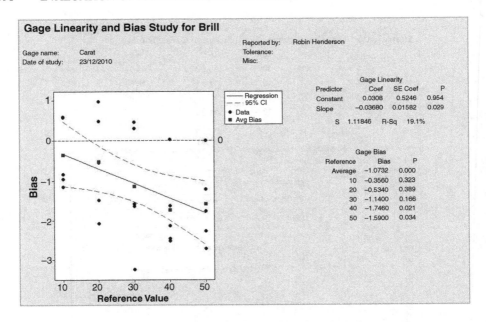

Figure 9.6   Carat linearity and bias analysis.

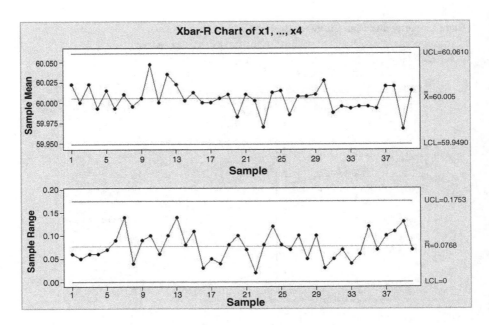

Figure 9.7   Control charts for rod length data.

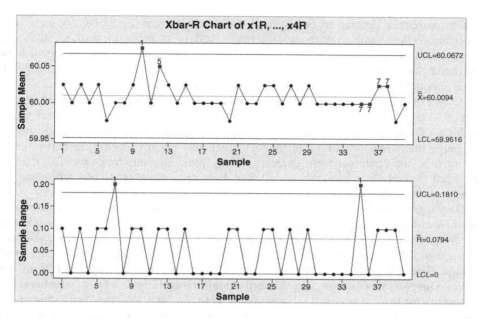

Figure 9.8 Control charts for rounded rod length data.

The process standard deviation estimated from the first range chart is approximately 0.04 mm, so for the second set of charts the measurement unit exceeds the estimated process standard deviation. As a result a process that was actually stable and predictable appeared to be subject to special cause variation.

In terms of a control chart for sample ranges, based on samples of up to 10 measurements, a measurement unit in excess of the process standard deviation generally leads to the occurrence of five or fewer values of range within the control limits. Wheeler and Lyday (1989, p. 8) comment:

The measurement unit borders on being too large when there are only 5 possible values within the control limits on the range chart. Four values within the limits will be indicative of inadequate measurement units, and fewer than four values will result in appreciable distortion of the control limits.

In the above example the fact that there are only two values of range within the control limits with the rounded data signals the inadequacy of the measurement unit. In some cases the problem of inadequate measurement units may simply be a result of failure to record enough significant figures from the measuring tool. In other cases it may be the result of a measurement system being incapable of detecting the variation in the product characteristic being measured.

9.2 Gauge repeatability and reproducibility studies

Gauge repeatability and reproducibility (R&R) studies may be used to estimate the components of variation that contribute to the overall variation in the measurements obtained from a measurement process. As a consequence, such studies enable the discriminatory power of a measurement process to be assessed. Czitrom (1997, p. 3) wrote: 'A gage study is performed to characterize the measurement system in order to help identify areas of improvement in the measurement process and to ensure that the process signal is not obscured by noise in measurement.' In performing such studies it is usual to assume that the gauge is both bias-free and performing in a stable manner.

As an example we consider again a data set used in Chapter 8 giving the height (mm) of a set of 10 bottles measured twice using a height gauge by each of a group of three trainee Six Sigma Green Belts at Ardagh Glass Ltd., Barnsley. On each occasion the operators of the gauge measured the bottles in random order, and when measuring a bottle for the second time the operators were unaware of any previous measurements. The data are displayed in Table 9.3, available (in stacked format) in Heights.xls and reproduced by permission of Ardagh Glass Ltd., Barnsley.

Once a gauge R&R experiment has been completed, it can be informative to display the data before formal analysis is carried out. One form of display that may be used is the multi-vari chart. However, Minitab provides a special type of run chart for use with data from gauge R&R experiments. It is available using **Stat > Quality Tools > Gage Study > Gage Run Chart...**. The dialog is shown in Figure 9.9. Specify **Part Numbers:** Bottle, **Operators:** Operator and **Measurement data:** Height. (**Trial numbers:** and **Historical mean:** are optional.) Under **Gage Info...** details of the measurement tool used, study date etc. may be inserted and **Options...** enables a customized title for the chart to be created if desired.

The chart is displayed in Figure 9.10. Each panel of the display corresponds to a bottle. The two measurements on a bottle obtained by an operator are plotted and linked by a line segment. Note that Minitab arranges the operators in alphabetical order in the display as indicated by the key in the box to the right of the chart. Horizontal or near horizontal segments indicate good repeatability. Widely separated segments in panels would indicate poor reproducibility. Scrutiny of the chart suggests that Paul had the best performance in terms of repeatability.

Formal analysis involves estimation of the components of variance. The variance components obtained in Section 8.1.4 using ANOVA are shown in Panel 9.1. (It was concluded that the operator–bottle component of variation was zero.)

Table 9.3 Bottle height data.

Bottle no.	Neil		Lee		Paul	
	Test 1	Test 2	Test 1	Test 2	Test 1	Test 2
1	214.82	214.83	214.84	214.89	214.84	214.84
2	214.61	214.64	214.65	214.62	214.60	214.58
3	214.52	214.53	214.51	214.57	214.51	214.51
4	214.57	214.61	214.61	214.62	214.61	214.61
5	214.64	214.64	214.64	214.65	214.64	214.65
6	214.72	214.73	214.72	214.73	214.72	214.72
7	214.60	214.61	214.63	214.63	214.61	214.62
8	214.73	214.75	214.74	214.76	214.73	214.73
9	214.70	214.67	214.70	214.67	214.66	214.68
10	214.80	214.78	214.81	214.81	214.78	214.79

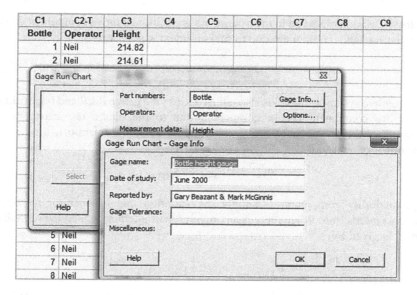

Figure 9.9   Dialog for creation of gauge run chart.

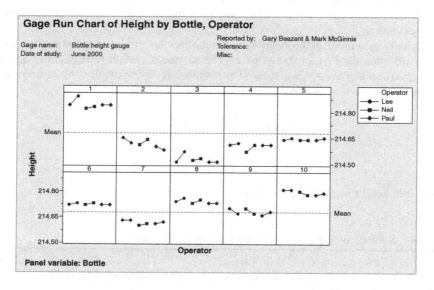

Figure 9.10   Gauge run chart.

	Estimated
Source	Value
Bottle	0.00934
Operator	0.00009
Error	0.00023

Panel 9.1   Components of variance for height gauge R&R study.

The total variation observed in measurements is partitioned into a component due to part-to-part variation and a component due to measurement system variation:

$$\sigma_{\text{Tot}}^2 = \sigma_{\text{Part}}^2 + \sigma_{\text{MS}}^2.$$

The measurement system variation is also referred to as total gauge R&R and is partitioned into a repeatability component (i.e. a component due to the gauge or measurement tool) and a reproducibility component (i.e. a component due to the operators, the users of the gauge or measurement tool):

$$\sigma_{\text{MS}}^2 = \sigma_{\text{Repeat}}^2 + \sigma_{\text{Reprod}}^2.$$

The reproducibility component is simply the variation due to operator if there is no operator–part interaction. When there is an operator–part interaction then the reproducibility component is given by

$$\sigma_{\text{Reprod}}^2 = \sigma_{\text{Oper}}^2 + \sigma_{\text{Oper} \times \text{Part}}^2.$$

The sources of variation and these formulae have been tabulated in Table 9.4. The variance components obtained directly from the ANOVA have been entered in bold, and the reader should check the calculation of the other entries in the column. The final column gives the standard deviations.

One widely used index in the evaluation of measurement processes is the proportion of the total variation that may be attributed to the measurement system (repeatability and reproducibility) variation. Variation is measured by standard deviation in this context, and it is usual to express the proportion as *percentage gauge R&R* (%R&R):

$$\%\text{R\&R} = \frac{\sigma_{\text{MS}}}{\sigma_{\text{Tot}}} \times 100$$

$$= \frac{0.0179}{0.0983} \times 100 = 18\%.$$

for the bottle height measurement system.

Table 9.4 Sources of variation for height gauge R&R study.

Source	Symbols and formulae	Variance component	Standard deviation
Part-to-part	σ_{Part}^2	**0.009 34**	0.0966
Operator	σ_{Oper}^2	**0.000 09**	0.0095
Operator × Part	$\sigma_{\text{Oper} \times \text{Part}}^2$	**0.000 00**	0.0000
Reproducibility	$\sigma_{\text{Reprod}}^2 = \sigma_{\text{Oper}}^2 + \sigma_{\text{Oper} \times \text{Part}}^2$	0.000 09	0.0095
Repeatability	σ_{Repeat}^2	**0.000 23**	0.0152
Total gauge R&R	$\sigma_{\text{MS}}^2 = \sigma_{\text{Repeat}}^2 + \sigma_{\text{Reprod}}^2$	0.000 32	0.0179
Total	$\sigma_{\text{Tot}}^2 = \sigma_{\text{Part}}^2 + \sigma_{\text{MS}}^2$	0.009 66	0.0983

The following guidelines for the acceptance of %R&R are given:

- Under 10% – generally considered to be an acceptable measurement system.

- 10% to 30% – may be acceptable based upon importance of application, cost of measurement device, cost of repair, etc.

- Over 30% – considered to be not acceptable – every effort should be made to improve the measurement system. (AIAG, 2002, p. 77)

Thus the bottle height measurement process falls into the middle of these three categories. These guidelines suggest that the measurement system could be improved. However, these guidelines are somewhat arbitrary and 'excessively conservative' according to Wheeler's (2003) critical review of them.

The determination of %R&R may be done directly in Minitab using **Stat > Quality Tools > Gage Study > Gage R&R Study (Crossed)....** The study is said to be crossed since every bottle was measured by every operator. The dialog is basically the same as in Figure 9.9; the default **Method of Analysis, ANOVA,** was accepted, as were the defaults under **Options....** In addition to the analysis in the Session window, a number of displays are provided – see Figure 9.11. (If desired the six plots may be displayed separately via **Options....**)

The height by bottle and height by operator plots are main effects plots that enable the mean measurements for each bottle and for each operator respectively to be compared. The operator–bottle interaction plot gives a visual indication of the presence or absence of an interaction effect. The Xbar chart by operator displays the mean result for each bottle for each operator. However, unlike a process monitoring control chart of means where signals of possible special cause variation are generally bad news, in this scenario the more points that plot outwith the chart limits the better the measurement system – it is desirable that a measurement system signals differences between parts! The R chart by operator highlights

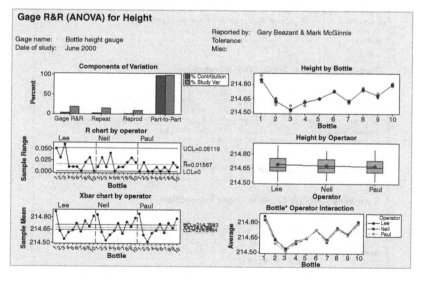

Figure 9.11   Displays from gauge R&R.

a suspicion that arose from scrutiny of the run chart in Figure 9.10, i.e. that Paul had the best repeatability. The signal for Lee provides evidence that he has significantly worse repeatability than the other two operators. Thus there may be a training issue that could be addressed.

The bar chart shows the components of variance and the matching standard deviations expressed as a percentage of total variance and total standard deviation, respectively. The second bar from the left represents the %R&R index. The ANOVA part of the Session window output is displayed in Panel 9.2. Minitab carries out the full ANOVA, with an interaction term. If the P-value for interaction exceeds 0.25 then the interaction term is removed from the model and the ANOVA re-calculated. (If desired the default value of 0.25 may be changed under **Options. . . .**) The ANOVA adopted is then used to compute the components of variance (as displayed in Panel 9.1) and to perform the calculations displayed in Table 9.4. The remaining portion of Session window output is shown in Panel 9.3.

The Total Gage R&R, expressed as a percentage of what Minitab refers to as study variation, and given as 17.96% in Panel 9.3, is the %R&R computed earlier as 18% from Table 9.4. (The reader should note that the percentages in the %Study Var column in the output do not sum to 100 whereas those in the %Contribution column do – this is because variances are additive but standard deviations are not.) At the foot of Panel 9.3 we are informed that the number of distinct categories that can be reliably differentiated by the measurement process is 7. This is another widely used index of measurement system performance and it is generally recommended that it should be 5 or more (AIAG, 2002, p. 45). The number of distinct categories is the rounded value of the discrimination ratio. Wheeler and Lyday (1989, pp. 54–59) give a detailed discussion of the discrimination ratio and its interpretation.

```
Gage R&R Study - ANOVA Method

Gage R&R for Height

Gage name:        Bottle height gauge
Date of study:    June 2000
Reported by:      Gary Beazant & Mark McGinnis
Tolerance:
Misc:


Two-Way ANOVA Table With Interaction

Source           DF      SS          MS          F         P
Bottle            9   0.506302   0.0562557   273.430   0.000
Operator          2   0.003863   0.0019317     9.389   0.002
Bottle * Operator 18  0.003703   0.0002057     0.863   0.621
Repeatability    30   0.007150   0.0002383
Total            59   0.521018


Alpha to remove interaction term = 0.25


Two-Way ANOVA Table Without Interaction

Source          DF      SS          MS          F         P
Bottle           9   0.506302   0.0562557   248.797   0.000
Operator         2   0.003863   0.0019317     8.543   0.001
Repeatability   48   0.010853   0.0002261
Total           59   0.521018
```

Panel 9.2   ANOVA tables from Minitab gauge R&R analysis.

```
Gage R&R

                                   %Contribution
Source                 VarComp     (of VarComp)
Total Gage R&R         0.0003114        3.23
  Repeatability        0.0002261        2.34
  Reproducibility      0.0000853        0.88
    Operator           0.0000853        0.88
Part-To-Part           0.0093383       96.77
Total Variation        0.0096497      100.00


                                   Study Var    %Study Var
Source                 StdDev (SD)   (6 * SD)      (%SV)
Total Gage R&R         0.0176462     0.105877       17.96
  Repeatability        0.0150370     0.090222       15.31
  Reproducibility      0.0092346     0.055408        9.40
    Operator           0.0092346     0.055408        9.40
Part-To-Part           0.0966347     0.579808       98.37
Total Variation        0.0982327     0.589396      100.00


Number of Distinct Categories = 7
```

Panel 9.3   Session window output from gauge R&R analysis.

Gage R&R Study (Crossed) would be used when each part is measured on more than one occasion by each operator. Minitab also provides, via **Stat** > **Quality Tools** > **Gage Study** > **Gage R&R Study (Nested)...**, analysis for situations in which each operator measures a specific set of parts unique to that operator. Skrivanek (2009) gives an example of the evaluation of the measurement system used to test the hardness of a reinforced plastic part of a prosthetic device. The test involved subjecting a sample of randomly selected parts to a specified force. A durometer was then used to measure the depth of the indentation in the material created by the force and the score recorded. The parts for the devices are supplied in lots of 50. Since the sampled parts are effectively destroyed during measurement, a nested analysis is appropriate. Three appraisers were selected at random. Five lots were selected to represent the full range of the manufacturing process. The data are available in Hardness.MTW and are reproduced by permission of MoreSteam.com LLC.

A gauge run chart may be created, and the nested analysis also provides a number of displays as in Figure 9.11 for the crossed case. Using the defaults for the nested analysis, the Session window output in Panel 9.4 was obtained. We have a percentage gauge R&R of 26.97 and five distinct categories, indicating a measurement process with marginal performance.

```
                                 Study Var    %Study Var
Source                StdDev (SD)  (6 * SD)      (%SV)
Total Gage R&R        0.93095       5.5857        26.97
  Repeatability       0.93095       5.5857        26.97
  Reproducibility     0.00000       0.0000         0.00
Part-To-Part          3.32415      19.9449        96.30
Total Variation       3.45205      20.7123       100.00


Number of Distinct Categories = 5
```

Panel 9.4   Session window output from Gage R&R analysis (nested).

Minitab also enables factors other than part and operator to be taken into account in assessing measurement system performance via **Stat > Quality Tools > Gage Study > Gage R&R Study (Expanded)....** Further information may be found at http://www.minitab.com/en-US/training/articles/articles.aspx?id=8900. Further general advice and guidance on the design and analysis of measurement systems capability may be found in the review paper by Burdick *et al.* (2003) and the book by Burdick *et al.* (2005).

9.3   Comparison of measurement systems

Situations arise where it is desirable to compare the performance of two measurement systems. For example, a supplier and customer may measure parts using their own systems and it may be of mutual benefit to compare the measurements obtained by each on a set of parts, or a new system might be under consideration for purchase and it could be of interest to compare its performance with that of the system currently in use.

The worksheet Outside_Diameters.MTW contains the diameters of 30 parts measured in random order by one appraiser, using both system A and system B. A useful initial display is a scatterplot of the measurement obtained from system B plotted against that obtained from system A. In the ideal situation both systems would be bias-free and the two measurements would be identical. Thus it can be informative to plot the line $y = x$ on the scatterplot. Once the basic scatterplot has been created the line may be added by right-clicking the graph and using **Add > Calculated Line....** Under **Coordinates** selection of the column containing the results from the first system as both the **Y column:** and the **X column:** yields the required line. The result is displayed in Figure 9.12.

The mouse pointer is shown located at the point corresponding to the first part. The point lies above the line, an indication that the measurement on that part from system B (47.956)

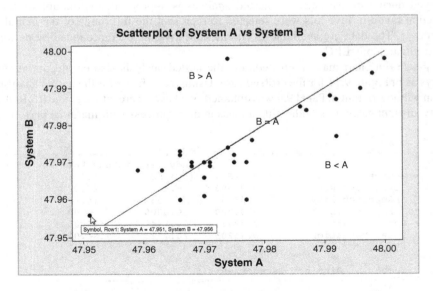

Figure 9.12   Scatterplot of measurements by two systems.

exceeds that from system A (47.951). Two points fall on the line, indicating that for parts 3 and 14 both systems gave the same measurement. Nearly twice as many points lie below the line as lie above it, suggesting the possibility of some relative bias for the two systems. Bland and Altman (1986) recommend plotting the difference between the two measurements on a part (A − B) versus the average of the two measurements (A + B)/2. They state: 'The plot of difference against mean also allows us to investigate any possible relationship between the measurement error and the true value. We do not know the true value, and the mean of the two measurements is the best estimate we have.' Such a plot is widely referred to as a Bland–Altman plot. They also advise display of the differences in a histogram – see Figures 9.13 and 9.14.

Bland–Altman plots are not available directly in Minitab. They may be created by adding reference lines to a scatterplot of difference against average or using a Minitab macro. Details of how to access and run the macro, which was created by Eli Walters, are given in Chapter 11 and at http://www.minitab.com/en-US/support/answers/answer.aspx?id=2504.

The mean of the differences is 0.000 10 and a reference line indicates this value, rounded to two decimal places. This value is close to zero and a formal t-test of the null hypothesis that the mean difference is zero yields a P-value of 0.951, so there is no evidence of relative bias for the two methods. Upper and lower limits of agreement, labelled ULA and LLA respectively, are also shown. The standard deviation of the differences is 0.008 89 and the limits are

$$\text{Mean} \pm 1.96 \times \text{Standard deviation}$$
$$= 0.000\,10 \pm 1.96 \times 0.008\,89$$
$$= (-0.017\,32, 0.017\,52),$$

or (− 0.017, 0.018) to three decimal places. If it is reasonable to consider the differences to be normally distributed then the calculated limits of agreement would be expected to contain

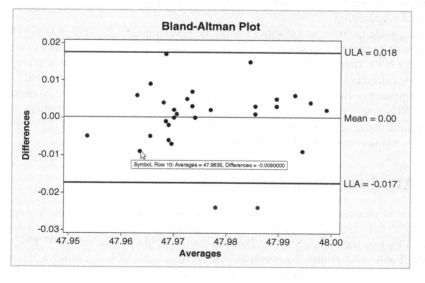

Figure 9.13 Bland–Altman plot of diameter measurements.

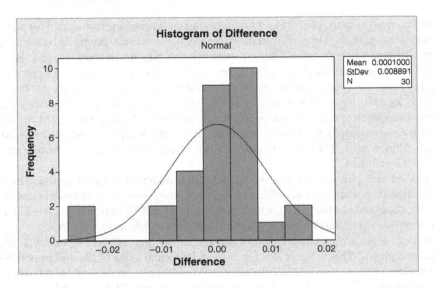

Figure 9.14   Histogram of the differences.

around 95% of the plotted points – in this case 93% of points fall between the limits. The relatively poor fit of the normal curve in Figure 9.14 casts doubt on this assumption – doubt that is confirmed by a normal probability plot. Ideally we would wish see the points in the Bland–Altman plot form a randomly distributed horizontal band of uniform width. A horizontal band implies that there is no relationship between difference in measurements and the size of the component, represented by the average of the two measurements. Correlation may be used to test formally for the presence of such a relationship. Here the correlation between difference and average is 0.015 with P-value 0.939, so there is no evidence of a linear relationship.

Setting aside the reservation about normality, the fundamental question is whether or not the limits of accuracy imply satisfactory repeatability for the two systems. Provided that differences in the range −0.017 to 0.018 are not of manufacturing or functional significance, the two measurement systems could be used interchangeably to make diameter measurements on the components. In the follow-up example at the end of the chapter an approximate method for the computation of confidence intervals for the limits of agreement is given. Further guidance is provided in the paper by Bland and Altman (1986) that is freely available at http://www-users.york.ac.uk/~mb55/meas//ba.htm.

9.4   Attribute scenarios

Specimens of silicon are inspected before dispatch to a customer. Twenty-five specimens, which have been classified as either accept or reject by senior inspectors, were assessed and classified by two trainee inspectors. The data are available in Silicon.MTW. Use of **Stat >** **Quality Tools > Attribute Agreement Analysis...** provides informative analyses and displays. A portion of the data and the dialog are shown in Figure 9.15. Column C1 contains the reference number of the specimens, C2 the classification made by the experts (the

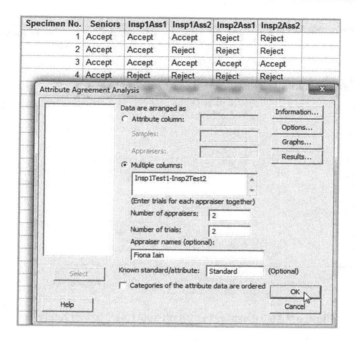

Specimen No.	Seniors	Insp1Ass1	Insp1Ass2	Insp2Ass1	Insp2Ass2
1	Accept	Accept	Accept	Reject	Reject
2	Accept	Accept	Reject	Reject	Reject
3	Accept	Accept	Accept	Accept	Accept
4	Accept	Reject	Reject	Reject	Reject

Figure 9.15   Data and dialog for attribute agreement analysis.

standard), C3 the first classification made by the first trainee, C4 the second classification made by the first trainee, and C5 and C6 contain the classifications made by the second trainee. For example, the senior inspectors deemed that the second specimen was acceptable; the first trainee deemed it as acceptable on the first assessment but as a reject on a second assessment, whereas the second trainee deemed it a reject on both his assessments.

The left-hand chart in Figure 9.16, labelled Within Appraisers, gives the proportion of assessments for each assessor where their first assessment agrees with their second. For example, Fiona's first assessment was the same as her second for 21 of the 25 specimens. This corresponds to 84% agreement, indicated by the solid circle. The line segment gives a confidence interval for the true proportion, for the first appraiser, of 64% to 95%. Similarly, Iain had 92% agreement with 95% confidence interval 74% to 99%. The corresponding Session window output for this first display is shown in Panel 9.5.

Fleiss' kappa statistic ranges from -1 to $+1$. The higher the value of kappa, the stronger is the agreement between the ratings. When kappa $= 1$ there is perfect agreement. When kappa $= 0$, then agreement is equivalent to that expected by chance. Minitab Help on this topic states that, as a general guide, kappa values less than 0.7 indicate that the measurement system requires improvement, while kappa values in excess of 0.9 are desirable. Dunn (1989, p. 37) discusses the benchmarks for the evaluation of kappa values displayed in Table 9.5. These benchmarks were proposed by Landis and Koch (1977), although Dunn comments that 'In the view of the present author these are too generous, but any series of standards such as these are bound to be subjective'.

The second chart displayed in Figure 9.16, labelled Appraisers vs. Standard, displays the proportions, with 95% confidence intervals, for agreement between appraisers and

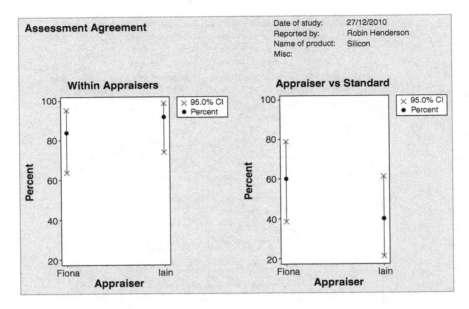

Figure 9.16   Displays from attribute agreement analysis.

```
Within Appraisers

Assessment Agreement

Appraiser  # Inspected  # Matched   Percent       95% CI
Fiona               25         21     84.00  (63.92, 95.46)
Iain                25         23     92.00  (73.97, 99.02)

# Matched: Appraiser agrees with him/herself across trials.


Fleiss' Kappa Statistics

Appraiser  Response     Kappa  SE Kappa        Z  P(vs > 0)
Fiona      Accept    0.652778       0.2  3.26389     0.0005
           Reject    0.652778       0.2  3.26389     0.0005
Iain       Accept    0.750000       0.2  3.75000     0.0001
           Reject    0.750000       0.2  3.75000     0.0001
```

Panel 9.5   Session window output for attribute agreement analysis.

Table 9.5   Benchmarks for the evaluation of kappa values.

Value of kappa	Strength of agreement
0.00	Poor
0.01–0.20	Slight
0.21–0.40	Fair
0.41–0.60	Moderate
0.60–0.80	Substantial
0.81–1.00	Almost perfect

```
Each Appraiser vs Standard

Assessment Agreement

Appraiser  # Inspected  # Matched   Percent       95% CI
Fiona               25          15     60.00   (38.67, 78.87)
Iain                25          10     40.00   (21.13, 61.33)

# Matched: Appraiser's assessment across trials agrees with the known standard.


Assessment Disagreement

              # Reject /            # Accept /
Appraiser      Accept   Percent      Reject   Percent  # Mixed   Percent
Fiona               4    22.22            2    28.57        4     16.00
Iain               13    72.22            0     0.00        2      8.00

# Reject / Accept:  Assessments across trials = Reject / standard = Accept.
# Accept / Reject:  Assessments across trials = Accept / standard = Reject.
# Mixed: Assessments across trials are not identical.


Fleiss' Kappa Statistics

Appraiser  Response       Kappa  SE Kappa           Z  P(vs > 0)
Fiona      Accept      0.255952  0.141421     1.80986     0.0352
           Reject      0.255952  0.141421     1.80986     0.0352
Iain       Accept     -0.129079  0.141421    -0.91273     0.8193
           Reject     -0.129079  0.141421    -0.91273     0.8193
```

Panel 9.6   Session window output for attribute agreement analysis.

the assessment made by the experts recorded in the column named Seniors. For 15 of the 25 specimens Fiona agreed with the experts on both tests. Thus her proportion for agreement with the standard was 60%, with 95% confidence interval 39% to 79%. Iain's proportion was 40%, with 95% confidence interval 21% to 61%. The corresponding Session window output for the second display is shown in Panel 9.6.

In addition to the agreement proportions and confidence intervals shown in the graphical display, an analysis of disagreement with the standard is given in the Session window output for each appraiser. In the case of Fiona, there were four specimens where she made the decision to reject on both tests, when the standard was to accept. The shorthand '# Reject/Accept' indicates the number of specimens for which the decision made by the appraiser was reject across all trials given that the standard specified the specimen as accept. The '/' symbol may be interpreted as the phrase 'given that' since what is being quoted are estimates of conditional probabilities. There were two specimens where she made the decision to accept on both tests when the standard was to reject. There were four specimens where she made the decision to reject on one of the tests and to accept on the other, regardless of the standard. Minitab refers to these cases as being mixed. Both kappa values for Fiona are less than 0.7, thus providing further evidence of an inadequate measurement system.

Windsor (2003) reports a case study of the measurement phase of a Six Sigma Black Belt project for which the type of analysis described above led to annual savings of $400,000. Some refer to attribute agreement analysis as 'attribute gage R&R (short method)'. He concludes that 'an attribute gage R&R can normally be performed at very low cost with little impact on the process' and that 'significant benefits can be gained from looking at even our most basic processes'.

Minitab also provides **Attribute Gage R&R (Analytic Method)**.... This may be used to assess the performance of measurement systems such as 'go/no-go' plug gauges used to measure the dimensions of machined components. The method will not be considered in this book. Further information may be obtained from Minitab Help or from AIAG (2002, pp. 125–140).

9.5   Exercises and follow-up activities

1. The worksheet Bands.MTW gives data for a bias and linearity study of a micrometer system carried out by a manufacturer of band saw blades. In the study five measurements were made of the width of each of ten reference pieces of steel band with known width of 5, 10, 15, 20, 25, 30, 35, 40, 45 and 50 mm. Use Minitab to assess the bias and linearity of the system.

2. If possible, obtain a digital micrometer and a set of ten parts – diameters of coins of the same denomination may be measured, for example. Set up a gauge R&R study involving at least two operators and at least two measurements by each operator of each part. Plan your experiment carefully after some initial trials with the equipment and the chosen parts. The author has used measurements of coin diameter with training groups in the past. Use Minitab to assess the performance of the measurement process.

3. The worksheet ARGageR&R.MTW contains data form a gauge repeatability and reproducibility study of a digital micrometer carried out by the author and his wife (who had never used a digital micrometer before the day on which the experiment was carried out). The measurements were of the heights of cylindrical wooden beads.

   (i) Display the data in a gauge run chart.

   (ii) Assess the performance of the measurement system.

   (iii) Unstack the data and assess each operator separately.

4. Bland and Altman (1986) provide data on the peak expiratory flow rates (PEFR) for 17 subjects measured using two different meters – LWPFM and SWPFM – which are provided in PEFR.MTW and reproduced by permission of *The Lancet*. Display the data as for the example given earlier and confirm that the limits of agreement are -78 to 74.
   The authors state that the standard error of the limits is given approximately by $\sqrt{3s^2/n}$, where s is the standard deviation of the differences and n is the number of parts measured. Calculate approximate 95% confidence intervals for the limits of agreement.

5. Following further training, the appraisers Fiona and Iain, referred to in Section 9.3, assessed the same 5 specimens twice in a second attribute agreement study. The data are available in Reassess.MTW. Perform an attribute agreement analysis and comment on the effectiveness of the further training.

10

Regression and model building

Statisticians, like artists, have the bad habit of falling in love with their models. (Attributed to George Box)

Overview

Regression has already been encountered in earlier chapters. In this chapter regression modelling is examined in more detail. In terms of the process model shown earlier in Figure 1.3, regression methods enable the models to be built in terms of linking process inputs (Xs) to process performance measures (Ys) via functional relationships of the form $\mathbf{Y} = f(\mathbf{X})$. The links between regression models and design of experiments will be established. Scenarios in which the response variable is categorical will be dealt with under the heading of logistic regression. The Minitab facilities for the creation, analysis and checking of regression models will be exemplified.

10.1 Regression with a single predictor variable

In Section 3.2.1 reference was made to data on the diameter (Y, mm) of machined automotive components and the temperature (X, °C) of the coolant supplied to the machine at the time of production. Given that the target diameter is 100 mm, a scatterplot indicated the possibility of improving the process through controlling the coolant temperature, thereby leading to less variability in the diameter of the components. Use of **Graph > Scatterplot...** and the **With Regression** option yielded the scatterplot in Figure 10.1, with the addition of the least squares regression line modelling the linear relationship between diameter and temperature. (Use of **Data View...** and the **Regression** tab indicates that the default is to fit a linear model as displayed, but quadratic and cubic models may also be fitted.) The data are available in Diameters.MTW.

Six Sigma Quality Improvement with Minitab, Second Edition. G. Robin Henderson.
© 2011 John Wiley & Sons, Ltd. Published 2011 by John Wiley & Sons, Ltd.

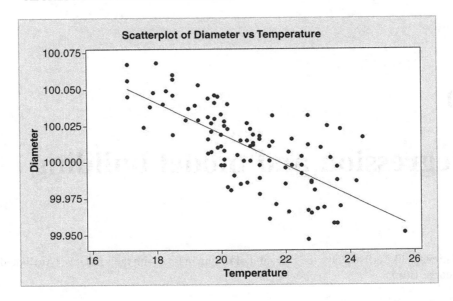

Figure 10.1   Scatterplot of diameter against temperature of coolant.

The equation of the least squares regression line may be found using **Stat > Regression > Regression....** In the dialog **Response:** Diameter and **Predictors:** Temperature must be specified. Thus for a functional relationship $Y = f(X)$, Y is a response and X is a predictor in the terminology used by Minitab. Given that the data are recorded in time order, it is appropriate to check **Four in one** under **Graphs....** Defaults were chosen elsewhere. Various aspects of what is known as *simple linear regression* will now be discussed with reference to the output.

The first portion of the Session Window output is shown Panel 10.1. The equation of the least squares regression line fitted to the data is

$$\text{Diameter} = 100.234 - 0.010\,726 \times \text{Temperature}$$

or, alternatively,

$$y = 100.234 - 0.010\,726x.$$

Regression Analysis: Diameter versus Temperature

```
The regression equation is
Diameter = 100 - 0.0107 Temperature


Predictor          Coef    SE Coef         T      P
Constant        100.234      0.022   4475.56  0.000
Temperature   -0.010726   0.001069    -10.04  0.000
```

Panel 10.1   Regression analysis for diameter versus temperature.

In dealing with regression situations statisticians often make use of the *linear model*

$$Y_i = \alpha + \beta x_i + \varepsilon_i,$$

where α is the intercept parameter, β is the slope parameter and ε_i is the random error, a random variable with mean 0 and standard deviation σ. It follows that

$$E(Y_i) = E(\alpha) + E(\beta x_i) + E(\varepsilon_i) = \alpha + \beta x_i,$$
$$\mathrm{var}(Y_i) = \mathrm{var}(\varepsilon_i) = \sigma^2,$$

where σ^2 is a constant. This means that if Y is observed repeatedly for a particular value of x, then the resulting population of Y-values will have a statistical distribution with mean $\alpha + \beta x$ and variance σ^2.

The further assumptions that the random errors are *independent* and *normally distributed* are also frequently made. One consequence of these assumptions is that the population of Y-values that could be observed for a given x has the normal distribution with mean $\alpha + \beta x$ and variance σ^2.

The fitted least squares regression line may be written as

$$y = a + bx$$

where $a = 100.234$ and $b = -0.010\,726$ respectively estimate the model parameters α and β.

The P-values shown in Panel 10.1 are for two t-tests of hypotheses for the model in which $Y_i \sim N(\alpha + \beta x_i, \sigma^2)$. The first test concerns the intercept parameter, α, in the model:

$$H_0 : \alpha = 0, \quad H_1 : \alpha \neq 0.$$

With P-value 0.000, to three decimal places, we have very strong evidence that the intercept parameter is nonzero.

The second test concerns the slope parameter, β, in the model:

$$H_0 : \beta = 0, \quad H_1 : \beta \neq 0.$$

With P-value 0.000, to three decimal places, we have very strong evidence that the slope parameter is nonzero. This second test may be considered as a test of whether or not there is a linear relationship between Y and x, i.e. between diameter and temperature. (The t-test performed here is equivalent to the test the that the correlation coefficient is zero.)

Whenever a least squares line has been fitted to a set of bivariate data, residuals may be calculated. For any statistical model fitted to data,

$$\mathrm{Data} = \mathrm{Fit} + \mathrm{Residual},$$

as the reader will recall from Chapter 7. The residual may be thought of as that 'component' of the data that remains when the model has performed its task of 'explaining' the data:

$$\mathrm{Residual} = \mathrm{Data} - \mathrm{Fit} = y_i - \hat{Y}_i = y_i - (a + bx_i).$$

The symbol $\hat{Y}_i$ is used for the fitted value of y. For example, for the first data point we have $x_1 = 20.9$ so the fitted value is $\hat{Y}_1 = 100.234 - 0.010\,726 \times 20.9 = 100.234 - 0.224 = 100.010$, to three decimal places. Hence, the first residual is calculated to be

$y_1 - \hat{Y}_1 = 100.001 - 100.010 = -0.009$. The reader is invited to check that the residual for the last data point is 0.026.

Checking **Residuals** and **Fits** under **Storage** during the **Regression** dialog enables columns of the residuals and fitted values to be created. The software gives the first residual as $-0.008\,975\,5$ which rounds to -0.009. Having fitted a linear model to data using least squares, it is standard practice to plot the residuals in various ways. The **Four in one** facility in Minitab yields the following plots:

- histogram of residuals;

- normal probability plot of residuals;

- residuals versus fits;

- residuals versus order.

The four plots for the regression of diameter on temperature are displayed in Figure 10.2.

The shape of the histogram and the linear normal probability plot support the assumption of normality in the model. The fact that the plot of residuals against fitted values has the appearance of a horizontal band of randomly distributed points supports the assumption of constant variability in the model. Finally, the absence of any patterns or trends in the plot of residuals against order suggests that there have been no time-related factors influencing the process. Thus we may consider the residual plots to be satisfactory in this case.

The next portion of the Session window output is shown in Panel 10.2. The value of s provides an estimate of the standard deviation, σ, in the model. The value of R^2 (R-Sq), in this case where there is a single predictor variable, is r^2, the coefficient of determination between diameter and temperature expressed as a percentage – see Section 3.2.2. It indicates the proportion of the variation in diameter that can be attributed to its linear dependence on

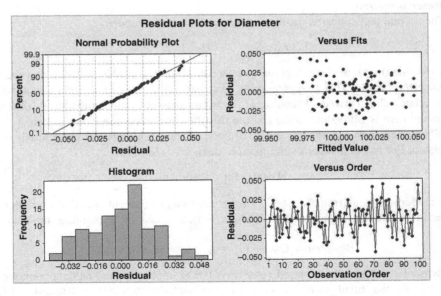

Figure 10.2   Residual plots for regression of diameter on temperature.

```
S = 0.0192354    R-Sq = 50.7%    R-Sq(adj) = 50.2%
```

Panel 10.2 Further output from regression analysis.

temperature. R^2 may also be calculated as the square of the correlation between the observed and fitted values of diameter, the response.

We may summarize the model as follows. The expected value of diameter is linearly related to temperature by the equation

$$\text{Diameter} = 100.234 - 0.010\,726 \times \text{Temperature}$$

For any specific temperature, diameter is normally distributed, with mean estimated by

$$100.234 - 0.010\,726 \times \text{Temperature}$$

and with standard deviation estimated to be 0.0192. Just over half the variation in diameter may be explained through its linear dependence on temperature.

An enhanced version of the plot in Figure 10.1 may be obtained using **Stat > Regression > Fitted Line Plot. . .**, specifying Diameter as the response and Temperature as the predictor and accepting defaults otherwise. The plot is shown in Figure 10.3. This plot is annotated by Minitab with the equation of the regression line, the estimated standard deviation of the random errors and the value of R^2. (The adjusted value of R^2, denoted by R-Sq(adj) will be discussed later in the chapter.) This plot would clearly complement the written summary of the model in, for example, a project report. (The first and last data points have been indicated by square symbols and vertical line segments drawn from the regression line to these data points. The magnitudes of these segments are the magnitudes of the residuals. The first data point lies below the regression line corresponding to its associated negative residual; the last data point lies above the line

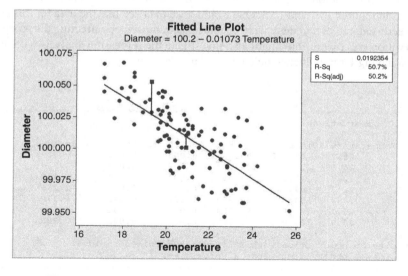

Figure 10.3 Fitted line plot for diameter on temperature.

```
Analysis of Variance

Source       DF       SS          MS         F       P
Regression    1   0.0372752   0.0372752   100.74   0.000
Error        98   0.0362602   0.0003700
Total        99   0.0735354
```

Panel 10.3   ANOVA for regression analysis.

corresponding to its associated positive residual. The fitted least squares regression line is such that the sum of the squares of all 100 residuals is a minimum.)

The Session window output also includes the ANOVA output displayed in Panel 10.3. The ANOVA provides an alternative method of testing the null hypothesis that the slope parameter, β, is zero. In fact the F-statistic of 100.74 quoted is the square, allowing for rounding, of the t-statistic of -10.04 quoted in Panel 10.1. The P-value of 0.000, to three decimal places, provides strong evidence that null hypothesis $H_0 : \beta = 0$ should be rejected in favour of the alternative hypothesis $H_1 : \beta \neq 0$.

The calculation of residuals has already been detailed. Minitab also computes standardized residuals by dividing the residuals by their standard deviation. The final part of the Session window output alerts the user to any observations in the data set that yield a standardized residual with an absolute value in excess of 2. Details of such observations are listed under the heading Unusual Observations and identified by a letter R at the end of the line – see Panel 10.4. With a standard normal distribution the probability of obtaining values exceeding 2, in absolute value, is approximately 5%. Thus being alerted to six observations in this category, when five would be expected from a sample of 100 observations, need not be a major concern.

The user is also alerted to any observations in the data set that have x-values that give them large influence. An X at the end of the line of output indicates such observations. Observation number 31 is adjudged to have a temperature value which gives it large influence in the analysis. This observation corresponds to the point in the bottom right-hand corner of the scatterplots in Figures 10.1 and 10.3. One would wish to check that the data for this point had been correctly entered or that there were no unusual circumstances affecting the process when the observation was made. One might also investigate the regression analysis with all points flagged as having large influence deleted from the data set.

```
Unusual Observations

Obs  Temperature  Diameter      Fit   SE Fit  Residual  St Resid
 31         25.7    99.952    99.958    0.005    -0.006    -0.35 X
 59         21.5    99.961   100.004    0.002    -0.043    -2.22R
 69         22.7   100.032    99.991    0.003     0.041     2.17R
 71         22.7    99.947    99.991    0.003    -0.044    -2.29R
 76         24.4   100.017    99.972    0.004     0.045     2.37R
 86         23.3   100.025    99.984    0.003     0.041     2.15R
 99         23.7   100.023    99.980    0.004     0.043     2.28R

R denotes an observation with a large standardized residual.
X denotes an observation whose X value gives it large leverage.
```

Panel 10.4   List of unusual observations from regression analysis.

The target diameter was 100 mm. Substitution of this value for diameter into the fitted linear equation yields the following:

$$\text{Diameter} = 100.234 - 0.010\,726 \times \text{Temperature}$$
$$100 = 100.234 - 0.010\,726 \times \text{Temperature}$$
$$0.010\,726 \times \text{Temperature} = 0.234$$
$$\text{Temperature} = \frac{0.234}{0.010\,726} = 21.8.$$

Thus the model suggests that the temperature of the coolant should be controlled at 21.8 °C.

The model may be used to make predictions of diameter for any specific temperature of interest. Predictions may be obtained by using **Stat** > **Regression** > **Regression. . .** again, with **Response:** Diameter and **Predictors:** Temperature, specified as previously. In order to predict for temperature 21.8, for example, the value 21.8 must be entered under **Options. . .** by specifying **Prediction intervals for new observations:** 21.8. (More than one temperature value may be entered if required.) The resulting output in the Session window is shown in Panel 10.5.

The order of presentation is arguably not the best. The second section informs the user that diameter has been predicted for temperature 21.8 °C. The first section indicates that the predicted diameter is 100.000 (under the heading Fit), thus confirming that the calculation performed earlier was correct! Two intervals are given. The 95% CI of (99.996,100.005) is a 95% confidence interval for the *mean* diameter obtained when the process is operated with temperature 21.8 °C. (The SE Fit value of 0.002 is the estimated standard deviation required in the computation of the confidence interval.) The 95% PI of (99.962,100.039) is a prediction interval in which we can have 95% confidence that an *individual* diameter, obtained when the process is operated with temperature 21.8 °C, will fall. Note that the prediction interval is wider than the confidence interval.

Instead of listing one or more temperature values under **Options. . .** in **Prediction intervals for new observations:**, one can create a column of temperature values of interest and insert the column name instead of a list. This was done in order to create Table 10.1. Temperatures from 17 °C to 26 °C, at intervals of 1 °C, were selected in order to cover the range of temperatures encountered in the investigation.

Note that the width of the intervals varies, with the narrowest intervals occurring for temperature 21 °C. In fact the narrowest possible intervals occur when temperature equals the mean of all 100 temperatures recorded in the given data set, i.e. 20.9 °C. In order to display

```
Predicted Values for New Observations

New Obs      Fit   SE Fit          95% CI               95% PI
      1  100.000    0.002  (99.996, 100.005)  (99.962, 100.039)


Values of Predictors for New Observations

New Obs   Temperature
      1          21.8
```

Panel 10.5 Predictions of diameter from model.

Table 10.1 Confidence and prediction intervals for diameter.

Temperature	Predicted diameter	95% confidence limits for diameter		95% prediction limits for diameter	
		Lower	Upper	Lower	Upper
17	100.052	100.043	100.061	100.013	100.091
18	100.041	100.034	100.048	100.002	100.080
19	100.030	100.025	100.036	99.992	100.069
20	100.020	100.015	100.024	99.981	100.058
21	100.009	100.005	100.013	99.971	100.047
22	99.998	99.994	100.003	99.960	100.037
23	99.987	99.982	99.993	99.949	100.026
24	99.977	99.969	99.984	99.938	100.016
25	99.966	99.956	99.976	99.927	100.005
26	99.955	99.944	99.967	99.915	99.995

these intervals use was made of **Stat > Regression > Fitted Line Plot...**, specifying diameter as the response and temperature as the predictor, and under **Options...** checking both **Display confidence interval** and **Display prediction interval** as **Display Options**. The resultant plot is shown in Figure 10.4. The arrowed vertical line segment that has been added to the plot indicates the 95% prediction interval corresponding to temperature 21.8 °C – the temperature predicted by the model to yield the desired mean diameter of 100 mm. From a process improvement point of view, the modelling has demonstrated that there is the potential to reduce the variability of diameter and hence to increase process capability through control of the coolant temperature.

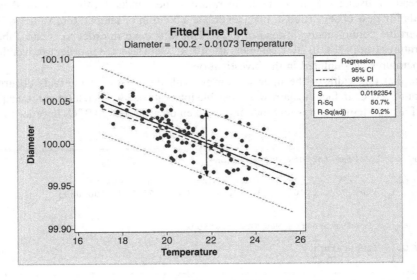

Figure 10.4 Fitted line plot with confidence and prediction intervals.

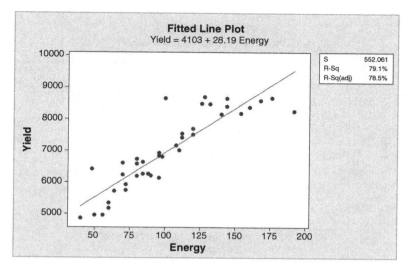

Figure 10.5   Fitted line plot for lactation data.

As a second example, consider the data available in Lactation.MTW. It gives milk yield (kg) and energy intake (MJ/d) for a sample of 40 cows. The fitted line plot for the linear regression of yield on energy is displayed in Figure 10.5. The R^2 value indicates that the linear relationship between yield on energy explains around 80% of the variation in yield. (The reader is invited to verify that the P-value for the slope parameter is 0.000, to three decimal places. Thus there is strong evidence of a linear relationship.)

The plot of residuals against fitted values is shown in Figure 10.6. The fact that this plot has an arched appearance (indicated by the broad arc superimposed on the plot) rather than the

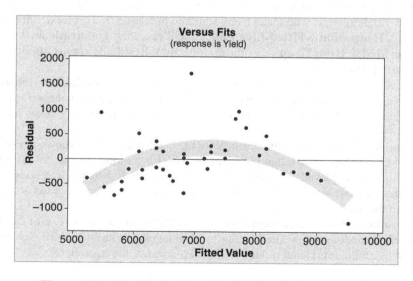

Figure 10.6   Residuals versus fitted values for linear model.

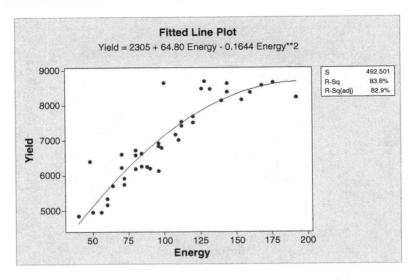

Figure 10.7   Quadratic model fitted to lactation data.

appearance of a horizontal band of randomly distributed points suggests that the linear model is inadequate. The plot indicates that there is still structure of a curvilinear nature in the data that might be utilized in order to improve the model. The simplest way to introduce curvature into the model is to make use of the *quadratic model*

$$Y_i = \alpha + \beta_1 x_i + \beta_{11} x_i^2 + \varepsilon_i,$$

where α is the intercept parameter, β_1 is the slope parameter, β_{11} is the quadratic parameter and ε_i is the random error, with mean 0 and standard deviation σ. This model may be fitted using **Stat > Regression > Fitted Line Plot. . .** and checking **Quadratic** as the **Type of Regression Model**. Using **Graphs. . .** the option to plot **Residuals versus fits** was checked under **Residual Plots**.

The fitted line plot is shown in Figure 10.7. The quadratic relationship between yield and energy explains around 84% of the variation in yield – the linear model explained around 80% of the variation, so the addition of the quadratic term has yielded a modest increase in explanatory power.

The plot of residuals against fitted values for the quadratic model is shown in Figure 10.8 and is a more satisfactory residual plot than the one for the previous model. The Session window output is displayed in Panel 10.6. The heading Polynomial Regression Analysis indicates that the software has fitted a polynomial curve, in this case a quadratic curve or parabola, to the data.

The equation of the quadratic curve is given together with the estimate, s, of the standard deviation of the random errors. The R^2 value is given together with an analysis of variance for the overall quadratic regression model. It provides a test of the hypotheses

$$H_0 : \beta_1 = \beta_{11} = 0, \quad H_1 : \text{Not all } \beta s \text{ are zero}.$$

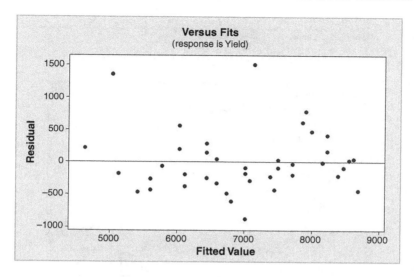

Figure 10.8 Residuals versus fitted values for quadratic model.

With *P*-value 0.000, to three decimal places, the null hypothesis cannot be rejected. The sequential analysis of variance, with *P*-values less than 0.01 for both the linear and quadratic terms, provides evidence that both terms are worth including in the model.

In order to carry out an alternative analysis it is necessary to use **Calc** > **Calculator** to calculate a new column, named Energy**2, say, containing the squares of the energy values. The quadratic regression may be found using **Stat** > **Regression** > **Regression....** In the dialog **Response:** Yield and **Predictors:** Energy and 'Energy**2' must be specified as indicated in Figure 10.9. Residual plots may be created using **Graphs...** as before.

```
Polynomial Regression Analysis: Yield versus Energy

The regression equation is
Yield = 2305 + 64.80 Energy - 0.1644 Energy**2


S = 492.501    R-Sq = 83.8%    R-Sq(adj) = 82.9%


Analysis of Variance

Source        DF        SS          MS        F       P
Regression     2    46355973    23177986    95.56   0.000
Error         37     8974605      242557
Total         39    55330577


Sequential Analysis of Variance

Source        DF        SS        F        P
Linear         1    43749257   143.55   0.000
Quadratic      1     2606716    10.75   0.002
```

Panel 10.6 Session window output for the quadratic regression model.

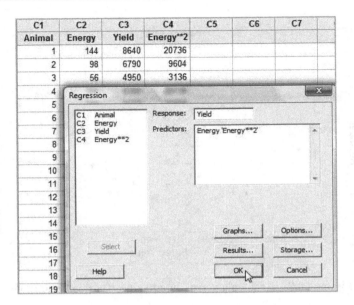

Figure 10.9 Dialog for fitting quadratic regression model.

Part of the Session window output is shown in Panel 10.7. Here the P-values provide evidence that the constant term in the model is nonzero and that the coefficients of the energy term and the energy squared term are both nonzero. The latter two P-values are identical to those given in the sequential analysis of variance in Panel 10.6.

From the point of view of quality improvement the creation of a quadratic model can assist with the determination of optimum conditions under which to run a process. The fitted quadratic curve, extrapolated for energy values up to 250, is shown in Figure 10.10. The model suggests to animal scientists that yield of milk could potentially be maximized by targeting energy intake levels of around 200 for the cows. Further investigation would be needed to confirm that yield could be expected to decrease for energy levels beyond 200.

The data for the final example in this section are reproduced from Gorman and Toman (1966) by permission of the American Society for Quality and are discussed by Hogg and

---

Regression Analysis: Yield versus Energy, Energy**2

```
The regression equation is
Yield = 2305 + 64.8 Energy - 0.164 Energy**2


Predictor      Coef   SE Coef       T      P
Constant      2305.3    593.7    3.88  0.000
Energy         64.80    11.36    5.70  0.000
Energy**2    -0.16438  0.05014  -3.28  0.002


S = 492.501    R-Sq = 83.8%    R-Sq(adj) = 82.9%
```

Panel 10.7 Session window output for the quadratic regression model.

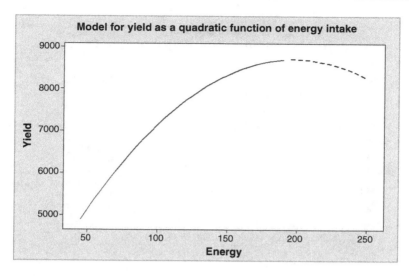

Figure 10.10   Dialog for quadratic regression model.

Ledolter (1992, pp 393–398). They are available in Rut.MTW and give change in rut depth (y) and viscosity of the asphalt (x) for experimental sections of pavement. Scrutiny of the scatterplot of the data in Figure 10.11 suggests that the relationship is nonlinear.

Rather than fit a polynomial model, such as a quadratic or cubic, Hogg and Ledolter applied a logarithmic transformation to both x and y. This transformation may be carried out directly using **Calc > Calculator** to calculate new columns containing the logarithms to base 10 of both x and y. The linear regression of $\log_{10}y$ on $\log_{10}x$ may then be obtained using **Stat > Regression > Regression....** Alternatively, use may be made of **Stat > Regression > Fitted**

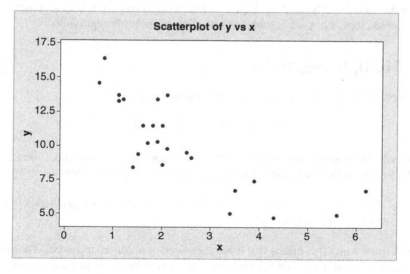

Figure 10.11   Plot of change in rut depth against asphalt viscosity.

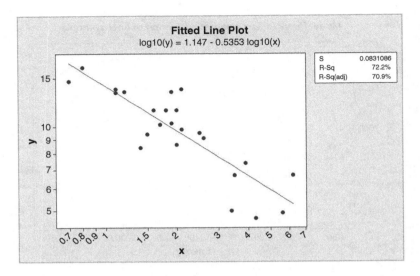

Figure 10.12   Regression of $\log_{10} y$ on $\log_{10} x$.

Line Plot..., specifying y as the response and x as the predictor. Under **Options...**, **Transformations** it is necessary to check **Logten of Y** and **Display logscale for Y variable** together with **Logten of X** and **Display logscale for X variable**. The output is shown in Figure 10.12. This model has an R^2 of 72%, whereas a simple linear regression model fitted to the data yields an R^2 of 64%. The residual plots are satisfactory, so the logarithmic transformations have yielded an adequate model of the situation.

So far we have considered situations where we have created models of the form $Y = f(X)$ for a single response, Y, in terms of a single factor or predictor, X. In the next section we consider models of the form $Y = f(\mathbf{X})$ or $Y = f(X_1, X_2, \ldots)$ for a single response, Y, in terms of a series of factors or predictors, X_1, X_2, ... under the heading of multiple regression.

10.2 Multiple regression

The simplest multiple regression model is the *linear model*

$$Y_i = \beta_0 + \beta_1 x_{1i} + \beta_2 x_{2i} + \varepsilon_i,$$

where β_0 is the intercept parameter, β_1 and β_2 are the (partial) regression coefficients, and ε_i is the random error, with mean 0 and standard deviation σ. It follows that

$$E(Y_i) = E(\beta_0) + E(\beta_1 x_{1i}) + E(\beta_2 x_{2i}) + E(\varepsilon_i) = \beta_0 + \beta_1 x_{1i} + \beta_2 x_{2i},$$
$$\operatorname{var}(Y_i) = \operatorname{var}(\varepsilon_i) = \sigma^2,$$

where σ^2 is a constant. This means that if Y is measured or observed repeatedly for a particular pair of values of x_1 and x_2, then the resulting values will have a statistical distribution with mean $\beta_0 + \beta_1 x_{1i} + \beta_2 x_{2i}$ and variance σ^2.

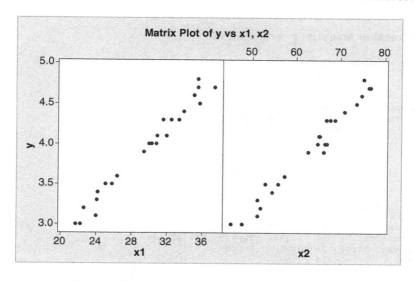

Figure 10.13   Scatterplots of Y versus x_1 and x_2.

The further assumptions that the random errors are *independent* and *normally distributed* are also frequently made. A consequence of these assumptions is that the population of Y values that could be observed for a given pair of values of x_1 and x_2 has the normal distribution with mean $\beta_0 + \beta_1 x_{1i} + \beta_2 x_{2i}$ and variance σ^2.

The next example concerns a market research company wishing to predict weekend circulation of daily newspapers in market areas. It ascertained circulation (Y, in thousands) in a sample of 25 market areas, together with total retail sales (x_1, in millions of dollars) and population density (x_2, adults per square mile). The data are available in Circulation.MTW and are from W. Daniel and J. Terrell, *Business Statistics for Management and Economics*, 5th edition, p. 531, © 1989 by Houghton Mifflin Company and used with permission. As a first step the data may be displayed in the form of scatterplots of y versus each of the two x variables as displayed in Figure 10.13. These may be obtained using **Graph > Matrix Plot...** and selecting **Each Y versus each X** and **Simple**.

These plots indicate linear relationships between both Y and x_1 and Y and x_2. A multiple regression model may be fitted using **Stat > Regression > Regression...**, specifying Y as the response and both x_1 and x_2 as predictors. As always, diagnostic plots of the residuals should be selected under **Graphs....**

Part of the Session window output is shown in Panel 10.8. The multiple regression equation is stated initially. The ANOVA table provides a test of the null hypothesis that both β_1 and β_2 are zero. The P-value of 0.000, to three decimal places, provides strong evidence that the null hypothesis should be rejected in favour of the alternative hypothesis that not both not both of the partial regression coefficients are zero.

In the earlier section of the output the results of individual t-tests on the βs are given. With P-values of 0.004, 0.003 and 0.026 respectively, the null hypothesis that $\beta_0 = 0$, the null hypothesis that $\beta_1 = 0$ and the null hypothesis that $\beta_2 = 0$ would all be rejected at the 5% level of significance. In addition, the standard deviation of the random error is estimated to be 0.0822 and the R^2 value is 98%. Thus 98% of the variation in circulation (Y) can be explained by its

```
Regression Analysis: y versus x1, x2

The regression equation is
y = 0.382 + 0.0678 x1 + 0.0244 x2


Predictor      Coef   SE Coef      T      P
Constant     0.3822    0.1203   3.18  0.004
x1          0.06779   0.02006   3.38  0.003
x2          0.02443   0.01021   2.39  0.026


S = 0.0821991   R-Sq = 98.0%   R-Sq(adj) = 97.8%


Analysis of Variance

Source          DF       SS       MS       F       P
Regression       2   7.3818   3.6909  546.25   0.000
Residual Error  22   0.1486   0.0068
Total           24   7.5304
```

Panel 10.8   Session window output for multiple regression.

linear dependence on both total retail sales (represented by x_1) and population density (x_2). Both a normal probability plot of the residuals and a plot of the residuals against fitted values appear to be satisfactory. Thus we can be confident that we have an adequate model of the situation.

Suppose that the company wished to predict circulation for a market area with total retail sales of \$25 million and population density 50 adults per square mile. Predictions may be obtained by using **Stat > Regression > Regression...** again, with **Response:** y and **Predictors:** $x_1 x_2$, specified as previously. In order to predict for $x_1 = 25$ and $x_2 = 50$, under **Options...** specify **Prediction intervals for new observations:** 25 50. Note that order of entry is important here – the values of the predictor variables must be entered in the same order as the variables were entered under predictors. More than one pair of values for x_1 and x_2 may be entered if required, or alternatively, the names of two columns containing matching pairs of values for x_1 and x_2 may be entered.

The resulting Session window output is shown in Panel 10.9. Thus for the market area of interest the model predicts circulation of 3.2984 thousands, i.e. of 3298 to the nearest whole number. The 95% confidence interval converts to (3198, 3399). We can be 95% confident that

```
Predicted Values for New Observations

New Obs    Fit  SE Fit       95% CI              95% PI
     1  3.2984  0.0487  (3.1975, 3.3993)  (3.1003, 3.4965)


Values of Predictors for New Observations

New Obs    x1     x2
     1   25.0   50.0
```

Panel 10.9   Prediction using the multiple regression model.

MULTIPLE REGRESSION   429

the *mean* circulation, for market areas with total retail sales of $25 million and population density of 50, will lie in this interval. The 95% prediction interval converts to (3100, 3496). We can be 95% confident that the *individual* circulation, for an individual market area with total retail sales of $25 million and population density of 50, will lie in this interval. Such predictions could be of value to the company in terms of making improvements to production scheduling and distribution.

As a second example, data on percentage elongation of 24 specimens of a steel alloy will be investigated. The data are available in Elongation.MTW and are reproduced by permission of Oxford University Press Inc., New York. The data, available in Elongation.MTW are from p.426 of *Fundamental Concepts in the Design of Experiments*, 5th edition by Charles R. Hicks and Kenneth V. Turner, Jr, copyright © 1964, 1973, 1982, 1993, 1999 and used by permission of Oxford University Press, Inc., New York. Elongation (Y) and the percentages of five specific chemical elements (x_1, x_2, x_3, x_4 and x_5) were determined for each specimen. In terms of creating a multiple regression model here, which is linear in the predictor variables x_1, x_2, x_3, x_4 and x_5, we have two choices for each predictor – either omit or include. Thus there are $2^5 = 32$ possible models, 31 if we discount the trivial model that involves none of the predictors. Initial exploration of potential models may be made using **Stat** > **Regression** > **Best Subsets...** with y specified as **Response:**, x_1, x_2, x_3, x_4 and x_5 as **Free Predictors** and defaults accepted otherwise.

The Session window output is shown in Panel 10.10. By default, summary information is provided in the output for the 'best' two models involving 1, 2, 3 and 4 predictors and the single model involving all 5 predictors. In addition to R^2 and the adjusted R^2, Mallow's C_p statistic is listed for each of the nine models. (The C_p statistic here should not be confused with the capability index C_p.) For example, the fourth row of information (highlighted in bold) indicates that the multiple regression involving the two predictors x_2 and x_5 (indicated by the crosses vertically below the 2 and 5 in the headings) has $R^2 = 49.2\%$, adjusted $R^2 = 44.4\%$ and Mallow's C_p statistic of 0.8. Hicks and Turner (1999, p. 425) comment: 'Even though many independent variables may be used, simpler models are more appealing and easier to interpret. Thus, we try to identify the smallest subset of the independent variables that will provide an adequate model.' Many statisticians compare the adjusted R^2 values for candidate models.

```
Best Subsets Regression: y versus x1, x2, x3, x4, x5

Response is y

                        Mallows        x x x x x
Vars  R-Sq  R-Sq(adj)      Cp     S    1 2 3 4 5
   1  31.4       28.3     5.4  2.3077  X
   1  28.3       25.1     6.5  2.3590    X
   2  49.2       44.4     0.8  2.0328  X       X
   2  47.3       42.3     1.5  2.0703  X X
   3  50.7       43.3     2.3  2.0529  X X     X
   3  49.8       42.3     2.6  2.0710  X   X X
   4  51.2       40.9     4.1  2.0952    X X X X
   4  51.0       40.7     4.1  2.0988  X X X   X
   5  51.4       37.9     6.0  2.1477  X X X X X
```

Panel 10.10   Session window output from best subsets regression.

With a simple linear regression model, i.e. one involving a single predictor variable x, and a response y, R^2 is the coefficient of determination given by the square of the correlation coefficient, r, between x and y. With two or more predictor variables, R^2 may be considered as the square of the correlation coefficient between the observed data value, y, and the fitted value $\hat{Y}$. Every new predictor added to a multiple regression model will lead to an increase in the value of R^2. The adjusted R^2, denoted by R-Sq (adj) in Minitab, is given by

$$R_{\text{adj}}^2 = 1 - \frac{n-1}{n-p-1}(1 - R^2),$$

where n is the number of observations and p is the number of predictor variables in the model. The statistic R^2 is computed from sample data and may therefore be considered as an estimate of a population value. The value of adjusted R^2 provides a better estimate of this population value than does R^2.

A plot of the 'best' adjusted R^2 values for each number of predictors is displayed in Figure 10.14. This points to the model involving the two predictors x_2 and x_5 as being worthy of further investigation. This may be done using **Stat > Regression > Regression....** The usual residual plots should be examined together with plots of residual against predictor (x) variables not currently in the model. The residual for the 23rd specimen is unusually large. In situations such as this, assuming no data input error has been made, analysis of the data with that observation excluded can be informative. Pattern or structure in residual plots can indicate the need to introduce predictors that are the squares of x variables (curvature terms) or the products of pairs of x variables (interaction terms).

Some statisticians use Mallow's C_p statistic as an aid to model selection. 'The C_p statistic appeals nicely to common sense and is developed from considerations of the proper compromise between excessive bias incurred when one underfits (chooses too few model terms) and excessive prediction variance produced when one overfits (has redundancies in the

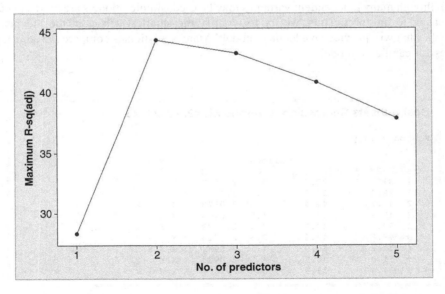

Figure 10.14   Plot of maximum adjusted R^2 versus number of predictors.

model)' (Walpole and Myers, 1989, p. 447). If p denotes the number of model parameters then '$C_p > p$ indicates a model that is biased due to being an underfitted model, while $C_p \approx p$ indicates a reasonable model' (Walpole and Myers, 1989, p. 448).

Minitab provides another major tool for the development of regression models – stepwise regression methods. It will not be considered in this book. Hicks and Turner (1999, p. 428) comment that 'these procedures may miss some models considered by the all-subsets procedure' and 'use of all-subsets regression is recommended when adequate computing facilities are available'. Readers who wish to learn more about variable selection in regression model building might find it beneficial to consult the book by Montgomery *et al.* (2006).

10.3 Response surface methods

The creation of response surface models essentially involves fitting multiple regression models to experimental data. To introduce response surface experimental designs, we consider an experiment carried out to investigate the tensile strength (Y, g/cm) of film used in the food industry. Customers of the manufacturer of the film had been experiencing problems due to the film tearing during food packaging operations. The manufacturer set up a Six Sigma project in order to determine if changes to manufacturing process settings would yield stronger film. The project team decided that the factors seal temperature (°C) and the amount of a plastic additive (%) should be investigated. The settings currently used in production were 140 °C and 4% respectively, and mean tensile strength was stable and predictable with a mean around 63 g/cm. Phase 1 of the experimentation involved a 2^2 factorial design (replicated twice) with low and high levels of 120 °C and 160 °C for temperature and 2% and 6% for amount of additive, supplemented by four runs carried out with the current settings of 140 °C and 4%. The five factor–level combinations involved are displayed in Figure 10.15. There is no simple rule for

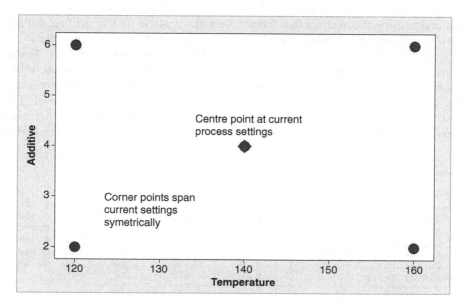

Figure 10.15 Factor–level combinations for 2^2 factorial experiment with centre points.

Table 10.2   Phase 1 data for 2^2 factorial experiment with centre points.

Temperature	Additive	y
160	2	64.1
140	4	61.6
160	2	60.8
140	4	62.4
120	2	52.1
120	6	60.6
160	6	69.6
120	2	53.3
120	6	61.7
140	4	63.1
140	4	62.5
160	6	70.5

selection of the low and high levels in such scenarios – the project team would need to consider the selection carefully, taking into account the views of people with knowledge and experience of running the process.

Use was made of **Stat > DOE > Factorial > Create Factorial Design...** to carry out the design and create a pro forma for recording results. With **Number of factors:** 2, under **Designs...** the specifications made were **Number of center points per block:** 4, **Number of replicates for corner points:** 2 and **Number of blocks:** 1. Randomization was used. The key data are shown in Table 10.2, with the factor levels in the random order obtained from the software, and the full data set is available in Phase1.MTW. (If the reader wishes to re-create the design and perform the analysis that follows then the response data in the final column of Table 10.2 will have to be entered into his/her worksheet in the appropriate order.)

The initial ANOVA obtained using **Stat > DOE > Factorial > Analyze Factorial Design...** is shown in Panel 10.11. It provides no evidence of any interaction or of any curvature (P-values 0.598 and 0.263, respectively). Before creating a contour plot of the response surface one may therefore remove the interaction term from the model. This is achieved by using **Stat > DOE > Factorial > Analyze Factorial Design...** again using **Terms...** to remove the interaction (AB) term from the **Selected Terms:** window. One must

```
Analysis of Variance for y (coded units)

Source                  DF   Seq SS   Adj SS   Adj MS      F      P
Main Effects             2  302.712  302.712  151.356  127.42  0.000
  Temperature            1  173.911  173.911  173.911  146.41  0.000
  Additive               1  128.801  128.801  128.801  108.43  0.000
2-Way Interactions       1    0.361    0.361    0.361    0.30  0.598
  Temperature*Additive   1    0.361    0.361    0.361    0.30  0.598
  Curvature              1    1.760    1.760    1.760    1.48  0.263
Residual Error           7    8.315    8.315    1.188
  Pure Error             7    8.315    8.315    1.188
```

Panel 10.11   ANOVA for Phase 1 experiment.

Factorial Fit: y versus Temperature, Additive

```
Estimated Effects and Coefficients for y (coded units)

Term            Effect      Coef   SE Coef        T       P
Constant                  61.858    0.3109   198.99   0.000
Temperature      9.325     4.662    0.3807    12.25   0.000
Additive         8.025     4.013    0.3807    10.54   0.000
```

Panel 10.12   Revised model for Phase 1.

also uncheck **Include center points in the model** in order to create a contour plot as a follow-up to the model revision. With no evidence of interaction or curvature a plane surface will provide an adequate model in the region of the design space explored in Phase 1, so omitting the centre points is justified. Of particular interest for this revised model is the output displayed in Panel 10.12. The P-values for temperature and amount of additive provide very strong evidence of important main effects of both.

Having fitted the revised model, one may proceed to use **Stat > DOE > Factorial > Contour/Surface Plots...** to create the required display. **Contour plot** was checked and **Setup...** involved specifying **Response:** C7 y. In addition, under **Contours...** the option to **Use defaults** was accepted for **Contour Levels** and, under **Data Display**, both **Contour Lines** and **Symbols at design points** were checked, but not **Area**. The plot is displayed in Figure 10.16.

Note the solid circle symbols indicating the five FLCs, or design points, used in the experimentation so far and previously displayed in Figure 10.15. A line from the centre point of the design, in the direction of increasing tensile strength, y, at right angles to the contour lines, has been added. This is the line of steepest ascent, and Phase 2 of the experimentation involved duplicate process runs at a series of FLCs along the line of steepest ascent. (A method

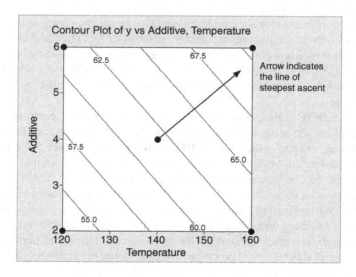

Figure 10.16   Contour plot for Phase 1.

Table 10.3   Phase 2 data from exploration along line of steepest ascent.

Temperature	Additive	y_1	y_2	Mean
170	6.6	74.8	74.3	74.55
190	8.3	78.8	77.5	78.15
210	10.0	80.1	79.2	79.65
230	11.7	76.3	77.5	76.90
250	13.5	74.4	74.6	74.50

for determining FLCs along a line of steepest ascent is detailed in one of the follow-up exercises.) The data are displayed in Table 10.3 and available in Phase2.MTW.

Note that, as the investigation moved further away from the original centre point, tensile strength began to increase and then decreased again. It peaked with temperature 210 °C and 10% additive, so Phase 3 of the experimentation involved a response surface design, with two replications, centred at this point.

In order to create such a design use may be made of **Stat > DOE > Response Surface > Create Response Surface Design. . . .** The default **Central Composite** type of design was accepted, with **Number of factors:** 2. Clicking on **Designs. . .** reveals the two available designs for the case of two factors. The default **Full** design involves 13 runs (or FLCs) and one block. This design was adopted with **Number of replicates:** 2 specified and defaults accepted otherwise. Under **Factors:** the default that **Levels Define Cube points** (corners of the square defining the 2^2 factorial that forms the basis of the design in terms of coded units) was accepted and then the factors temperature and amount of additive with the levels selected by the project team (200 °C and 220 °C for temperature and 9% and 11% for amount of additive) were specified in the usual way. The reader should note that these FLCs for corner points are symmetrically placed relative to the centre point levels of 210 °C for temperature and 10% for amount of additive identified in Phase 2 as being a region of the design space worth exploring. The data are displayed in Figure 10.17 and available in Phase3.MTW.

In terms of coordinates in the temperature–additive space, the centre point for the design was (210, 10), with the four corner points (200, 9), (220, 9), (200, 11), (220, 11). In addition, the central composite response surface design involved the FLCs corresponding to the axial points (195.9, 10), (210, 11.4), (224.1, 10), (210, 8.6). Note that Minitab uses codes in the PtType column of the worksheet to indicate centre points (code 0), corner (or cube) points (code 1) and axial points (code − 1).

Use of **Stat > DOE > Response Surface > Analyze Response Surface Design. . .** is required to analyse the data. Once this had been done the contour plot displayed in Figure 10.18 was created using **Stat > DOE > Response Surface > Contour/Surface Plots. . . .** The contour plot indicates that maximum strength, y, of around 79 g/cm can be expected with the FLC of temperature 206 °C and additive 9.8%. Thus, given that with current operating conditions mean strength was around 63 g/cm, the Six Sigma project indicates that a 25% increase to a mean of around 79 g/cm appears feasible. The peak on the response surface has been indicated by the triangular symbol and the nine FLCs for the response surface design are indicated by the solid circles. Hogg and Ledolter (1992, pp. 409–410) provide a calculus-based procedure for calculating the factor levels corresponding to an optimum point and for determining the nature of the optimum point. The solid circles indicate the FLCs used in the response surface design.

↓	C1	C2	C3	C4	C5	C6	C7
	StdOrder	RunOrder	PtType	Blocks	Temperature	Additive	y
1	23	1	0	1	210.0	10.0	78.9
2	4	2	1	1	220.0	11.0	77.9
3	12	3	0	1	210.0	10.0	78.4
4	9	4	0	1	210.0	10.0	79.4
5	1	5	1	1	200.0	9.0	78.3
6	22	6	0	1	210.0	10.0	79.3
7	8	7	-1	1	210.0	11.4	77.5
8	13	8	0	1	210.0	10.0	78.0
9	19	9	-1	1	224.1	10.0	78.4
10	16	10	1	1	200.0	11.0	79.0
11	18	11	-1	1	195.9	10.0	78.2
12	5	12	-1	1	195.9	10.0	79.8
13	6	13	-1	1	224.1	10.0	79.3
14	20	14	-1	1	210.0	8.6	78.2
15	26	15	0	1	210.0	10.0	79.8
16	11	16	0	1	210.0	10.0	78.5
17	15	17	1	1	220.0	9.0	77.1
18	3	18	1	1	200.0	11.0	77.6
19	14	19	1	1	200.0	9.0	77.8
20	17	20	1	1	220.0	11.0	77.8
21	21	21	-1	1	210.0	11.4	77.5
22	2	22	1	1	220.0	9.0	79.1
23	25	23	0	1	210.0	10.0	79.2
24	7	24	-1	1	210.0	8.6	79.4
25	24	25	0	1	210.0	10.0	78.7
26	10	26	0	1	210.0	10.0	78.0

Figure 10.17 Response surface design experiment worksheet.

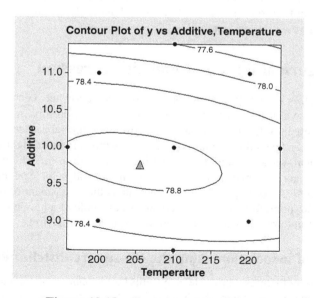

Figure 10.18 Contour plot for Phase 3.

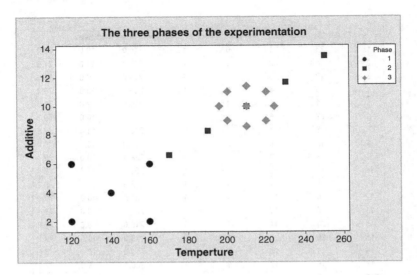

Figure 10.19   Display of all FLCs used in the three phases of the experimentation.

The FLCs used in all three phases of the experimentation are displayed in Figure 10.19. The display highlights the iterative nature of the experimentation. The factorial experiment performed in Phase 1 indicated a promising direction in which to carry out further investigation. Having found an FLC at which yield reached a peak from Phase 2, a response surface design centred in the region of this combination provided Phase 3 and indicated a likely optimal FLC.

Minitab enables central composite response surface designs to be created and analysed for up to 10 factors. In many situations process teams have to consider a number of responses, and it may not be possible to optimize all of them simultaneously. Montgomery (2005a, 2009) gives comprehensive details and examples.

10.4   Categorical data and logistic regression

The use of correlation to investigate relationships or association between continuous random variables and the use of least squares regression to model relationships in which the response was continuous were introduced in Chapter 3 and developed further earlier in this chapter. The rest of this chapter is devoted to the introduction of methods for investigation of association between categorical variables and for modelling relationships in which the response is categorical. There are two main types of measurement scales for categorical variables – *ordinal* and *nominal*. An example of an ordinal scale is the assessment of the condition of a road surface as bad, poor, fair, good or excellent. An example of a nominal scale is the plant where a model of automobile was assembled, e.g. Linwood or Ryton.

10.4.1   Tests of association using the chi-square distribution

A company that manufactures marine radar systems builds scanners using main bearings from two suppliers, A and B. One year after installation of systems, engineers from the company

Table 10.4   Contingency table of scanner data.

Classification of sample of scanners by bearing supplier and bearing state		Bearing state		
		Sound	Worn	
Bearing supplier	A	32	3	35
	B	48	17	65
		80	20	100

service the scanners and check the main bearings for wear. From a random sample of 100 service reports the scanners were categorized according to the bearing supplier and the state of the main bearing, yielding Table 10.4.

The null hypothesis of interest here is that there is no association between bearing supplier and bearing state, with the alternative hypothesis being that there is an association, i.e. that the bearing state is *contingent* on the supplier. Tables such as Table 10.4 are known as *contingency tables*. If the null hypothesis is true then the events that a bearing was supplied by A and is sound are independent. From the table we may then carry out the calculations displayed in Box 10.1. The reader is invited to calculate, using the method outlined in Box 10.1, the expected counts for the other three cells – all four expected counts are shown in Table 10.5. In the case of a 2×2 contingency table as we have here, once one expected frequency has been obtained the others may be obtained by ensuring that the correct marginal totals are obtained. Thus a 2×2 contingency table has one degree of freedom – an $r \times c$ contingency table, with r rows and c columns, has degrees of freedom given by $(r - 1) \times (c - 1)$.

The test statistic for the hypothesis test is:

$$\chi^2 = \sum \frac{(O_i - E_i)^2}{E_i} = \frac{(32 - 28)^2}{28} + \frac{(3 - 7)^2}{7} + \frac{(48 - 52)^2}{52} + \frac{(17 - 13)^2}{13}$$
$$= 0.571 + 2.286 + 0.308 + 1.231$$
$$= 4.396, \text{ with 1 degree of freedom.}$$

An estimate for the probability that a bearing was supplied by A is $P(A) = 35/100$. An estimate for the probability that a bearing was sound is $P(S) = 80/100$. If the null hypothesis is true then an estimate of $P(A \cap S)$ is

$$P(A \cap S) = P(A) \times P(S) = \frac{35}{100} \times \frac{80}{100}.$$

Hence, the expected frequency of scanners supplied by A with sound bearings is

$$100 \times \frac{35}{100} \times \frac{80}{100} = \frac{35 \times 80}{100} = \frac{\text{Row total} \times \text{Column total}}{\text{Sample size}}.$$

Box 10.1   Calculation of an expected frequency.

Table 10.5 Contingency table showing observed (O_i) and expected (E_i) counts.

Classification of sample of scanners by bearing supplier and bearing state		Bearing state		
		Sound	Worn	
Bearing supplier	A	32 (28)	3 (7)	35
	B	48 (52)	17 (13)	65
		80	20	100

Use of **Graph > Probability Distribution Plot... > View Probability** yields the P-value. Under the **Distribution** tab, **Distribution** Chi-Square is specified from the drop-down menu with **Degrees of freedom:**1. Under the **Shaded Area** tab **X Value** is checked, **X Value:** 4.396 specified and **Right Tail** selected. On clicking **OK** the display reveals the P-value to be 0.036, correct to three decimal places. Since this is less than 0.05 the null hypothesis that there is no association between bearing supplier and bearing state would be rejected in favour of the alternative there is an association, at the 5% level of significance. This test is often referred to as *Pearson's chi-square test* in honour of the statistician Karl Pearson.

This result may be obtained directly from Minitab using **Stat > Tables > Chi-Square Test (Two-Way Table in Worksheet)...** having set up the four counts from the contingency table in, say, columns C1 and C2 of a worksheet. Having specified **Columns containing the table:** C1 C2, clicking **OK** yields the Session window output in Panel 10.13 which confirms all the calculations performed earlier.

From the contingency table we have an estimate of the probability of wear for a bearing from supplier A, $p_A = 3/35 = 0.0857$, and from supplier B, $p_B = 17/65 = 0.2615$. Thus we could say that we estimate that a bearing from supplier A is approximately one third as likely to

```
Chi-Square Test: C1, C2

Expected counts are printed below observed counts
Chi-Square contributions are printed below expected counts

          C1      C2   Total
   1      32       3      35
        28.00    7.00
        0.571   2.286

   2      48      17      65
        52.00   13.00
        0.308   1.231

Total     80      20     100

Chi-Sq = 4.396, DF = 1, P-Value = 0.036
```

Panel 10.13 Output for chi-square test for association.

show signs of wear as a bearing from supplier B. One measure of association for a two-way contingency table is the *relative risk* defined as the ratio of the two probabilities: here the relative risk (A to B) is $p_A/p_B = 0.0857/0.2615 = 0.338$; similarly, the relative risk (B to A) is the reciprocal of this, 3.051. For an event with probability p the odds are defined as $p/(1 - p)$. Hence, we have that the odds for wear in a bearing from supplier A are

$$\frac{p_A}{1 - p_A} = \frac{0.0857}{0.9143} = 0.094,$$

and in a bearing from supplier B are

$$\frac{p_B}{1 - p_B} = \frac{0.2615}{0.7385} = 0.354.$$

Another measure of association for a two-way contingency table is the *odds ratio*, defined as the ratio of the two odds. Thus the odds ratio (A to B) is

$$\frac{\text{Odds}_A}{\text{Odds}_B} = \frac{0.0937}{0.3541} = 0.26,$$

and similarly the odds ratio (B to A) is the reciprocal of that, 3.78.

Intuitively, the *relative risk* is an easier measure of association to interpret than the odds ratio. The two are related by the equation

$$\text{Relative risk(A to B)} = \text{Odds ratio(A to B)} \times \frac{1 - p_A}{1 - p_B}.$$

Odds ratios occur in binary logistic regression analysis and will be referred to again in the next section.

The raw data extracted from the scanner service reports is provided in the worksheet Bearings.MTW. The first column contains scanner service reference numbers, the second the supplier of the main bearing, and the third indicates whether or not signs of wear were found. In order to cross-tabulate the data to form the contingency table and to carry out the chi-square test of association one may use **Stat > Tables > Cross Tabulation and Chi-Square...** as indicated in Figure 10.20. Supplier was allocated to rows and Wear to columns, counts checked under **Display, Fisher's exact test for 2 × 2 tables** checked under **Other Stats...** and Chi-Square analysis checked under **Chi-Square....**

The reader is invited to check that the resultant Session window output includes the key output in Panel 10.13 together with a P-value of 0.027 for an alternative chi-square test and a P-value of 0.039 for Fisher's exact test for association. The latter test is named in honour of the statistician Ronald Fisher. All three tests lead to the same conclusion, i.e. that the null hypothesis of no association would be rejected at the 5% level of significance.

The Pearson chi-square test involves a degree of approximation. Minitab displays the number of cells that have expected counts less than 5. Minitab Help (**Stat > Tables > Cross Tabulation and Chi-Square... > Help > see also > Methods and formulas > Chi-Square test**) states: 'Some statisticians hesitate to use the χ^2 test if more than 20% of the cells have expected counts below five, especially if the p-value is small and these cells give

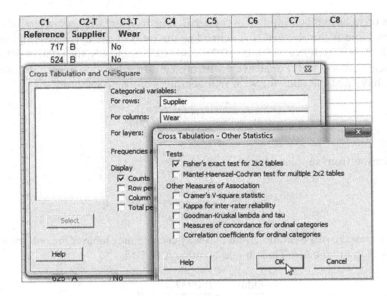

Figure 10.20 Dialog for cross-tabulation and chi-square test of association.

a large contribution to the total χ^2 value'. If the expected counts for some are small it may be possible to carry out an analysis by combining or omitting row and/or column categories. On the other hand, as the name implies, the Fisher test is exact and may be used with confidence when expected counts in a 2×2 contingency table are low.

Duncan (1959) gives the data in Table 10.6 on causes of rejection of metal casting by week of manufacture. The data, from Hunt (1948), reproduced by permission of the American Society for Quality, are available in Castings.MTW. The reader is invited to use **Stat** > **Tables** > **Chi-Square Test (Two-Way Table in Worksheet)...** to carry out the test and verify that the Session window output includes the statement 'Chi-Sq = 45.597, DF = 12, P-Value = 0.000'. No expected counts less than 5 were obtained so the analysis provides very strong evidence (P-value < 0.001) that there is a difference in the distribution of rejects from week to week.

Table 10.6 Causes of rejection of metal castings.

Cause of rejection	Week 1	Week 2	Week 3
Sand	97	120	82
Misrun	8	15	4
Shift	18	12	0
Drop	8	13	12
Corebreak	23	21	38
Broken	21	17	25
Other	5	15	19

Table 10.7   Space Shuttle launch O-ring data.

Temperature x	Launches	O-ring failure-free launches	$P(\text{Success}) = p$
53	1	0	0.0
57	1	0	0.0
58	1	0	0.0
63	1	0	0.0
66	1	1	1.0
67	3	3	1.0
68	1	1	1.0
69	1	1	1.0
70	4	2	0.5
72	1	1	1.0
73	1	1	1.0
75	2	1	0.5
76	2	2	1.0
78	1	1	1.0
79	1	1	1.0
81	1	1	1.0

10.4.2 Binary logistic regression

In binary logistic regression the response variable has two categories. In order to introduce the topic, we consider the data in Table 10.7 and Space_Shuttle.MTW from Dalal *et al.* (1989) and reprinted with permission from the *Journal of the American Statistical Association*, all rights reserved. It gives temperature (°F) at the time of launch for 23 Space Shuttle missions and a classification of each mission as either a success or failure in terms of O-ring performance. The classification was based on examination of O-rings that became available for inspection following launches. The temperature column gives the air temperature at the time of launch. For example, for the two launches that took place with air temperature 75 °F, there was one where O-ring failure is known to have occurred. Thus at air temperature 75 °F the estimated probability of a launch free of O-ring failures is $P(\text{Success}) = p = 1/2 = 0.5$.

The question of interest is whether or not there is any relationship between p and x. If the answer is in the affirmative then the question arises of whether or not the relationship can be modelled. As a first step a display was created in the form of a plot of p against x with a Lowess smoother applied. Locally weighted scatterplot smoothing, LOWESS, may be used to explore the relationship between two variables without fitting a specific model and may be added to a scatterplot by right-clicking the graph and selecting **Add** > **Lowess. . . .** The author used **Degree of smoothing:** 0.45 and **Number of steps:** 2 in creating Figure 10.21.

The fit from the smoother has the approximate appearance of an S-shaped or sigmoid curve. The logistic function is one mathematical function that may be used to model sigmoid curves and is described in Box 10.2.

The logit of p is a linear function of x so we could attempt to estimate the parameters α and β of the logistic model by fitting a straight line to a scatterplot of the logit of p versus x. We encounter an immediate problem with this approach for the current data set in that the logit

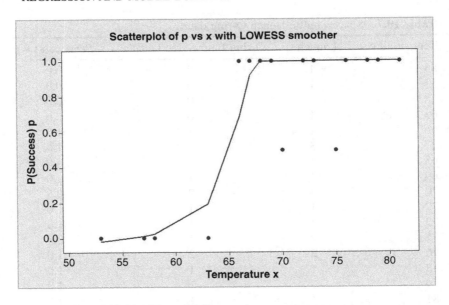

Figure 10.21   Plot of P(Success), p versus temperature, x.

function of p is undefined for $p = 0$ and for $p = 1$. In practice, the method of maximum likelihood is used to estimate the parameters α and β of the logistic model. It may be implemented in Minitab using **Stat > Regression > Binary Logistic Regression....** The dialog required is shown in Figure 10.22. **Response in event/trial format** was checked as the response data appear in the worksheet as two columns – one containing the number of O-ring failure-free launches and the other containing the number of launches. Thus **Number of events:** O-ring failure free launches and **Number of trials:** Launches were specified. Under **Model:** Temperature x was entered. No entries were required under **Factor:** in this case. Under **Storage...**, **Event probability** was checked under **Characteristics of Estimated Equation** in order to obtain fitted values of the probability of an O-ring failure-free launch for subsequent plotting.

A logistic function linking p to x may be written in the form

$$p = \frac{\exp(\alpha + \beta x)}{1 + \exp(\alpha + \beta x)},$$

where α and β are the function parameters. The formula may be conveniently rearranged in the form

$$\ln \frac{p}{1 - p} = \alpha + \beta x.$$

The function $\ln(p/(1 - p))$ is the logit of p.

Box 10.2   The logistic function.

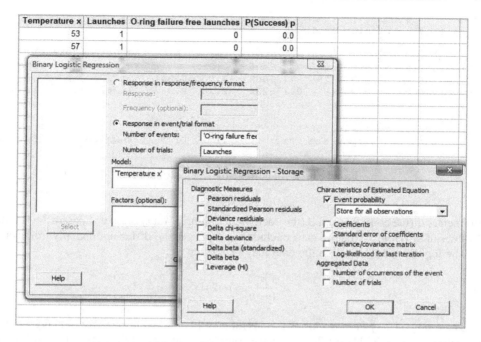

Temperature x	Launches	O-ring failure free launches	P(Success) p
53	1	0	0.0
57	1	0	0.0

Figure 10.22   Dialog for binary logistic regression.

Summary information given in the first section of the Session window output is shown in Panel 10.14. In simple linear regression, the expected value of the response, Y, is given by the linear function $\alpha + \beta x$, i.e. the expected value of the response is related directly to the linear function. In the form of logistic regression applied here the logit function of the expected probability of success is a linear function $\alpha + \beta x$. Thus in this case the logit function provides the link between the expected response of interest, probability of success, and the linear function of x. Hence the statement 'Link Function: Logit' in Panel 10.14. (Other link functions may be used but will not be considered in this book.) The Response Information summary displayed in Panel 10.14 may be readily checked from Table 10.7.

The next portion of the output is displayed in Panel 10.15. The method of maximum likelihood has provided the estimates of -15.04 and 0.2322 respectively for the logistic

---

Binary Logistic Regression: O-ring failu, Launches versus Temperature

```
Link Function: Logit


Response Information

Variable                        Value        Count
O-ring failure free launches    Event           16
                                Non-event        7
Launches                        Total           23
```

Panel 10.14   Summary information from logistic regression analysis.

```
Logistic Regression Table


                                                 Odds      95% CI
Predictor          Coef   SE Coef     Z      P  Ratio  Lower  Upper
Constant       -15.0429   7.37862  -2.04  0.041
Temperature x   0.232163  0.108236  2.14  0.032  1.26   1.02   1.56



Log-Likelihood = -10.158
Test that all slopes are zero: G = 7.952, DF = 1, P-Value = 0.005
```

Panel 10.15   The logistic regression table.

parameters α (the constant) and β (the coefficient of temperature x or slope parameter). With P-values of 0.041 and 0.032 we can conclude that both α and β differ significantly from zero.

The logistic model fitted to the data may be written as

$$\ln \frac{p}{1-p} = \alpha + \beta x.$$

The ratio $p/(1 - p)$ is the odds. The odds ratio in this context is the value of $\exp(\beta)$ and is estimated by $\exp(0.2322) = 1.26$, with 95% confidence interval (1.02, 1.56). The odds ratio may be interpreted as the factor by which the odds increase for a unit increase in x, i.e. for an increase of 1 °F in temperature,

The results from goodness-of-fit tests of the logistic model are provided in the next portion of Session window output in Panel 10.16, the null hypothesis being that a logistic model provides a good fit to the data. Since none of the three P-values are small the null hypothesis cannot be rejected, so we conclude that the logistic model provides a potential model of the situation.

The remaining Session window output comprises a table of observed and expected frequencies and a table of measures of association. The observed and fitted probabilities, computed using **Storage...**, from the model are plotted against temperature in Figure 10.23. The default name for the fitted event probability column, EPR01, was changed to pfit. Initially a scatterplot of p versus x was created. On right-clicking the plot **Add > Calculated Line...** was selected in order to plot the fitted logistic regression curve. Under **Coordinates** the selections **Y column:** pfit and **X column:** Temperature x were made.

Some statisticians quote the value of x that corresponds to probability 0.5. The dotted lines added to the plot indicate that, for the fitted model, the value of x for which p is 0.5 is

```
Goodness-of-Fit Tests

Method            Chi-Square  DF     P
Pearson            11.1303    14  0.676
Deviance           11.9974    14  0.607
Hosmer-Lemeshow    11.0395     8  0.199
```

Panel 10.16   Goodness-of-fit tests for the logistic regression model.

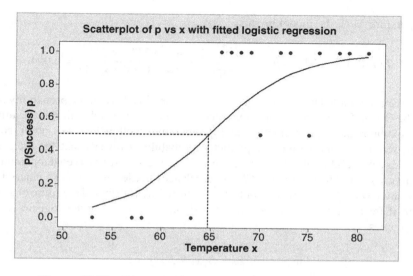

Figure 10.23   Binary logistic regression model fitted to data.

approximately 65. Thus the model predicts that there is probability of 0.5 that a Space Shuttle launch will be O-ring failure-free at temperature 65 °F. The calculation is as follows. We have $p = 0.5$, so

$$\ln \frac{0.5}{1 - 0.5} = \ln 1 = 0 = \alpha + \beta x$$

so

$$x = -\frac{\alpha}{\beta}.$$

Thus in this case the required temperature is estimated by

$$-\frac{-15.04}{0.2322} = 64.8.$$

The modelling carried out has established a relationship between the probability of a Space Shuttle launch being O-ring failure-free and air temperature. One may think of the random variable Y that takes the value 0 for a Space Shuttle launch with O-ring failures and the value 1 for a launch that was O-ring failure-free. The fitted logistic model enables prediction of the value of Y for a given value of x in the sense that prediction may be made of the conditional probability

$$P(Y = 1 | \text{Temperature} = x) = p = \frac{\exp(\alpha + \beta x)}{1 + \exp(\alpha + \beta x)}.$$

For temperature 31 °F the model predicts

$$P(Y = 1|x = 31) = p = \frac{\exp(-15.04 + 0.2322 \times 31)}{1 + \exp(-15.04 + 0.2322 \times 31)} = 0.004.$$

Thus for a Space Shuttle launch taking place at 31 °F the model predicts a probability of 0.004 for it to be O-ring failure-free. (Such a prediction should be viewed with some caution as it involves extrapolation, i.e. prediction at a temperature well below any observed value. However, scrutiny of the column of predicted probabilities reveals that the fitted logistic model yields a probability of 0.06 for a launch at 53 °F, the lowest observed temperature, to be O-ring failure free.) At the time of the *Challenger* Space Shuttle launch on 28 January 1986 the air temperature was 31 °F. The catastrophe that occurred was attributed to failure of an O-ring.

When there are two or more predictor variables x_1, x_2, ..., the logistic regression model becomes

$$\ln \frac{p}{1-p} = \beta_0 + \beta_1 x_1 + \beta_2 x_2 + \ldots.$$

As an example, we will consider data discussed by Everitt (1994, pp. 54–66) on patients suffering from acute myeloblastic leukaemia. The patients were given a course of treatment and the binary response recorded was whether or not a patient responded to treatment. The data are from a paper by Hart (1977), available in Leukaemia.MTW and reproduced by permission of John Wiley and Sons Inc., New York.

Six variables were recorded for each patient prior to treatment: age at diagnosis, x_1; smear differential percentage of blasts, x_2; percentage of absolute marrow leukaemia infiltrate, x_3; percentage labelling index of the bone marrow leukaemia cells, x_4; absolute blasts, x_5; and highest temperature prior to treatment, x_6. The responses recorded were: the response to treatment, y_1, where $0 =$ Fails to respond to treatment, $1 =$ Responds to treatment; survival time from diagnosis (months), y_2; and status, y_3, where $0 =$ Dead, $1 =$ Still alive. (No reference will be made to y_2 and y_3 in this book.)

To begin the analysis **Stat > Regression > Binary Logistic Regression...** was used to fit a model involving all six of the candidate predictor variables recorded prior to treatment. Here **Response: y1** was specified on checking **Response in response/frequency format** as the raw data are in raw binary format, and **Model: x1 x2 x3 x4 x5 x6** was entered.

Key Session window output is displayed in Panel 10.17. The Response Information indicates that 24 of the 51 patients in the sample responded to the treatment. The test of the null hypothesis that all slopes are zero,

$$H_0 : \beta_1 = \beta_2 = \beta_3 = \beta_4 = \beta_5 = \beta_6 = 0,$$

yields *P*-value 0.000, to three decimal places. Thus the null hypothesis would be rejected in favour of the alternative hypothesis that at least one of the βs is nonzero. None of the goodness-of-fit tests has a small *P*-value, so there is no evidence of lack of fit. However, scrutiny of the *P*-values in the Logistic Regression Table suggests that x_2, x_3 and x_5 do not contribute to the model. Thus a second model, involving only x_1, x_4 and x_6, was fitted. In addition, under **Graphs...** the two diagnostic plots of **Delta chi-square versus probability** and **Delta chi-square versus leverage** were selected. An explanation of the background to these plots

```
Binary Logistic Regression: y1 versus x1, x2, x3, x4, x5, x6

Link Function: Logit


Response Information

Variable  Value  Count
y1          1      24    (Event)
            0      27
          Total    51


Logistic Regression Table

                                                   Odds      95% CI
Predictor        Coef     SE Coef      Z      P    Ratio  Lower  Upper
Constant      98.5236    40.8532    2.41   0.016
x1           -0.0602925   0.0272871  -2.21  0.027   0.94   0.89   0.99
x2           -0.0047997   0.0410745  -0.12  0.907   1.00   0.92   1.08
x3            0.0362132   0.0393374   0.92  0.357   1.04   0.96   1.12
x4            0.398447    0.132773    3.00  0.003   1.49   1.15   1.93
x5            0.0134344   0.0578199   0.23  0.816   1.01   0.90   1.14
x6           -0.102229    0.0418088  -2.45  0.014   0.90   0.83   0.98


Log-Likelihood = -20.030
Test that all slopes are zero: G = 30.465, DF = 6, P-Value = 0.000


Goodness-of-Fit Tests

Method            Chi-Square  DF      P
Pearson             40.3923   44   0.627
Deviance            40.0599   44   0.641
Hosmer-Lemeshow      5.3804    8   0.716
```

Panel 10.17   Initial logistic regression analysis of leukaemia data.

is beyond the scope of this book, but values of delta chi-square of around 4 or higher flag the possible presence of unusual observations in the data, observations that merit further scrutiny.

The Logistic Regression Table for the second model is shown in Panel 10.18. The estimated value of the coefficient β_1 of age (x_1) in the model is -0.0585, and the corresponding odds ratio is 0.94. Consider two patients similar in all respects except that one is 1 year older than the other. The model predicts that the odds in favour of the older patient responding to the treatment are 94% of the odds in favour of the younger patient responding to the treatment. The estimated value of the coefficient β_4 of percentage labelling index of the bone marrow leukaemia cells (x_4) in the model is 0.3849, and the corresponding odds ratio is 1.47. Consider two patients similar in all respects except that one has percentage labelling index of the bone marrow leukaemia cells that is 1% higher than the other. The model predicts that the odds in favour of the patient with the higher percentage responding to the treatment are 147% of the odds in favour of the patient with the lower percentage responding to the treatment.

Binary Logistic Regression: y1 versus x1, x4, x6

```
Link Function: Logit


Response Information

Variable  Value  Count
y1        1      24     (Event)
          0      27
          Total  51


Logistic Regression Table

                                                 Odds      95% CI
Predictor         Coef     SE Coef     Z      P  Ratio  Lower  Upper
Constant       87.3880     35.4581  2.46  0.014
x1           -0.0585016  0.0255764 -2.29  0.022  0.94   0.90   0.99
x4            0.384926   0.121518   3.17  0.002  1.47   1.16   1.86
x6           -0.0889732  0.0360684 -2.47  0.014  0.91   0.85   0.98


Log-Likelihood = -21.633
Test that all slopes are zero: G = 27.259, DF = 3, P-Value = 0.000


Goodness-of-Fit Tests

Method           Chi-Square  DF     P
Pearson            41.1411   47  0.713
Deviance           43.2654   47  0.628
Hosmer-Lemeshow     8.6631    8  0.372
```

Panel 10.18   Logistic regression table for second model.

The plot of delta chi-square versus leverage is displayed in Figure 10.24. Brushing (introduced in Chapter 3) was used to identify the row numbers, which match the patient numbers, for those observations yielding values of delta chi-square in excess of 4. It certainly suggests that patient number 47 is unusual in some respect, and possibly also patients 48 and 50.

Consider again the data in Table 10.5 in the previous section presented in the form displayed in the worksheet in Figure 10.25. Also shown is the dialog required to analyse the data via binary logistic regression. Note that Supplier is specified both for **Model:** and **Factors (optional):**. The relevant Session window output is shown in Panel 10.19. Note that it gives the odds ratio (B to A) as 3.78, as obtained in the previous section, and that in addition it gives the 95% confidence interval (1.02, 13.95) for this ratio. An odds ratio of 1 corresponds to bearing wear being independent of supplier. The fact that the confidence interval does not include 1 provides evidence, at the 5% level of significance, that bearing wear is dependent on supplier. Since all values in the interval exceed 1 the conclusion from the data is that the odds of wear being present in the bearings are significantly greater for supplier B.

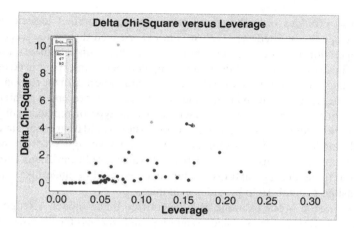

Figure 10.24   Delta chi-square versus leverage plot.

C1-T	C2-T	C3	C4	C5	C6	C7	C8	C9	C10	C11
Supplier	Wear	Count								
A	No	32								
A	Yes	3								
B	No	48								
B	Yes	17								

Binary Logistic Regression

• Response in response/frequency format
Response: Wear
Frequency (optional): Count

○ Response in event/trial format
Number of events:
Number of trials:

Model:
Supplier

Factors (optional):
Supplier

Select

Options...   Prediction...
Graphs...   Results...   Storage...
Help
OK   Cancel

Figure 10.25   2 × 2 contingency table set up for analysis using binary logistic regression.

```
Logistic Regression Table

                                          Odds      95% CI
Predictor     Coef     SE Coef      Z       P  Ratio  Lower  Upper
Constant   -2.36712   0.603807  -3.92   0.000
Supplier
  B          1.32914   0.666513   1.99   0.046   3.78   1.02  13.95
```

Panel 10.19   Logistic regression table for 2 × 2 contingency table.

Minitab also provides ordinal logistic regression and nominal logistic regression. If patients in a hospital classified the level of pain experienced during recovery from a knee replacement operation as being mild, moderate or severe then ordinal logistic regression would be appropriate in an investigation of the relationship between pain experienced and predictors such as gender and age. This is the case as there is a natural ordering in the response – moderate represents a greater level of pain than mild and, in turn, severe represents a greater level of pain than moderate. In addition to predictors, factors such as type of replacement joint may be introduced into the modelling. Nominal logistic regression would be appropriate where there is no natural ordering in possible values for the response. An investigation of preferred methods of payment for supermarket customers, with possible values credit card, bank card, cheque and cash, in relation to gender, age and disposable income could be undertaken using nominal logistic regression. Minitab Help provides examples.

Logistic regression is one example of a generalized linear model. Montgomery (2005a, p. 563) provides further details and examples and makes the following comments. 'Generalized linear models have found extensive application in biomedical and pharmaceutical research and development. As more software packages incorporate this capability, it will find widespread application in the general industrial research and development environment.'

10.5   Exercises and follow-up activities

1. Table 10.8 gives the mass (tonnes) and fuel usage (kilometres/litre) for a sample of 10 vehicles.

    (a) Obtain the least squares regression of y on x and store the residuals and fitted values.

    (b) Give an interpretation of the slope of the regression line.

    (c) Give an interpretation of the value of R^2.

    (d) Check the values of some fitted values and residuals.

    (e) Perform diagnostic checks of the model using residual plots.

2. In an investigation of the shelf life of a cereal, data were obtained on shelf time, x (days), and percentage moisture content, y. The data are given in Table 10.9.

    (a) Investigate the regression of y on x.

Table 10.8   Mass and fuel usage for a sample of vehicles.

Mass x	1.27	1.68	1.63	1.45	1.86	1.18	1.63	1.54	1.72	1.22
Fuel Usage y	6.1	5.3	5.5	5.8	5.2	6.3	5.6	5.5	5.5	6.0

Table 10.9   Moisture content and shelf time.

x	0	3	6	8	10	13	16	20	24	27	30	34	37	41
y	2.4	2.6	2.7	2.8	3.0	3.0	3.1	2.7	3.4	3.6	3.7	3.9	4.0	4.5

(b) Obtain a 95% prediction interval for the moisture content of an individual box of the cereal that has been stored on a shelf for 30 days.

Consumer testing indicated that the cereal is unacceptably soggy when the moisture content is greater than 4.0. On the basis of the prediction interval you have calculated, would you recommend to a supermarket manager that he continue to stipulate a shelf life of 30 days for this brand of cereal? Does analysis of the data following removal of the unusual data value alter your conclusion?

3. In the manufacture of glass bottles gobs of molten glass are poured from the furnace into the moulds in which the containers are formed by the action of compressed air. The gob temperature is of major importance and the manufacturer was interested in being able to predict gob temperature from the temperature obtained from a sensor located in the fore-hearth of the furnace. An experiment was conducted from which a series of 35 values of gob temperature (y) and fore-hearth temperature (x) were obtained. The data are available in the worksheet Gob.MTW and reproduced by permission of Ardagh Glass Ltd., Barnsley.

   (a) Carry out a regression analysis of y on x with diagnostic checks of the residuals.

   (b) Display the data, with fitted line and 95% prediction interval curves.

   (c) Obtain a 99% prediction interval for gob temperature when fore-hearth temperature is 1165 and state, with justification, whether or not you would advise the furnace supervisor to pour glass for a container requiring a target gob temperature of 1150 when fore-hearth temperature is displayed as 1165 on the furnace control panel.

   (d) State the value of fore-hearth temperature that would yield the narrowest prediction interval for gob temperature.

4. Table 10.10 gives systolic blood pressure (SBP, y) measured in millimetres of mercury and the age (x) measured in years for a sample of women considered to be in good health.

   Obtain the regression line of y on x and show that the slope differs significantly from zero.

   (a) Plot residuals and comment on the adequacy of the model.

   (b) Obtain 95% prediction intervals for the systolic blood pressure of women aged 50 years and 20 years, respectively.

5. During the development of a biocide for use in hospitals, a microbiologist carried out an experiment using a trial solution. Twelve beakers were set up, each containing 50 ml of a nutrient broth with microbial spores in suspension. At the start of the experiment a

Table 10.10 SBP and age for a sample of women.

Age (x)	71	53	40	42	47	49	74	43	50	49	67	37
SBP (y)	158	139	126	131	128	141	167	116	128	126	148	121

Table 10.11   Data from biocide experiment.

Time (minutes)	Spore count (organisms per ml)
10	16 518
20	15 177
30	10 084
40	8 533
50	8 823
60	6 690
70	6 042
80	4 890
90	4 065
100	3 166
110	3 012
120	1 852

fixed amount of the biocide was added simultaneously to each beaker. At 10-minute intervals thereafter, a beaker was selected at random and the spore count determined by a method that meant that the beaker could no longer be part of the experiment. The data in Table 10.11 were obtained and are provided in Biocide.MTW.

(a) Fit a least squares regression line of spore count on time.

(b) Explain how the plot of residuals confirms the poor fit of the model.
In order to improve the model she considered the equation

$$N = N_0 e^{-kt},$$

where N represents the spore count and t the time.

(c) Express $\log_e N$ as a linear function of t and obtain the least squares regression of $\log_e N$ on t.

(d) Perform diagnostic checks of the revised model.

(e) Use the revised model to estimate the decimal reduction time D, i.e. the time predicted by the model for the biocide to reduce the number of spores to one-tenth of its initial value.

6. The Minitab worksheet EXH_Regr.MTW, available in the Minitab Sample Data folder supplied with the software, contains in columns C9 and C10 values of energy consumption (y) and setting (x) for a type of machine. Use **Stat > Regression > Fitted Line Plot. . .** to fit a quadratic model to the data. Is the model improved by making a logarithmic transformation of the response? Create a column containing the values of x^2, fit the models using **Stat > Regression > Regression. . .** and carry out diagnostic checks of the residuals.

7. The Minitab worksheet Trees.MTW, available in the Minitab Sample Data folder supplied with the software, gives diameter, height and volume for a sample of 31 black

cherry trees from Allegheny National Forest in the USA. Diameter (feet) was measured 4.5 feet above ground. Height was measured in feet and volume in cubic feet.

Suppose that the forest management team wish to develop a model to enable prediction of timber production, i.e. a model that may be used to predict volume, which is difficult to measure, from diameter and height measurements. Explore models involving the three variables and also the three variables after logarithmic transformation of all three. Which model would you recommend?

8. The Minitab worksheet EXH_Regr.MTW (see Exercise 6 above) contains in columns C3 to C8 data, from a project on solar thermal energy, on total heat flux from homes. It is desirable to ascertain whether total heat flux can be predicted from insolation, the position of the focal points in the east, south, and north directions and from time.

   (i) Use best subsets regression to select a potential regression model to predict heat flux from a subset of the five candidate predictor variables.

   (ii) Fit your selected model and carry out diagnostic checks of the residuals.

9. The worksheet BHH1.MTW contains data from Phase 1 of an illustration of response surface methodology given by Box *et al.* (1978, pp. 514–525). (All the data in this exercise are reproduced by permission of John Wiley & Sons, Inc., New York.) The design used was a single replication of a 2^2 factorial with the addition of three centre points. The response was yield (g) from a laboratory-scale chemical production process and the factors were time (low 70 minutes, high 80 minutes) and temperature (low 127.5 °C, high 132.5 °C). The centre point (75, 130) represented current operating factor levels prior to the experimentation.

   (i) Verify, from scrutiny of the ANOVA from **Stat > DOE > Factorial > Analyze Factorial Design...**, that there is no evidence of curvature and no evidence of interaction.

   (ii) In order to create a contour plot, repeat the analysis having, under **Terms...**, removed the interaction term from the model and unchecked **Include center points in the model.** You should obtain the Session window output displayed in Panel 10.20 – providing evidence that both time and temperature appear to influence yield and indicating that the equation of the fitted model, in terms of the **coded** units, is $y = 62.0 + 2.35x_1 + 4.50x_2$ (with the constant and the coefficients quoted to three significant figures).

```
Estimated Effects and Coefficients for y (coded units)

Term            Effect    Coef  SE Coef       T      P
Constant                62.014   0.6011  103.16  0.000
Time            4.700    2.350   0.7952    2.96  0.042
Temperature     9.000    4.500   0.7952    5.66  0.005
```

Panel 10.20   Part of Session window output.

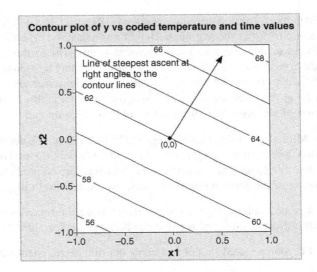

Figure 10.26   Line of steepest ascent.

(iii) Use **Stat > DOE > Factorial > Contour/Surface Plots...** to create a contour plot. Choose **Contour Plot** and under **Setup...** check **Display plots using: Coded units**. Under **Contours...** insert **Values: 56 58 60 62 64 66 68** as **Contour Levels**. Under **Data Display** select **Contour lines** and **Symbols at design points**. This will create the plot in Figure 10.26, but without the annotation. The point $(0, 0)$ in coded units represents the centre point of the design.

   The line of steepest ascent is the line through $(0,0)$ in the above plot at right angles to the contours in the direction in which yield increases. Some mathematics is presented in Box 10.3 that enables FLCs on the line of steepest ascent to be calculated.

(iv) Set up, in a Minitab worksheet, columns named x1, x2, Time and Temperature and assign values 0, 1, 2, 3, 4 and 5 to x1. Use **Calc > Calculator** to compute x2 from the equation of the line of steepest ascent and to decode the values in x1 and x2 to give the time and temperature values. The values you should obtain are shown in Table 10.12.

   The first row corresponds to the centre point (a useful cross-check on the computations) and the average yield for the three runs at this FLC is readily verified to be 62.3. The process team decided to carry out a further run with the FLC (80, 135). This gave the encouraging yield of 73.3, so the next run was carried out with the combination (100, 154). The result was a disappointing yield of 58.2. Then use of (90, 144) gave yield 86.8, representing substantial improvement.

(v) Add the yield column to your worksheet and display the Phase 2 data. (The data are available in BHH2.MTW)

   As a result of the knowledge gained from Phase 2, the Phase 3 experimentation involved a central composite design centred on time 90 minutes and temperature 145 °C. The data are available in worksheet BHH3.MTW.

In terms of the coded units, the contour lines are specified by the model equation

$$y = 62.0 + 2.35x_1 + 4.50x_2.$$

In particular, the contour for yield 62.0 is specified by

$$62.0 = 62.0 + 2.35x_1 + 4.50x_2.$$

This equation may be written as

$$x_2 = -\frac{2.35}{4.50}x_1.$$

To obtain the equation of the line of steepest ascent it is necessary to change the sign on the right-hand side and to invert the ratio which forms the coefficient of x_1. Thus the equation of the line of steepest ascent is given by

$$x_2 = +\frac{4.50}{2.35}x_1 = 1.91x_1.$$

The coding equations are

$$x_1 = \frac{\text{Time} - 75}{5} \quad \text{and} \quad x_2 = \frac{\text{Temperature} - 130}{2.5}.$$

They may be rearranged to give

$$\text{Time} = 5x_1 + 75 \quad \text{and} \quad \text{Temperature} = 2.5x_2 + 130.$$

Box 10.3   Determination of the line of steepest ascent.

Table 10.12   Data from Phase 2 exploration along line of steepest ascent.

		Phase 2 Experimentation on the line of steepest ascent		
x_1	x_2	Time	Temperature	Yield
0	0.00	75	130.000	62.3
1	1.91	80	134.775	**73.3**
2	3.82	85	139.550	
3	5.73	90	144.325	**86.8**
4	7.64	95	149.100	
5	9.55	100	153.875	**58.2**

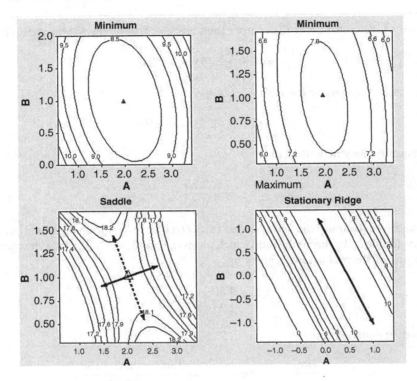

Figure 10.27   Examples of response surfaces.

(vi) Analyse the data and display the response surface as a contour plot, with time on the horizontal axis and temperature on the vertical axis.

Box *et al*. conclude the illustration by stating that the surface 'represents an oblique rising ridge with yields increasing from the lower right to the top left corner of the diagram, that is, yield increases as we progressively *increase* temperature and simultaneously *reduce* reaction time'. Were it desirable to increase yield still further then 'subsequent experimentation would have followed and further explored this rising ridge'.

10. The contour plots in Figure 10.27 display four types of response surface encountered in modelling process behaviour. In the cases of the saddle and the stationary ridge, what recommendations would you make, given that the goal was to maximize the responses?

11. Open the Pulse.MTW worksheet provided in the Minitab Sample Data folder. Use **Help > ? Help** and the **Search** tab to perform a search for Pulse. Double clicking on PULSE.MTW then reveals a description of and key to the data set.

  (i) Form a 2×2 contingency table that classifies the sample by smoking habit and gender and test for association using both the chi-square and Fisher's exact test.

  (ii) Form a contingency table that classifies the sample by smoking habit and level of activity and test for association. Note that a 2×4 contingency table results for which the chi-square analysis involves three expected frequencies less than 5, so

Table 10.13   Contingency table classifying defects in wafers.

Shift	Type of defect			
	A	B	C	D
1	15	21	45	13
2	26	31	34	5
3	33	17	49	20

Minitab issues a warning in the Session window output that the chi-square analysis is probably invalid.

Code activity levels 0 and 1 as low, 2 as moderate and 3 as high, and form and analyse the corresponding contingency table.

12. NIST/SEMATECH (2005) gives an industrial example in which 309 wafer defects were recorded and the defects were classified according to type (A,B,C, or D) and according to the production shift at time of manufacture of the wafer (1, 2, or 3). The data are given in Table 10.13.

   Emphasis has been placed on the display of data in this book, so prior to formal analysis the reader is invited to set up the contingency table as shown in Figure 10.28 and use **Graph > Bar Chart**. . . with **Bars represent:** Values from a table, clicking on **Cluster** under **Two-way table**, clicking on **OK**, then selecting **Graph variables:** A-D, **Row labels:** Shift and accepting defaults otherwise to display the data. Repeat with the second option for Table Arrangement. Do you consider one display to be more informative than the other? Carry out a formal test for association.

13. Cox (1970, p. 86) provides data on the duration of heating, T, for ingots and on the numbers not ready for rolling, R, summarized in Table 10.14 and reproduced by permission of Taylor & Francis Group. Model the relationship between readiness for rolling and duration of heating. On a scatterplot of the observed probability of readiness for rolling versus duration of heating superimpose the curve giving the fitted probability of readiness for rolling as a function of duration of heating.

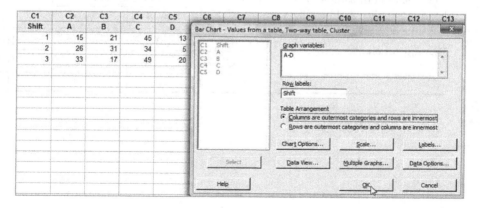

Figure 10.28   Contingency classifying defects in wafers.

Table 10.14   Data on ingot readiness for rolling.

Duration of heating T	No. not ready for rolling R	No. tested N
7	0	55
14	2	157
27	7	159
51	3	16

14. The worksheet ChemProc.MTW gives data on the reaction time and catalyst used in a chemical production process and an indication of whether or not the final product was of prime quality. Use binary logistic regression to fit a model involving both reaction time and catalyst. Catalyst has to be specified as a factor as well as being included in the model. Store the event probabilities (for all observations) and display them plotted against reaction time using **Graph > Scatterplot...**, selecting **With Connect and Groups** and specifying Catalyst under **Categorical variables for grouping**. State recommendations that can be made with regard to running the process.

15. Obtain the odds ratio with a 95% confidence interval for each of the scenarios Tables 10.15 and 10.16. Comment.

Table 10.15   Scenario 1.

Classification of sample of scanners by bearing supplier and bearing state		Bearing state		
		Sound	Worn	
Bearing supplier	P	5	5	10
	Q	7	3	10
		12	8	20

Table 10.16   Scenario 2.

Classification of sample of scanners by bearing supplier and bearing state		Bearing state		
		Sound	Worn	
Bearing supplier	P	50	50	100
	Q	70	30	100
		120	80	200

11

Learning more and further Minitab

Practice does not end here. (Ellis Ott, in Ott et al., 2000, p. 537)

Overview

To conclude the book, this final chapter highlights ways in which one can learn more about Minitab (the author invariably learns something new almost every time he uses the software!) and how one can obtain help with Minitab. Reference will be made to the creation of macros and their application. Brief reference will also be made to further features of Minitab not covered in this book and to Minitab Quality Companion.

11.1   Learning more about Minitab and obtaining help

11.1.1   Meet Minitab

The *Meet Minitab 16* guide (Minitab Inc., 2010), is available as a PDF at http://www.minitab.com/uploadedFiles/Shared_Resources/Documents/MeetMinitab/EN16_MeetMinitab.pdf. It deals with the following topics:

- Getting started
- Graphing data
- Analyzing data
- Assessing quality
- Designing an experiment

Six Sigma Quality Improvement with Minitab, Second Edition. G. Robin Henderson.
© 2011 John Wiley & Sons, Ltd. Published 2011 by John Wiley & Sons, Ltd.

- Using Session commands

- Generating a report

- Preparing a worksheet

- Customizing Minitab

- Getting help

- Reference.

Its comprehensive index makes it a very useful resource for Minitab users, particularly beginners.

11.1.2   Help

Some reference has already been made to the Help facility in Chapter 2. It may be accessed in a variety of ways:

- Click Help in any dialog box.

- Click the Help icon ⟦?⟧ on the tool bar.

- Select **Help** > **Help** from the menus.

General information on navigation for Help is available via **Help** > **Help** > **How to Use Help**. The reader is urged to explore! The author believes that each Minitab user will develop his/her own approach to the utilization of the Help facilities.

As an example of how one might seek assistance with a particular topic, consider the need to analyse quarterly sales data for a company, i.e. a time series of data values. The data are stored in Halcro.MTW as a single column of quarterly sales figures for the years 2006 to 2010 inclusive. On clicking the Help icon a window appears with two panes as displayed in Figure 11.1.

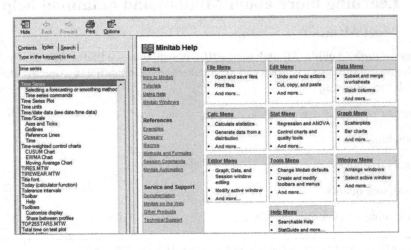

Figure 11.1   Help window.

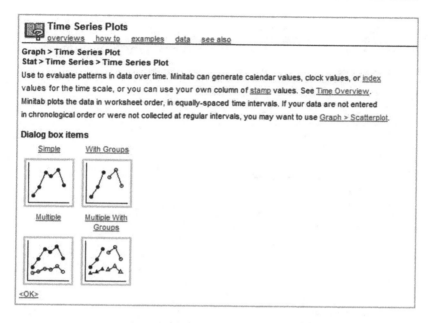

Figure 11.2   Help on time series plots.

Note the Hide icon in the top left-hand corner on the toolbar. On clicking on it the navigation pane on the left is hidden and the icon becomes a Show icon. Clicking on the **Index** tab and entering 'time series' reveals headings Time Series and Time Series Plot. Bearing in mind that display of data is wise whenever possible, on selecting Time Series Plot and clicking the **Display** button, a Topics Found window is displayed with links to **Time Series Plot (Character Graph menu)** and **Time Series Plot (Graph and Stat menus)**. This indicates that a time series plot may be created via both the **Graph** menu and the **Stat** menu or via the archaic **Character Graph** menu. In order to access the Help information on the topic one can highlight Time Series Plot (Graph and Stat menus) and double-click on it, or click the **Display** button below the navigation pane. Alternatively, one can click on a menu link in the window on the right and then click the required topic in the list that appears. Thus the sequence **Stat Menu** > **Time Series** > **Time Series Plot** will also reveal the information displayed in Figure 11.2.

Underneath the display heading there are links to:

- <u>overviews</u> of the topic;

- instructions on <u>how to</u> complete the dialog;

- <u>examples</u> of use of the command;

- an explanation of how <u>data</u> should be structured;

- <u>see also</u> links to other relevant topics and information.

Exploration of all of these links can prove very informative. Clearly the information on completion of dialogs and the examples provided can be very helpful, especially when one is using a particular command for the first time. The links to other relevant topics in the

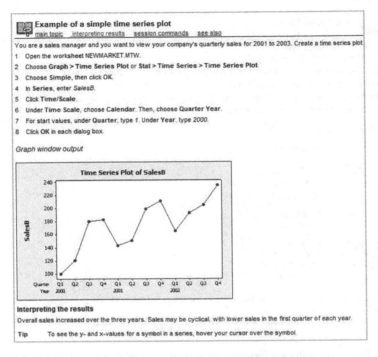

Figure 11.3   Help example of a simple time series plot.

majority of instances may be used to access methods, formulae and references to textbooks and journal articles.

With the display in Figure 11.2 on screen, clicking on <u>examples</u> and selecting **Simple** from the menu that appears yields the display in Figure 11.3.

The reader is invited to minimize the Help screen and to use the information displayed in Figure 11.3 to create the required time series plot. In step 1 you will need to to open Halcro. MTW, in step 4 the name (Sales £000) of the column containing the data must be entered, and in step 7 you will need to enter 1 and 2006. As a further exercise the reader is also invited to delete the plot, to maximize the Help screen, to use the **Back** arrow to return to the previous Help screen and to find out how to add a Lowess smoother to the plot by clicking on the **Data View...** button in the main dialog for the simple time series plot and then using the **Smoother** tab. The final plot is shown in Figure 11.4.

11.1.3   StatGuide

Having used a command, one may use StatGuide™ to access guidance on interpretation on the output, whether it be Session window output or graphical output. For example, having created the plot displayed in Figure 11.4, clicking the StatGuide icon 🔲 (immediately to the right of the Help icon) yields the output displayed in Figure 11.5 in the right-hand window. The left-hand window provides **Contents**, **Index** and **Search** tabs for Help.

There are three panes in the window. The first pane provides general information on the command, a **Topics** button giving access to a list of other associated topics and arrow keys that

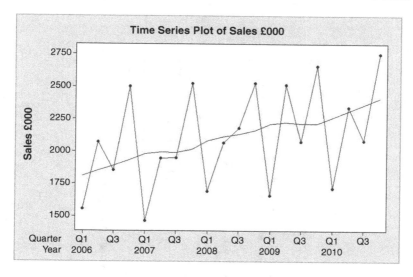

Figure 11.4   Time series plot of sales with smoother.

enable access to these. The second gives example output from the command (in this case a plot) and a **Data** button giving access to a description of the data used. The third pane provides interpretation of the output together with a **More** button that provides access to further information that is shown in Figure 11.6. The reader should note that clicking on any word or

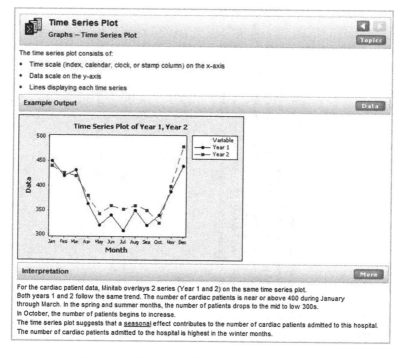

Figure 11.5   StatGuide for time series plot.

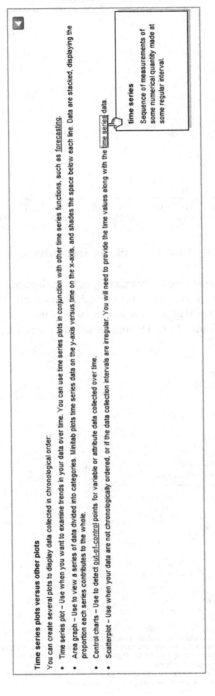

Figure 11.6 StatGuide information on time series plots versus other plots.

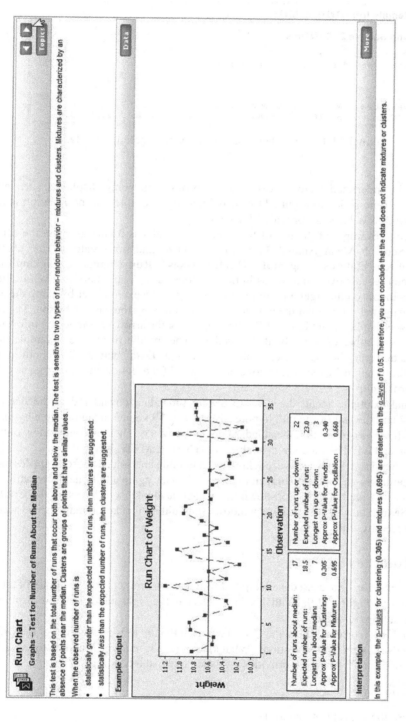

Run Chart

Graphs — Test for Number of Runs About the Median

This test is based on the total number of runs that occur both above and below the median. The test is sensitive to two types of non-random behavior – mixtures and clusters. Mixtures are characterized by an absence of points near the median. Clusters are groups of points that have similar values.

When the observed number of runs is

- statistically *greater* than the expected number of runs, then mixtures are suggested.
- statistically *less* than the expected number of runs, then clusters are suggested.

Example Output

Run Chart of Weight

Number of runs about median:	17	Number of runs up or down:	22
Expected number of runs:	18.5	Expected number of runs:	23.0
Longest run about median:	7	Longest run up or down:	3
Approx P-Value for Clustering:	0.305	Approx P-Value for Trends:	0.340
Approx P-Value for Mixtures:	0.695	Approx P-Value for Oscillation:	0.660

Interpretation

In this example, the P-values for clustering (0.305) and mixtures (0.695) are greater than the α-level of 0.05. Therefore, you can conclude that the data does not indicate mixtures or clusters.

Figure 11.7   StatGuide information on test for number of runs about the median.

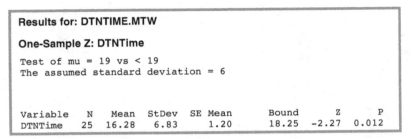

Results for: DTNTIME.MTW

One-Sample Z: DTNTime

```
Test of mu = 19 vs < 19
The assumed standard deviation = 6
```

Variable	N	Mean	StDev	SE Mean	Bound	Z	P
DTNTime	25	16.28	6.83	1.20	18.25	-2.27	0.012

Panel 11.1   Hypothesis test for door to needle time data.

phrase that is underlined leads to explanatory information being displayed such as the definition of a time series in Figure 11.6. It is interesting to note that no mention is made of run charts! Arrow keys are provided for navigation.

As a second example of the use of StatGuide, the reader is invited to open the Minitab worksheet Scottish_Mean_Annual_Temperatures.MTW and to create a run chart of mean annual temperatures using **Stat > Quality Tools > Run Chart . . . .** Clicking on the StatGuide icon reveals two windows as in the previous case. A summary is given in the top pane of the window on the right and further topics may be accessed either by using the arrow keys or by using the drop-down menu that appears on clicking the **Topics** button. The right-hand window is displayed in Figure 11.7 following use of the arrow keys to locate information on the test for number of runs about the median. The middle pane gives example output and information on the associated data set via the **Data** button. The bottom pane provides interpretation of the output together with a **More** button that provides access to further information.

As a final example, recall the door to needle time data considered at the beginning of Chapter 7 and the test of the null hypothesis H_0: $\mu = 19$ versus the alternative hypothesis H_1: $\mu < 19$. It was assumed that the standard deviation was 6. The Session window output is shown in Panel 11.1. The data are available in DTNTime.MTW.

With the Session window active, clicking on StatGuide and selecting from **Topics** the one entitled Hypotheses yields the display in Figure 11.8. Once again additional information may be explored using the buttons, drop-down menus and arrow keys or by using the **Contents**, **Index** and **Search** tabs in the left-hand window. StatGuide information may also be accessed using **Help > StatGuide.**

11.1.4   Tutorials

A set of tutorials may be accessed via **Help > Tutorials.** The contents of the library of tutorials are displayed in Figure 11.9. With P Chart selected the right-hand window displays the information in Figure 11.10.

Note the **Uses** (current), **Data** and **How To** tabs in the top left-hand corner of the window. **Data** provides answers to the questions 'What kind of data is required?' and 'What should my worksheet look like?'. **How To** provides guidance under the headings:

- Scenario

- Choose the appropriate analysis

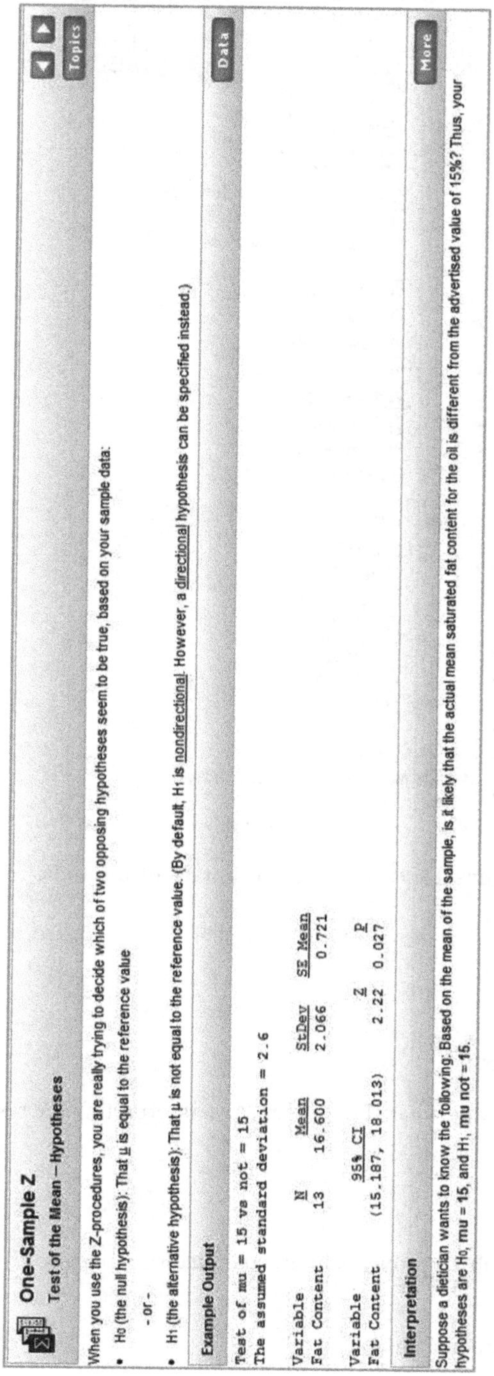

One-Sample Z

Test of the Mean – Hypotheses

When you use the Z-procedures, you are really trying to decide which of two opposing hypotheses seem to be true, based on your sample data:

- H0 (the null hypothesis): That μ is equal to the reference value

- or -

- H1 (the alternative hypothesis): That μ is not equal to the reference value. (By default, H1 is nondirectional. However, a directional hypothesis can be specified instead.)

Example Output

```
Test of mu = 15 vs not = 15
The assumed standard deviation = 2.6

Variable        N      Mean    StDev    SE Mean
Fat Content    13    16.600    2.066      0.721

Variable          95% CI              Z       P
Fat Content    (15.187, 18.013)    2.22   0.027
```

Interpretation

Suppose a dietician wants to know the following: Based on the mean of the sample, is it likely that the actual mean saturated fat content for the oil is different from the advertised value of 15%? Thus, your hypotheses are H0, mu = 15, and H1, mu not = 15.

Figure 11.8 StatGuide information on hypotheses for a one-sample z-test.

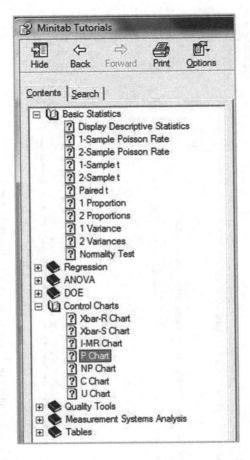

Figure 11.9   Contents of library of tutorials.

- Enter your data

- Specify options

- Interpreting the output

with the final one stating 'Now what? For guidance on interpreting the results of this analysis, see StatGuide'!

11.1.5 Assistant

This new facility in Minitab 16 provides interactive decision trees for users to follow on the topics following:

- Measurement system analysis

- Capability analysis

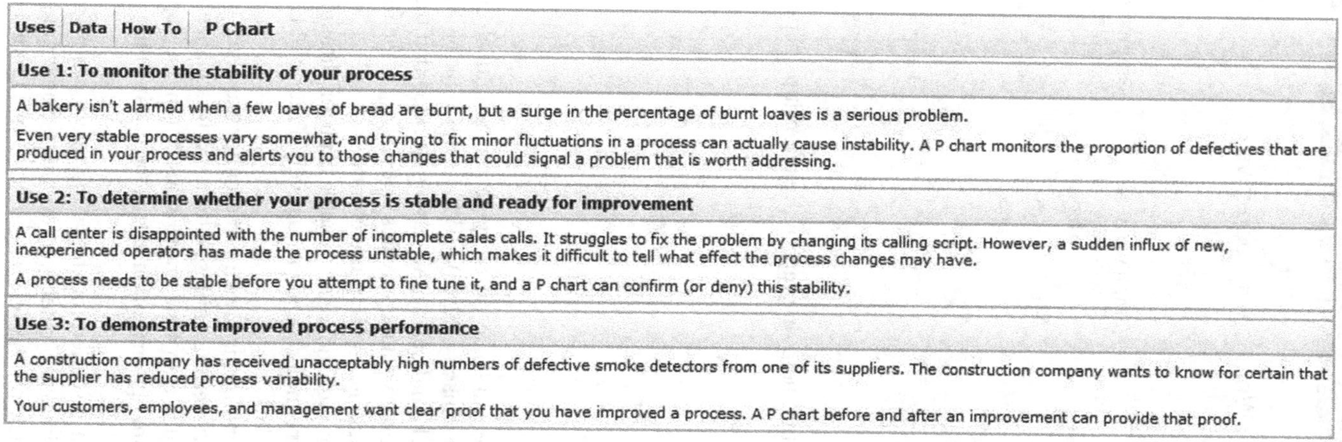

Uses	Data	How To	P Chart

Use 1: To monitor the stability of your process

A bakery isn't alarmed when a few loaves of bread are burnt, but a surge in the percentage of burnt loaves is a serious problem.

Even very stable processes vary somewhat, and trying to fix minor fluctuations in a process can actually cause instability. A P chart monitors the proportion of defectives that are produced in your process and alerts you to those changes that could signal a problem that is worth addressing.

Use 2: To determine whether your process is stable and ready for improvement

A call center is disappointed with the number of incomplete sales calls. It struggles to fix the problem by changing its calling script. However, a sudden influx of new, inexperienced operators has made the process unstable, which makes it difficult to tell what effect the process changes may have.

A process needs to be stable before you attempt to fine tune it, and a P chart can confirm (or deny) this stability.

Use 3: To demonstrate improved process performance

A construction company has received unacceptably high numbers of defective smoke detectors from one of its suppliers. The construction company wants to know for certain that the supplier has reduced process variability.

Your customers, employees, and management want clear proof that you have improved a process. A P chart before and after an improvement can provide that proof.

Figure 11.10   Part of tutorial on the P chart.

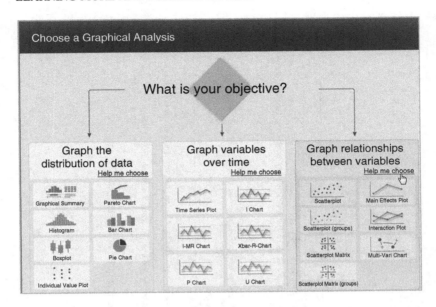

Figure 11.11   Assistant help on graphing relationships between variables.

- Graphical analysis

- Hypothesis tests

- Regression analysis

- Control charts.

Suppose that you would like guidance on carrying out graphical analysis of the data in Vehicles.MTW giving engine capacity (litres) and fuel consumption (litres per 100 km) for a sample of 20 cars. Figure 11.11 indicates how, having decided that the objective is to graph relationships between variables, Assistant provides further help in the form of the decision tree shown in Figure 11.12 by clicking on **Help me choose**.

One could select a path through the tree to reach a decision through visual scrutiny of the tree or one could **Click to start** and proceed interactively. Figure 11.13 shows the outcome of clicking on the first decision box.

In Figure 11.13 the mouse pointer is shown as the author was about to click on **Next** in order to select the **Association between variables** option and to advance to the next step. Thus **Association between variables** with **One continuous X** (engine capacity) and **No groups** leads the user to the **Scatterplot** option (bottom left of the tree in Figure 11.12) with the display in Figure 11.14.

The user is informed that a scatterplot shows the relationship between a continuous X (engine capacity) and a continuous Y (fuel consumption). Some guidelines are provided: to view the detail one must click on the relevant + sign. In Figure 11.14 the guidelines on examining the regression equation and the model fit are displayed. In order to create the scatterplot the user simply uses the **Click to Create Graph** icon. In the dialog box that appears

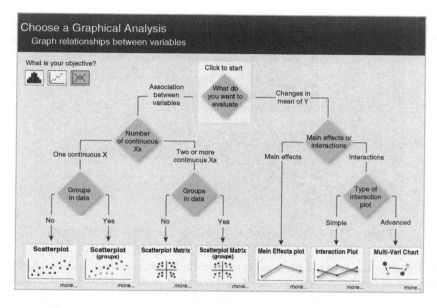

Figure 11.12   Decision tree for choosing a graphical analysis.

consumption is specified as the *Y* column and capacity as the *X* column. On clicking **OK** and moving the mouse pointer to the fitted line on the plot, the display in Figure 11.15 is obtained. (Note that the dialog for the creation of the scatterplot via the Assistant is simpler than that obtained via **Graph > Scatterplot ...**).

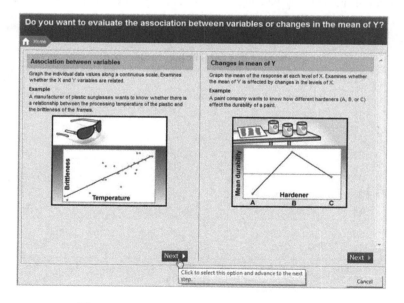

Figure 11.13   Assistance with first decision.

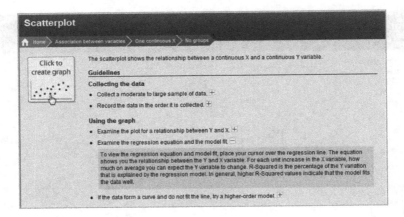

Figure 11.14   Scatterplot outcome via interactive decision tree.

The line provides a reasonably good fit to the data. The R^2 value indicates that 87.5% of the variation in fuel consumption may be attributed to engine capacity. Substituting capacity 3 into the equation yields consumption of 10.6. Thus the model fitted to the data predicts that, on average, cars with 3-litre engine capacity would consume 10.6 litres per 100 km.

Once again the reader is urged to explore use of the Assistant. As Minitab develops further in the future perhaps books such as this will become redundant!

11.1.6 Glossary, methods and formulas

Help > Glossary provides access to a statistical glossary. As indicated in Figure 11.16, a statistical term of interest may be sought either via the **Index** tab or using the **Search** tab.

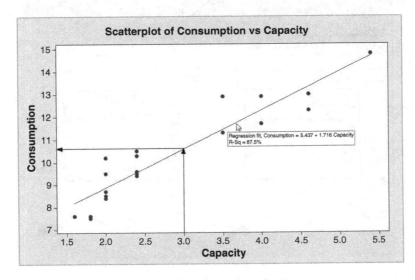

Figure 11.15   Scatterplot with fitted regression line.

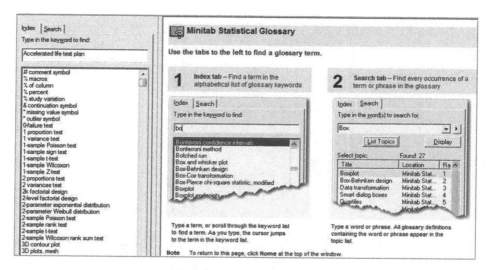

Figure 11.16   Statistical glossary.

For example, if one wished to obtain information on R^2 (R-squared) in regression then clicking on the **Search** tab, entering 'R' in the search window and clicking on **List Topics** reveals a list including R-squared. Double-clicking on this entry in the list reveals the explanation shown in Figure 11.17.

Should the details of how R-squared is calculated be of interest, then **Help > Methods and Formulas > Regression > General Regression > R^2** yields the required information shown in Figure 11.18.

Further information on methods and formulas in regression may be obtained by clicking on relevant items in the **General Regression** list shown.

11.1.7   Minitab on the web and Knowledgebase/FAQ

Either **Help > Minitab on the Web** or **Help > Knowledgebase/FAQ** accesses the Minitab website and a wealth of information. The home page displayed in Figure 11.19 was accessed using **Help > Minitab on the Web** and shows seven tabs: **Products, Training, Support, Academic, Store, Company** and **Theater**. The **Theater** tab provides access to a number of videos: Getting Started, Analysing Data, Assessing Capability, Data Manipulation, Graphing Data, Creating Control Charts, Using the Assistant Menu and Navigating Projects. The reader is urged to watch these in order to reinforce his or her understanding of the software and the statistical methods.

The Knowledgebase and answers to frequently asked questions may be accessed either using **Help > Knowledgebase/FAQ** or using the link at the bottom left of the home page. In Chapter 5 reference was made to the Deming funnel experiments. The displays in Figure 5.47 were created using a Minitab macro. Suppose that you wish to obtain a copy of the macro so that you can demonstrate the plots (in glorious colour!) to colleagues. Performing a search of the knowledgebase using the phrase Deming funnel led to the display in Figure 11.20.

The single result from the search performed has identification number ID 1259, and clicking on this leads to a display of the solution to the query – see Panel 11.2. The reader is

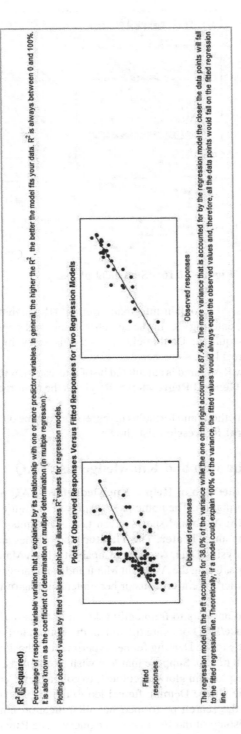

Figure 11.17 Glossary explanation of R-squared.

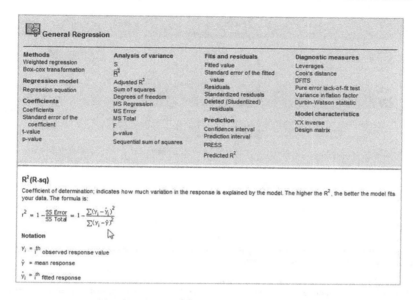

Figure 11.18   Formula for calculation of R-squared.

invited to access the link and to follow steps 2 and 3 so that the macro may be used as an example in the next section.

The Minitab Help menu also provides **Keyboard Map...**, **Check for Updates**, **Licensing...**, **Contact Us** and **About Minitab** which will not be referred to further. It should be noted that under the **Support** tab **Online Support** enables one to create an account with Minitab in order to seek help from the company's experts.

Figure 11.19   Minitab on the web.

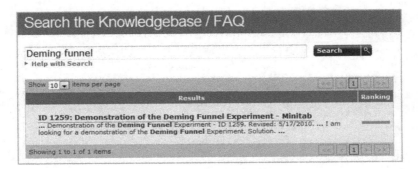

Figure 11.20 Knowledgebase search result.

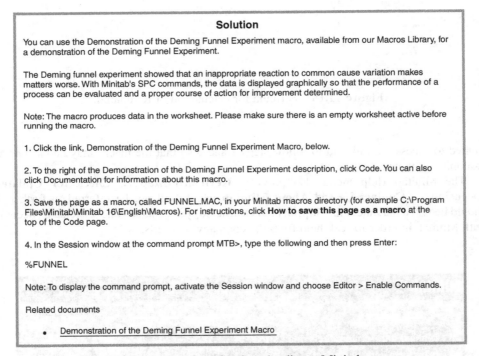

Panel 11.2 Instructions for downloading a Minitab macro.

11.2   Macros

11.2.1   Minitab Session commands

Each day the accident and emergency department at a hospital records the number of patients admitted (Patients) and the number of patients who wait for more than 2 hours to see a doctor (Number). The file December2010.xls includes data for December 2010. Each month an individuals (X) chart of the percentage of patients who waited for more than 2 hours to see a doctor is required, with tests for evidence of special causes 1, 2, 5 and 6 implemented.

```
MTB > WOpen "C:\Users\Public\Documents\Second Edition\Chapter
11_Dev\Ch11_Data\December2010.xls";
SUBC>    FType;
SUBC>      Excel.
Retrieving worksheet from file:
'C:\Users\Public\Documents\Second
Edition\Chapter 11_Dev\Ch11_Data\December2010.xls'
Worksheet was saved on 09/01/2011


Results for: Sheet1

MTB > Name C4 'Percentage'
MTB > Let 'Percentage' = 'Number' / 'Patients' * 100
MTB > IChart 'Percentage';
SUBC>    Test 1 2 5 6.

I Chart of Percentage

MTB >
```

Panel 11.3   Session window output for control chart of percentages.

The reader is invited to set up a new Minitab project and, with the Session window active, to use **Editor** to select **Enable Commands**. The MTB > prompt will appear in the Session window. Open the file December2010.xls. Use **Calc > Calculator,** with **Store result in variable:** Percentage and **Expression:** 'Number'/'Patients' * 100 specified using **Select** and the keyboard in the dialog box, to compute the required column of percentages, and **Stat > Control Charts > Variables Charts for Individuals > Individuals...** to create the required chart.

The Session window will appear as shown in Panel 11.3. Note the command prompt MTB > and the subcommand prompt SUBC >. As an alternative to using the menus and dialog boxes in Minitab one may perform tasks by typing session commands and subcommands directly into the Session window (once commands have been enabled as described above.) For example, typing the command Runchart c3 1 into the Session window, and pressing the enter key, creates a run chart of the data. Alternatively, Runc 'Percentage' 1 may be used; many commands may be abbreviated to the first four letters and variable names in quotes may be used instead of column numbers.

Information on Session commands is available via **Help > Help > Session Commands**. (The first version of Minitab that the author used was Release 7. It was entirely driven using session commands, and on occasion he finds it quicker to carry out work with a series of typed commands rather than using the menus. However, many whose first introduction to Minitab is via Release 16 may never use session commands.)

Suppose that we had omitted to activate the third test for evidence of special causes on the control chart. Use of **Window > Project Manager** and clicking on the **History** folder or clicking on the History icon [≥] reveals the Session commands generated as the tasks described earlier were performed. The pair of commands requiring editing may be highlighted as shown in Figure 11.21 – a command and any associated subcommands must all be highlighted for editing. When highlighting more than one line, keep the control key pressed after highlighting the first and click on the remaining lines requiring highlighting.

On selecting **Edit > Command Line Editor** the final line may be edited to read Test 1 2 3 5 6. Clicking on **Submit Commands** causes the individuals chart to be re-created with the additional test implemented. The History folder now contains two additional lines. The two

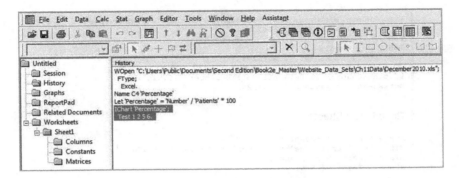

Figure 11.21   Session commands in History folder.

lines highlighted in Figure 11.21 may be deleted together with the first three lines that are the commands for reading the data from one particular file. A right click reveals a menu. Selection of **Save History As . . .** enables the commands for the creation of the control chart of the monthly data to be saved as an exec file.

The reader is invited to select **Save History As . . .** and to choose **Save as type:** Exec Files (*.MTB) and to save the commands in a file named Monthly.MTB. On opening the file November 2010.xls in Minitab the commands stored in the exec file may be implemented using **File > Other Files > Run an Exec . . . .** Accept one, the default number of times to execute the set of stored commands, select the file Monthly.MTB and click **Open**. The commands will be executed, the percentages calculated and displayed on the control chart. Of course the set of tasks being carried out by the exec in this case is small, but the example serves to illustrate the potential value of an exec for carrying out repetitive routine tasks.

11.2.2 Global and local Minitab macros

An exec is one form of Minitab macro. In addition, one can create global and local macros. An exec file may be converted into a global macro by using Notepad and adding three lines of text to the set of commands stored in the exec file as indicated in bold in Panel 11.4.

The first and last additional lines *must* be gmacro and endmacro, and the second additional line is the macro name. The file may then be stored as IChart.MAC, with the extension. MAC indicating that it contains a Minitab macro. Instructions for this may be found by carrying out the Knowledgebase search displayed in Figure 11.20 and following the guidelines thereby

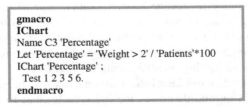

```
gmacro
IChart
Name C3 'Percentage'
Let 'Percentage' = 'Weight > 2' / 'Patients'*100
IChart 'Percentage' ;
  Test 1 2 3 5 6.
endmacro
```

Panel 11.4   Global macro for creation of control chart of percentages.

obtained that are displayed in Panel 11.2. If it is stored in the Macros folder in the Minitab 16 folder then it may be invoked simply by typing the command %IChart in the Session window. Alternatively, the file may be stored as a text file named Ichart.txt and invoked by typing the command %Ichart.txt in the Session window.) Readers are encouraged to try this out for themselves. If the file is not stored in the Macros folder then the path must be specified. Readers are also encouraged now to try out the Deming funnel experiment macro they were invited to download in Section 11.1.5. The file FunnelBU.txt is supplied for the benefit of any reader unable to make the download.

In creating a local macro one is essentially writing a program that can involve variables and control statements such as **if** and **goto**, in addition to Minitab commands. Global macros operate directly on the current worksheet. Local macros are more complicated and hence more difficult to write. The term 'local macro' is used because this type of macro creates a 'local worksheet' during the running of the macro that is erased from the computer memory once the macro has been executed. Arguments and variables may be passed between the global and local worksheets during the running of a local macro. Details of the creation of local macros will not be covered in this book. Readers wishing to learn more about local macros will find comprehensive information available via **Help > ? Help** and a single click on **Macros** under the **References** heading in the **Minitab Help** window.

Some macros require information from the user. For example, the Minitab macro for the creation of the Bland–Altman plot in Figure 9.13 requires the user to indicate the two columns that contain the measurements. Thus if the macro is stored as blandalt. MAC in the Minitab Macros folder it would be invoked using %blandalt c2 c3 to create the plot in Figure 9.13 as the second and third columns of the supplied worksheet Outside_Diameters.MTW contain the measurements. The reader is invited to re-create the plot.

An example of a macro that requires numerical information to be provided via the keyboard is the macro blankch for the creation of a blank Xbar or individuals (X) control chart. As an exercise the interested reader could use the macro to create a blank chart for the Xbar chart component of Figure 5.13. The rod cutting process had historical mean and standard deviation of 60.00 and 0.02 respectively, and the samples were of size 4.

11.3 Further features of Minitab

In reliability and survival statistics interest often centre on the time to failure of a system, T. The survival or cumulative failure function is an alternative name for the cumulative distribution function of the random variable T. **Stat > Reliability/Survival** provides facilities for the display and analysis of data for time to failure of systems. Breyfogle (2003) devotes four chapters to reliability topics.

In a paper on multivariate methods and Six Sigma, Yang (2004, p. 94) made the following comments.

> Multivariate statistical methods are powerful methods that are playing important roles in many fields. Data mining and chemometrics are among the successful applications of multivariate statistical methods in business and industry. However, multivariate statistical methods are seldom applied in Six Sigma practice as well as quality assurance practice in general ... multivariate statistical methods can play important roles in Six Sigma (DMAIC) practice.

The multivariate procedures provided by Minitab are available under **Stat** > **Multivariate**. Johnson and Wichern (2001) give a comprehensive account of applications of multivariate statistical methods.

Comprehensive facilities for the analysis and modelling of time series are also provided under **Stat** > **Time Series**. Of particular importance for quality improvement are models that take into account time series in which autocorrelation is present. Montgomery (2009, pp. 527–545) gives further details and references.

11.4   Quality Companion

Six Sigma product and process improvement invariably involves work being done by project teams. Minitab Quality Companion® Release 3.0 is a project management aid that includes a variety of tools that are widely used in Six Sigma quality improvement. Minitab claims that Quality Companion simplifies quality initiatives by providing the soft tools need to develop, organize and execute projects. With tools, files and data in a single project file, project teams are able to focus less on managing details and more on reaching goals. Using Quality Companion it is possible to:

- create a plan of action with a Roadmap™;

- centralize and share project data;

- standardize project deliverables with built-in forms;

- customize data and forms to meet the needs of the organization;

- map processes and assign data to any process step;

- establish the flow of value through the organization;

- organize ideas and challenges with brainstorming tools;

- create presentations to keep stakeholders informed;

- guide project teams with templates and coaches.

The default DMAIC project template provided by the software contains a Roadmap based on the Six Sigma stages: define, measure, analyse, improve and control. However, customization of the template is possible for use with quality improvement strategies other than Six Sigma. An overview is provided in *Getting Started: Quality Companion® 3* (Minitab Inc., 2009), available as a PDF at http://www.minitab.com/uploadedFiles/Shared_Resources/ Documents/Brochures/companion3getstarted.pdf. The software provides coaches to help the user select an appropriate statistical method to analyse process data captured, for example, before and after a process change. The analysis capture tools enable output from Minitab analyses to be added to a Quality Companion project and Minitab project files may readily be added as related documents. Figure 11.22 displays a Quality Companion project window with Minitab output. Note the Project Manager window on the left with components of DMAIC.

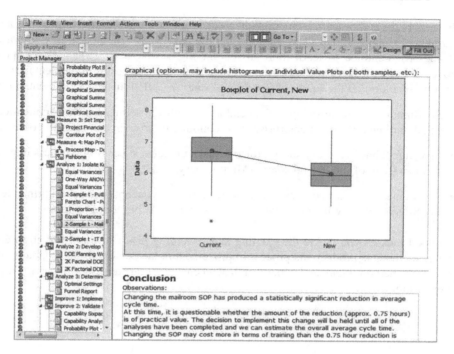

Figure 11.22   A Quality Companion project incorporating Minitab output.

11.5   Postscript

Hoerl and Snee (2005) claim that 'Six Sigma is the best way we have found to actually deploy statistical thinking broadly'. The author makes no apology for presenting for a second time in this book, in Box 11.1, the summary of the measure, analyse, improve and control phases of DMAIC given by Hoerl (1998). Statistical thinking has a major role to play in all four phases and Minitab is a powerful and versatile tool for the application of statistical tools that may be used to foster understanding of variation and to reduce it.

Measure – Based on customer input, select the appropriate responses (the *Y*s) to be improved and ensure that they are quantifiable and can be accurately measured.

Analyze – Analyze the preliminary data to document current performance or baseline process capability. Begin identifying root causes of defects (the *X*s or independent variables) and their impact.

Improve – Determine how to intervene in the process to significantly reduce the defect levels. Several rounds of improvement may be required.

Control – Once the desired improvements have been made, put some type of system into place to ensure the improvements are sustained, even though additional Six Sigma resources may no longer be focused on the problem.

Box 11.1   Description of key phases in applying Six Sigma methodology.

Invariably the author learns something new each time he uses Minitab – it provides a vast and growing resource. It is his earnest hope that the reader has benefited from study of this book and will continue to learn more about Six Sigma and other strategies for quality improvement and the statistical tools that are vital for their successful implementation.

In a presentation on the future of Six Sigma, Antony (2010) made the following points:

- Six Sigma is on the upswing in financial and healthcare services and its application in small and medium-sized enterprises will continue to grow in the years to come.

- 'Six Sigma has been very successful – perhaps the most successful business improvement strategy of the last 50 years' (Montgomery, 2005b).

- Six Sigma will continue to evolve. However, the key concepts and sound principles of Six Sigma will stay for many years whatever the 'next big thing' will be.

So good luck as you tackle the exciting and rewarding task of using Minitab to apply statistical methods to product and process improvement in the context of Six Sigma or other improvement methodologies and remember that, whenever you are dealing with data, 'Display is an obligation'!

Appendix 1

Sigma quality levels and number of nonconformities per million opportunities

The shaded cell indicates that a sigma quality level of 3.17 corresponds to 47 461 nonconformities per million opportunities.

SQL	0.00	0.01	0.02	0.03	0.04	0.05	0.06	0.07	0.08	0.09
1.50	501350	497317	493286	489256	485229	481205	477185	473167	469154	465144
1.60	461140	457140	453146	449157	445175	441199	437229	433267	429313	425366
1.70	421427	417498	413577	409665	405763	401871	397989	394118	390258	386409
1.80	382572	378747	374934	371134	367347	363573	359813	356067	352335	348618
1.90	344915	341228	337556	333900	330259	326636	323028	319438	315864	312308
2.00	308770	305250	301748	298264	294799	291352	287925	284517	281129	277761
2.10	274412	271084	267776	264489	261223	257977	254753	251550	248369	245209
2.20	242071	238956	235862	232791	229742	226716	223712	220732	217774	214839
2.30	211928	209040	206175	203333	200516	197722	194951	192205	189482	186783
2.40	184108	181457	178831	176228	173650	171095	168565	166059	163578	161120
2.50	158687	156278	153893	151533	149197	146885	144597	142333	140094	137878
2.60	135687	133519	131376	129256	127161	125089	123040	121016	119015	117037
2.70	115083	113152	111245	109360	107499	105660	103845	102052	100282	98534
2.80	96809	95106	93425	91767	90130	88515	86921	85350	83799	82270
2.90	80762	79275	77809	76363	74938	73534	72149	70785	69440	68116
3.00	66811	65525	64259	63011	61783	60573	59382	58210	57056	55920
3.10	54801	53701	52618	51553	50504	49473	48459	47461	46480	45515
3.20	44567	43634	42717	41816	40931	40060	39205	38364	37539	36728

Six Sigma Quality Improvement with Minitab, Second Edition. G. Robin Henderson.
© 2011 John Wiley & Sons, Ltd. Published 2011 by John Wiley & Sons, Ltd.

SQL	0.00	0.01	0.02	0.03	0.04	0.05	0.06	0.07	0.08	0.09
3.30	35931	35149	34380	33626	32885	32157	31443	30742	30055	29379
3.40	28717	28067	27429	26804	26190	25588	24998	24420	23852	23296
3.50	22750	22216	21692	21179	20675	20182	19699	19226	18763	18309
3.60	17865	17429	17003	16586	16178	15778	15386	15004	14629	14262
3.70	13904	13553	13209	12874	12546	12225	11911	11604	11304	11011
3.80	10724	10444	10170	9903	9642	9387	9138	8894	8656	8424
3.90	8198	7976	7760	7549	7344	7143	6947	6756	6569	6387
4.00	6210	6037	5868	5703	5543	5386	5234	5085	4940	4799
4.10	4661	4527	4396	4269	4145	4025	3907	3793	3681	3573
4.20	3467	3364	3264	3167	3072	2980	2890	2803	2718	2635
4.30	2555	2477	2401	2327	2256	2186	2118	2052	1988	1926
4.40	1866	1807	1750	1695	1641	1589	1538	1489	1441	1395
4.50	1350	1306	1264	1223	1183	1144	1107	1070	1035	1001
4.60	968	935	904	874	845	816	789	762	736	711
4.70	687	664	641	619	598	577	557	538	519	501
4.80	483	466	450	434	419	404	390	376	362	349
4.90	337	325	313	302	291	280	270	260	251	242
5.00	233	224	216	208	200	193	185	178	172	165
5.10	159	153	147	142	136	131	126	121	117	112
5.20	108	104	100	96	92	88	85	82	78	75
5.30	72	69	67	64	62	59	57	54	52	50
5.40	48	46	44	42	41	39	37	36	34	33
5.50	32	30	29	28	27	26	25	24	23	22
5.60	20.7	19.8	19.0	18.1	17.4	16.6	15.9	15.2	14.6	14.0
5.70	13.4	12.8	12.2	11.7	11.2	10.7	10.2	9.8	9.4	8.9
5.80	8.5	8.2	7.8	7.5	7.1	6.8	6.5	6.2	5.9	5.7
5.90	5.4	5.2	4.9	4.7	4.5	4.3	4.1	3.9	3.7	3.6
6.00	3.4	3.2	3.1	3.0	2.8	2.7	2.6	2.4	2.3	2.2

Appendix 2

Factors for control charts

Sample size	Hartley's constant	c_4 constant	For charts based on ranges			For charts based on standard deviations		
			Xbar chart limits	R chart limits		Xbar chart limits	S chart limits	
n	d_2	c_4	A_2	D_3	D_4	A_3	B_3	B_4
2	1.128	0.7979	1.880	*	3.267	2.659	*	3.267
3	1.693	0.8862	1.023	*	2.575	1.954	*	2.568
4	2.059	0.9213	0.729	*	2.282	1.628	*	2.266
5	2.326	0.9400	0.577	*	2.115	1.427	*	2.089
6	2.534	0.9515	0.483	*	2.004	1.287	0.030	1.970
7	2.704	0.9594	0.419	0.076	1.924	1.182	0.118	1.882
8	2.847	0.9650	0.373	0.136	1.864	1.099	0.185	1.815
9	2.970	0.9693	0.337	0.184	1.816	1.032	0.239	1.761
10	3.078	0.9727	0.308	0.223	1.777	0.975	0.284	1.716
11	3.173	0.9754	0.285	0.256	1.744	0.927	0.321	1.679
12	3.258	0.9776	0.266	0.283	1.717	0.886	0.354	1.646
13	3.336	0.9794	0.249	0.307	1.693	0.850	0.382	1.618
14	3.407	0.9810	0.235	0.328	1.672	0.817	0.406	1.594
15	3.472	0.9823	0.223	0.347	1.653	0.789	0.428	1.572
16	3.532	0.9835	0.212	0.363	1.637	0.763	0.448	1.552
17	3.588	0.9845	0.203	0.378	1.622	0.739	0.466	1.534
18	3.640	0.9854	0.194	0.391	1.608	0.718	0.482	1.518
19	3.689	0.9862	0.187	0.403	1.597	0.698	0.497	1.503
20	3.735	0.9869	0.180	0.415	1.585	0.680	0.510	1.490
21	3.778	0.9876	0.173	0.425	1.575	0.663	0.523	1.477
22	3.819	0.9882	0.167	0.434	1.566	0.647	0.534	1.466

Six Sigma Quality Improvement with Minitab, Second Edition. G. Robin Henderson.
© 2011 John Wiley & Sons, Ltd. Published 2011 by John Wiley & Sons, Ltd.

Sample size	Hartley's constant	c_4 constant	For charts based on ranges			For charts based on standard deviations		
			Xbar chart limits	R chart limits		Xbar chart limits	S chart limits	
n	d_2	c_4	A_2	D_3	D_4	A_3	B_3	B_4
23	3.858	0.9887	0.162	0.443	1.557	0.633	0.545	1.455
24	3.895	0.9892	0.157	0.451	1.548	0.619	0.555	1.445
25	3.931	0.9896	0.153	0.459	1.541	0.606	0.565	1.435

An asterisk indicates that the corresponding chart limit is set to zero.

When using ranges an estimate of process standard deviation is:

$$\hat{\sigma} = \frac{\bar{R}}{d_2}.$$

When using standard deviations a recommended estimate of process standard deviation, discussed in Chapter 5, is:

$$\hat{\sigma} = \frac{s_{\text{Pooled}}}{c_4}.$$

Appendix 3

Formulae for control charts

The constants in the following formulae may be found in Appendix 2

Charts based on variables (measurements)

Individuals chart

$$\text{UCL} = \bar{x} + 2.66\overline{MR}$$

$$\text{LCL} = \bar{x} - 2.66\overline{MR} \quad \text{These formulae apply with moving ranges of length 2.}$$

$$\text{CL} = \bar{x}$$

Moving range chart

$$\text{UCL} = 3.267\overline{MR}$$

LCL is effectively zero   The formula for UCL applies with moving ranges of length two.

$$\text{CL} = \overline{MR}$$

Mean or Xbar chart (range based)

$$\text{UCL} = \bar{\bar{x}} + A_2\bar{R}$$

$$\text{LCL} = \bar{\bar{x}} - A_2\bar{R}$$

$$\text{CL} = \bar{\bar{x}}$$

Six Sigma Quality Improvement with Minitab, Second Edition. G. Robin Henderson.
© 2011 John Wiley & Sons, Ltd. Published 2011 by John Wiley & Sons, Ltd.

Range chart

$$UCL = D_4\bar{R}$$
$$LCL = D_3\bar{R}$$
$$CL = \bar{R}$$

Mean or Xbar chart (standard deviation based)

$$UCL = \bar{\bar{x}} + A_3\hat{\sigma}$$
$$LCL = \bar{\bar{x}} - A_3\hat{\sigma}$$
$$CL = \bar{\bar{x}}$$

Standard deviation chart

$$UCL = B_4\hat{\sigma}$$
$$LCL = B_3\hat{\sigma}$$
$$CL = \bar{s}$$

EWMA chart

$$UCL = \bar{\bar{x}} + 3\hat{\sigma}\sqrt{\frac{\alpha}{2-\alpha}} \quad \text{in the long term}$$

$$CL = \bar{\bar{x}}$$

$$LCL = \bar{\bar{x}} - 3\hat{\sigma}\sqrt{\frac{\alpha}{2-\alpha}} \quad \text{in the long term}$$

where $\hat{\sigma}$ denotes an estimate of the process standard deviation and α denotes the smoothing constant.

Charts based on attributes (counts)

Proportion nonconforming (P) chart

$$UCL = \bar{p} + 3\sqrt{\frac{\bar{p}(1-\bar{p})}{n}}$$

$$LCL = \bar{p} - 3\sqrt{\frac{\bar{p}(1-\bar{p})}{n}}$$

$$CL = \bar{p}$$

Number nonconforming (NP) chart

$$UCL = n\bar{p} + 3\sqrt{n\bar{p}(1 - \bar{p})}$$

$$LCL = n\bar{p} - 3\sqrt{n\bar{p}(1 - \bar{p})}$$

$$CL = n\bar{p}$$

Count of nonconformities (P) chart

$$UCL = \bar{c} + 3\sqrt{\bar{c}}$$

$$LCL = \bar{c} - 3\sqrt{\bar{c}}$$

$$CL = \bar{c}$$

Mean number of nonconformities per unit (U) chart

$$UCL = \bar{u} + 3\sqrt{\frac{\bar{u}}{n}}$$

$$LCL = \bar{u} - 3\sqrt{\frac{\bar{u}}{n}}$$

$$CL = \bar{u}$$

Appendix 4

Tests for evidence of special causes on Minitab control charts

Six Sigma Quality Improvement with Minitab, Second Edition. G. Robin Henderson.
© 2011 John Wiley & Sons, Ltd. Published 2011 by John Wiley & Sons, Ltd.

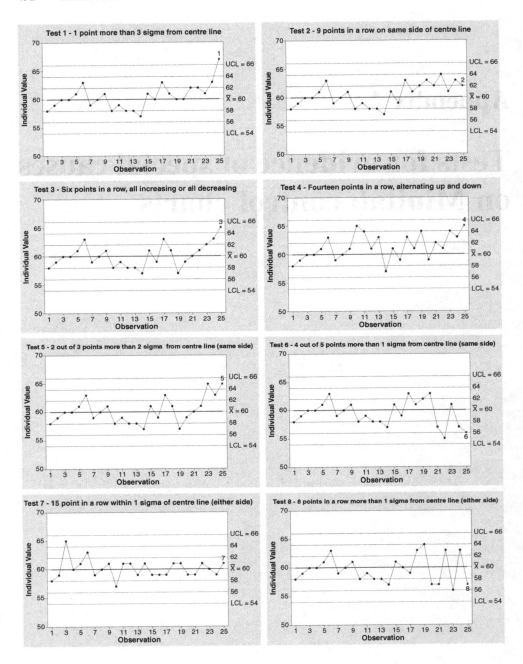

The default set of tests provided in Minitab is illustrated for the case of an individuals or X chart of a variable with mean 60 and standard deviation 2. Note that, in each case, a pattern of points that is the mirror image in the centre line of that displayed would provide a second example of a signal from the corresponding test.

References

Antony, J. (2010) The future of Six Sigma: Viewpoints from world leading practitioners and academics. Presentation given at an Aston Business School Seminar: Performance Measurement and Quality Management: http://www1.aston.ac.uk/aston-business-school/research/groups/oim/esrcseminars/past-seminars/#Seminar5 (accessed 28 January 2011).

Automotive Industry Action Group (2002) *Measurement Systems Analysis*, 3rd edn. Southfield, MI: Automotive Industry Action Group.

Berger, P.D. and Maurer, R.E. (2002) *Experimental Design with Applications in Management, Engineering and the Sciences*. Belmont, CA: Duxbury/Thomson Learning.

Bisgaard, S. and Kulachi, M. (2000) Finding assignable causes. *Quality Engineering*, 12(4): 633–640.

Bland, J.M. and Altman, D.G. (1986) Statistical methods for assessing agreement between two methods of clinical measurement. *The Lancet*, 1(8476): 307–310.

Box, G.E.P. (1990) George's column. *Quality Engineering*, 12(3): 365–369.

Box, G.E.P. (1999) Statistics as a catalyst to learning by scientific method – a discussion. *Journal of Quality Technology*, 31(1): 16–29.

Box, G.E.P. and Liu P.Y.T. (1999) Statistics as a catalyst to learning by scientific method – an example. *Journal of Quality Technology*, 31(1): 1–15.

Box, G.E.P. and Luceño, A. (1997) *Statistical Control by Monitoring and Feedback Adjustment*. New York: John Wiley & Sons, Inc.

Box, G.E.P. and Luceño, A. (2000) Six Sigma, process drift, capability indices, and feedback adjustment. *Quality Engineering*, 12(3): 297–302.

Box, G.E.P., Hunter W.G. and Hunter J.S. (1978) *Statistics for Experimenters – An Introduction to Design, Data Analysis, and Model Building*. New York: John Wiley & Sons, Inc.

Box, G.E.P., Bisgaard, S. and Fung, C. (1988) An explanation and critique of Taguchi's contributions to quality engineering. *Quality and Reliability Engineering International*, 4(2): 123–131.

Box, G.E.P., Hunter W.G. and Hunter J.S. (2005) *Statistics for Experimenters – Design, Innovation, and Discovery*, 2nd edn. Hoboken, NJ: John Wiley & Sons, Inc.

Breyfogle, F.W. (2003) *Implementing Six Sigma: Smarter Solutions Using Statistical Methods*. Hoboken, NJ: John Wiley & Sons, Inc.

Six Sigma Quality Improvement with Minitab, Second Edition. G. Robin Henderson.
© 2011 John Wiley & Sons, Ltd. Published 2011 by John Wiley & Sons, Ltd.

British Broadcasting Corporation (2004) Blood groups – an overview. http://www.bbc.co.uk/dna/h2g2/A2116621 (accessed 28 January 2011).

Burdick, R.K., Borror, C.M. and Montgomery, D.C. (2003) A review of methods for measurement systems capability analysis. *Journal of Quality Technology*, 35(4): 342–354.

Burdick, R.K., Borror, C.M. and Montgomery, D.C. (2005) *Design and Analysis of Gauge R&R Studies*. Philadelphia, PA: Society for Industrial and Applied Mathematics; and Alexandria, VA: American Statistical Association.

Caulcutt, R. (1995) *Achieving Quality Improvement*. London: Chapman & Hall.

Caulcutt, R. (2004) Managing by fact. *Significance*, 1(1): 36–38.

Clarke, R.D. (1946) An application of the Poisson distribution. *Journal of the Institute of Actuaries*, 72: 481.

Coleman, S.Y., Greenfield, T., Jones, R., Morris, C. and Puzey, I. (1996) *The Pocket Statistician – A Practical Guide to Quality Improvement*. London: Arnold.

Cox, D. R. (1970) *The Analysis of Binary Data*. London: Methuen.

Criqui, M.H. and Ringel, B.L. (1994) Does diet or alcohol explain the French paradox? *Lancet*, 344: 1719–1723.

Czitrom, V. (1997) Introduction to gauge studies. In V. Czitrom and P.D. Spagon (eds), *Statistical Case Studies for Industrial Process Improvement*. Philadelphia, PA: Society for Industrial and Applied Mathematics; and Alexandria, VA: American Statistical Association.

Dalal, S.R., Fowlkes, E.B. and Hoadley, B. (1989) Risk analysis of the Space Shuttle: Pre-Challenger prediction of failure. *Journal of the American Statistical Association*, 84(408): 945–957.

Daly, F., Hand, D.J., Jones, M.C., Lunn, A.D. and McConway, K.J. (1995) *Elements of Statistics*. London: Prentice Hall.

Daniel, W.W. and Terrell, J.C. (1989) *Business Statistics for Management and Economics*. Boston: Houghton Mifflin.

de Mast, J. (2008) A history of industrial statistics. In S.Y. Coleman, T. Greenfield, D. Stewardson, and D.C. Montgomery (eds), *Statistical Practice in Business and Industry*. Chichester: John Wiley & Sons, Ltd.

Deming, W.E. (1986) *Out of the Crisis*. Cambridge: Cambridge University Press.

Deming, W.E. (1988) A special seminar for statisticians. University of Nottingham, Nottingham.

Dodgson, J. (2003) A graphical method for assessing mean squares in saturated fractional designs. *Journal of Quality Technology*, 35(2): 206–212.

Douglass, J. and Coleman, S.Y. (2000) Improving product yield utilising a statistically designed experiment. In *Proceedings of the Industrial Statistics in Action 2000 International Conference*. Newcastle upon Tyne: University of Newcastle upon Tyne.

Duncan, A.J. (1959) *Quality Control and Industrial Statistics*. Homewood, IL: Richard D. Irwin.

Dunn, G (1989) *Design and Analysis of Reliability Studies*. New York: Oxford University Press.

Everitt, B.S. (1994) *A Handbook of Statistical Analyses Using S-PLUS*. London: Chapman & Hall.

Fisher, R.A. (1954) *Statistical Methods for Research Workers*, 12th edn. Edinburgh: Oliver and Boyd.

Gorman, J.W. and Toman, R.J. (1966) Selection of variables for fitting equations to data. *Technometrics*, 8: 27–51.

Harry, M. and Schroeder, R. (2000) *Six Sigma*. New York: Random House.

Hart, J.S., George, S.L., Frei, E., Bodey, G.P., Nickerson, R.C. and Freireich, E. (1977) Prognostic significance of pretreatment proliferative activity in adult acute leukemia. *Cancer*, 39(4): 1603–1617.

Hellstrand, C. (1989) The necessity of modern quality improvement and some experience, with its implementation in the manufacture of rolling bearings. *Philosophical Transactions of the Royal Society, Series A*, 327: 529–537.

Henderson, G.R. (2001) EWMA and industrial applications to feedback adjustment and control. *Journal of Applied Statistics*, 28(3–4): 399–407.

Henderson, G.R., Mead G.E., van Dijke M.L., Ramsay S., McDowall M.A. and Dennis M. (2008) Use of statistical process control charts in stroke medicine to determine if clinical evidence and changes in service delivery were associated with improvements in the quality of care. *Quality and Safety in Health Care*, 17: 301–306.

Henderson, G.R., Davies, R. and Macdonald, D. (2010) Bringing data to life with post-hoc CUSUM charts. *Case Studies in Business, Industry and Government Statistics*, 3(2).http://www.bentley.edu/csbigs/documents/Henderson.pdf (accessed 28 January 2011).

Hicks, C.R. and Turner, K.V. Jr (1999) *Fundamental Concepts in the Design of Experiments*, 5th edn. Oxford: Oxford University Press.

Hoerl, R.W. (1998) Six Sigma and the future of the quality profession. *Quality Progress*, 31(6): 35–44.

Hoerl, R.W. (2001) Six Sigma Black Belts: what do they need to know? *Journal of Quality Technology*, 33(4): 391–406.

Hoerl, R.W. and Snee, R. (2005) Six Sigma beyond the Factory Floor: Deployment Strategies for Financial Services, *Healthcare and the Rest of the Real Economy*. Upper Saddle River, NJ: Prentice Hall.

Hogg, R.V. and Ledolter, J. (1992) *Applied Statistics for Engineers and Physical Scientists*, 2nd edn. New York: Macmillan.

Hunt, G.A. (1948) A training program becomes a clinic. *Industrial Quality Control*, January: 26.

Hunter, J.S. (1989) A one-point equivalent to the Shewhart chart with Western Electric rules. *Quality Engineering*, 2: 13–19.

Iman, R.L. and Conover, W.J. (1989) *Modern Business Statistics*, 2nd edn. New York: John Wiley & Sons, Inc.

Johnson, R.A. and Wichern, D.W. (2001) *Applied Multivariate Statistical Analysis*, 5th edn. Upper Saddle River, NJ: Prentice Hall.

Kolarik, W.J. (1995) *Creating Quality*. New York: McGraw-Hill.

Landis, J.R. and Koch, G.G. (1977) The measurement of observer agreement for categorical data. *Biometrics*, 33: 159–174.

Lenth, R.V. (1989) Quick and easy analysis of unreplicated factorials. *Technometrics*, 31: 469–473.

López-Alvarez, T. and Aguirre-Torres, V. (1997) Improving field performance by sequential experimentation: a successful case study in the chemical industry. *Quality Engineering*, 9(3): 391–403.

Lynch, R.O. and Markle, R.J. (1997) Understanding the nature of variability in a dry etch process. In V. Czitrom and P.D. Spagon (eds), *Statistical Case Studies for Industrial Process Improvement*. Philadelphia: Society for Industrial and Applied Mathematics; and Alexandria, VA: American Statistical Association.

Mahmoud, M.A., Henderson G.R., Eprecht, E.K. and Woodall W.H. (2010) Estimating the standard deviation in quality-control applications. *Journal of Quality Technology*, 42(4): 348–357.

Mendenhall, W., Scheaffer, R.L. and Wackerly, D.D. (1986) *Mathematical Statistics with Applications*. Boston: Duxbury.

Minitab Inc. (2009) *Getting Started: Quality Companion 3*. http://www.minitab.com/uploadedFiles/Shared_Resources/Documents/Brochures/companion3getstarted.pdf (accessed 25 March 2011).

Minitab Inc., (2010) *Meet Minitab 16*. http://www.minitab.com/uploadedFiles/Shared_Resources/Documents/MeetMinitab/EN16_MeetMinitab.pdf (accessed 28 January 2011).

Montgomery, D.C. (2005a) *Design and Analysis of Experiments*, 6th edn. Hoboken, NJ: John Wiley & Sons, Inc.

Montgomery, D.C. (2005b) Generation III Six Sigma (Editorial). *Quality and Reliability Engineering International*, 21(6), iii–iv.

Montgomery, D.C. (2009) *Introduction to Statistical Quality Control*, 6th edn. Hoboken, NJ: John Wiley & Sons, Inc.

Montgomery, D.C. and Runger, G.C. (2010) *Applied Statistics and Probability for Engineers*, 4th edn. Hoboken, NJ: John Wiley & Sons, Inc.

Montgomery, D.C. and Woodall, W.H. (2008) An overview of Six Sigma. *International Statistical Review*, 76(3): 329–346.

Montgomery, D.C., Peck, E.A. and Vining, G.G. (2006) *Introduction to Linear Regression Analysis*, 4th edn. Hoboken, NJ: John Wiley & Sons, Inc.

Moore, D.S. (1996) *Statistics – Concepts and Controversies*, 4th edn. New York: W.H. Freeman & Co.

NIST/SEMATECH (2005) *e-Handbook of Statistical Methods*. http://www.itl.nist.gov/div898/handbook/ (accessed 28 January 2011)

Ott, E.R., Schilling, E.G. and Neubauer, D.V. (2000) *Process Quality Control – Troubleshooting and Interpretation of Data*, 3rd edn. New York: McGraw-Hill.

Page, E.S. (1954) Continuous inspection schemes. *Biometrics*, 41(1): 100–115.

Pande, P.S., Neuman, R.P. and Cavanagh, R.R. (2000) *The Six Sigma Way: How GE, Motorola and Other Top Companies Are Honing Their Performance*. New York: McGraw-Hill.

Perez-Wilson, M. (1999) *Six Sigma*. Scottsdale, AZ: Advanced System Consultants.

Pignatiello, J.J., Jr. and Ramberg, J.S. (1985) Off-line quality control, parameter design, and the Taguchi method – a discussion. *Journal of Quality Technology*, 17(4): 198–206.

Quinlan, J. (1985) Product improvement by application of Taguchi methods. In *Third Supplier Symposium on Taguchi Methods*. Dearborn, MI: American Supplier Institute.

Reynard, S. (2007) Making a profit from game-changing inventions. *iSigSigma Magazine,* January/February: 20–27. http://6sigmaexperts.com/presentations/JF07_Motorola_Corporate_Leader.pdf (accessed 9 March 2011).

Ryan, T.P. (2000) *Statistical Methods for Quality Improvement*, 2nd edn. New York: John Wiley & Sons, Inc.

Shewhart, W.A. (1931) *Economic Control of Quality of Manufactured Product*. New York: D. Van Nostrand. Also available in a 50th anniversary edition published in 1980 by the American Society for Quality, Milwaukee, WI.

Skrivanek, S. (2009) How to conduct an MSA when the part is destroyed during measurement. http://www.moresteam.com/whitepapers/nested-gage-rr.pdf (accessed 19 March 2011).

Snee, R.D. (2004) Six-Sigma: the evolution of 100 years of business improvement methodology. *International Journal of Six Sigma and Competitive Advantage*, 1(1): 4–20.

Spiegelhalter, D. (2002) Funnel plots for institutional comparison. *Quality and Safety in Healthcare*, 11: 390–391.

Steiner, S.H., Abraham, B. and MacKay, R.J. (1997) Understanding process capability indices. IIQP Research Report no. 2, University of Waterloo, Canada.

Sterne, A.C. and Davey Smith G. (2001) Sifting the evidence – what's wrong with significance tests? *British Medical Journal*, 322: 226–231.

Truscott, W.T, (2003) *Six Sigma: Continual Improvement for Business*. Oxford: Butterworth-Heinemann.

Tukey, J.W. (1977) *Exploratory Data Analysis*. Reading, MA: Addison-Wesley.

Tukey, J.W. (1986) Displaying results for people: static, dynamic or computer-selected (via cognostics). Paper presented to the Edinburgh Local Group of the Royal Statistical Society, 10 November.

United States Golf Association (2008) The rules of golf. Appendix III: The ball. http://www.usga.org/ bookrule.aspx?id=14324 (accessed 28 January 2011).

Vickers V. (2010) *What is a p-Value Anyway? 34 Stories To Help You Actually Understand Statistics.* Boston, MA: Addison-Wesley.

Walpole R.E. and Myers R.H. (1989) *Probability and Statistics for Engineers and Scientists.* New York: Macmillan.

Wasiloff, E. and Hargitt, C. (1999) Using DOE to determine AA battery life. *Quality Progress*, 32(3): 67–71.

Welch, J.with Byrne J.A. (2001) *Jack – Straight from the Gut.* London: Headline.

Wheeler, D.J. (1993) *Understanding Variation – The Key to Managing Chaos.* Knoxville, TN: SPC Press.

Wheeler, D.J. (2003) Good data, bad data, and process behaviour charts. http://www.spcpress.com/pdf/ good_data_%20bad_data.pdf (accessed 28 January 2011).

Wheeler, D.J. (2007) Shewhart, Deming, and Six Sigma. http://www.spcpress.com/pdf/DJW187.pdf (accessed 16 March 2011).

Wheeler, D.J. and Chambers, D.S. (1992) *Understanding Statistical Process Control*, 2nd edn. Knoxville, TN: SPC Press.

Wheeler, D.J. and Lyday R.W.D.S. (1989) *Evaluating the Measurement Process*, 2nd edn. Knoxville, TN: SPC Press.

Wheeler, D.J. and Poling S.R. (1998) *Building Continual Improvement.* Knoxville, TN: SPC Press.

Windsor, S.E. (2003) Attribute gage R&R. *Six Sigma Forum Magazine*, 2(4): 23–28.

Yang, K. (2004) Multivariate statistical methods and Six Sigma. *International Journal of Six Sigma and Competitive Advantage*, 1(1): 76–96.

Index

Six Sigma Quality Improvement with Minitab, Second Edition. G. Robin Henderson.
© 2011 John Wiley & Sons, Ltd. Published 2011 by John Wiley & Sons, Ltd.